# The Growth of Firms, Middle East Oil and Other Essays

# The Growth of Firms, Middle East Oil and Other Essays

**Edith Penrose**
Professor of Economics in the
University of London

FRANK CASS & CO. LTD.
1971

First Published in 1971 by
FRANK CASS AND COMPANY
67 Great Russell Street, London WC1B 3BT

Copyright © EDITH PENROSE

ISBN 0 7146 2772 0

Printed in Great Britain by
Clarke, Doble and Brendon Limited, Plymouth

# Contents

## SECTION I

GROWTH OF FIRMS AND PRIVATE FOREIGN INVESTMENT

| | | |
|---|---|---|
| I | Biological Analogies in the Theory of the Firm<br>With Comment by A. A. Alchian and Stephen Enke, and Rejoinder | 3 |
| II | Limits to the Growth and Size of Firms | 30 |
| III | The Growth of the Firm: A Case Study, The Hercules Powder Company | 43 |
| IV | Foreign Investment and the Growth of the Firm | 64 |
| V | Problems Associated with the Growth of International Firms | 82 |
| VI | International Economic Relations and the Large International Firm | 91 |
| VII | The State and the Multinational Enterprise in Less-Developed Countries | 119 |

## SECTION II

MULTINATIONAL FIRMS IN THE INTERNATIONAL PETROLEUM INDUSTRY

| | | |
|---|---|---|
| VIII | Middle East Oil: The International Distribution of Profits and Income Taxes | 139 |
| IX | Profit Sharing between Producing Countries and Oil Companies in the Middle East<br>With Reply by H. W. Page and Rejoinder | 151 |
| X | Monopoly and Competition in the International Petroleum Industry | 178 |
| XI | Vertical Integration with Joint Control of Raw Material Production: Crude Oil in the Middle East | 200 |
| XII | Government Partnership in the Major Concessions of the Middle East: The Nature of the Problem | 217 |

| XIII | OPEC and the Changing Structure of the International Petroleum Industry | 228 |
|---|---|---|
| XIV | The Role of OPEC in Changing Circumstances | 236 |

# SECTION III

## MIDDLE EAST ESSAYS

| XV | Economic Development and the State—A Review Article | 247 |
|---|---|---|
| XVI | Money, Prices, and Economic Expansion in the Middle East, 1952–1960 | 256 |
| XVII | Oil and State in Arabia | 281 |
| XVIII | Iran as a Pacemaker in Middle East Oil | 296 |
| XIX | A Note on Development Planning and the Role of the Enterprise, with Special Reference to Egypt | 302 |

# SECTION IV

Economics and the Aspirations of *Le Tiers Monde*—An Inaugural Lecture         319

# Preface

AT first glance it might seem that the three subjects dealt with in the essays written over the last twenty years and now collected in this volume could hardly be more diverse, beginning with the growth of the firm and moving from the international petroleum industry to the Middle East generally. Oddly enough, however, these subjects are connected by the same type of historical logic that characterizes the diversification of an industrial firm: the logic in the simple principle that one thing leads to another.

My studies of the growth of the firm, of which essays I, II and III are off-shoots, and which culminated in the publication of a book on that subject,[1] led naturally to a consideration of the growth of firms through expansion abroad—commonly known as direct private foreign investment. Essay IV was the first of a series on this aspect of the subject and was written during a stimulating year spent at the Australian National University in 1955.

Two years later I went with my husband to teach at the University of Baghdad for the academic years 1957/58 and 1958/59. If one is interested in the growth of firms, especially in their expansion abroad by direct investment, and one is in Baghdad, what can one do but look at the international oil industry? Much to my surprise I discovered that at that time no serious academic work on the economics of the international oil industry was available.[2] Economists had studied the industry in the United States and in Europe; many writers had examined the politics of oil; there were a few good histories of oil companies; but nothing on the economics of the international industry. My work in this field produced in its turn a book dealing with the economics of the international oil companies in developing countries.[3] Some of the essays in Section II are by-products of that book; others are additions to it.

Finally, my stay in the Middle East led to an enduring interest in the area. I have spent much time there since and am now engaged, with my husband, on a book on the development of modern Iraq. Section III of the present volume reflects this aspect of my work in recent years.

For good measure this collection of essays is rounded out with the inaugural lecture I delivered on taking the Chair of Economics with Reference to Asia at the School of Oriental and African Studies in the University of London in 1964–65. It deals with the importance of including economics in modern Asian studies as one of the aspects of the societies of modern Asia that cannot be neglected if one is to understand them.

The essays are arranged in rough chronological order within each section, although I have deviated from strict chronology on occasion in order to achieve a more rational ordering within the sections. There is a little repetition, especially in the exposition of the basic characteristics of the international oil industry, in the illustrations drawn from it, and in the presentation of relevant aspects of its history. I see no way of eliminating this repetition, for even though each essay deals with a distinct analytical problem, the relevant background of the industry had to be explained in each. I trust the reader will understand, and tolerantly skip when necessary. Fortunately the repetition is not serious—indeed it is much less serious than I had feared when I began putting the essays together for this volume. The very few changes I have made in the essays are designed only to improve exposition or to clarify the discussion.

EDITH PENROSE

School of Oriental and
African Studies
University of London
August 1970

## NOTES

1 *The Theory of the Growth of the Firm.* (Oxford: Basil Blackwell; New York: Wiley.) First published 1959. Translated into Japanese, French, Spanish, and Italian.
2 There have been some good books on the subject written since.
3 *The Large International Firm in Developing Countries: The International Petroleum Industry.* (London: Allen and Unwin; Cambridge, Mass.: MIT Press. 1968.)

# Acknowledgments

I AM indebted to the following for permission to reprint the papers contained in this volume:
*The American Economic Review*, published by the American Economic Association; *The Business History Review*, published by Harvard Graduate School of Business Administration; *The Economic Journal*, Quarterly Journal of the Royal Economic Society; *Tijdschrift voor vennootschappen verenigingen en stichtingen*; Frank Cass, Ltd; Professor John Dunning, University of Reading; *Economica*, published by the London School of Economics and Political Science; Stevens & Sons, Ltd., publisher of the *Yearbook of World Affairs*; Fuad Itayim, publisher of the *Middle East Economic Survey*; *Economic Development and Cultural Change*, published by the University of Chicago Press; *Rivista Internazionale di Scienze Economiche e Commerciali*, published under the auspices of the Università Commerciale Luigi Bocconi; and the School of Oriental and African Studies, University of London.

I am also grateful to the following individuals who were kind enough to permit me to re-publish their comments on two of the articles: Professor Armen A. Alchian, Professor Stephen Enke, and Mr. H. W. Page, Standard Oil Company (New Jersey).

# SECTION I

## GROWTH OF FIRMS AND PRIVATE FOREIGN INVESTMENT

Nearly twenty years separate the time of writing the first essays in this section from the last. The first four essays were written while I was preparing *The Theory of the Growth of the Firm* and are closely related to the problems dealt with in that book. The last of these four provides a transition to the rest of the essays in this section, which were written after my book on the international firm in developing countries appeared, and continued my work on that subject.

# I

## BIOLOGICAL ANALOGIES IN THE THEORY OF THE FIRM[*][1]

*Biological analogies are used in a variety of ways in the social sciences. Although this paper deals specifically with the theory of the firm, where such analogies are no longer prominent, the general arguments advanced are relevant to uses of biological analogies in other contexts.*

ECONOMICS has always drawn heavily on the natural sciences for analogies designed to help in the understanding of economic phenomena. Biological analogies in particular have been widely used in discussions of the firm. Probably the best known and most common of these analogies is that of the *life cycle*, in which the appearance, growth and disappearance of firms is likened to the processes of birth, growth, and death of biological organisms. Marshall's reference to the rise and fall of the trees in the forest is an oft-quoted example of this type of analogy. Recently, two additional biological analogies have been presented—a natural selection analogy, dubbed by one writer *viability analysis*, and the *homeostasis* analogy designed to explain some aspects of the behavior of firms. The former, like the life cycle analogy, is for use in long-run analysis only. The latter is exclusively for short-run analysis. Both are supposed to represent improvements on the existing theory of the firm at the core of which lies the chief target of attack—the assumption that firms attempt to maximize profits.

The purpose of this paper is to examine critically all three types of reasoning and to show that they lead in most cases to a serious neglect of important aspects of the problem that do not fit the particular type of analogical reasoning employed. The chief danger of carrying sweeping analogies very far is that the problems they are designed to illuminate become framed in such a special way that significant matters are frequently inadvertently obscured. Biological analogies contribute little either to the theory of price or to the theory of growth and development of firms and in general tend to confuse the nature of the important issues.

*The 'Life Cycle' Theory of the Firm*
  Implicit in the notion that firms have a 'life cycle' analogous to

[*] Published in *The American Economic Review*, Vol. XLII, No. 5, December, 1952.

that of living organisms is the idea that there are 'laws' governing the development of firms akin to the laws of nature in accordance with which living organisms appear to grow, and that the different stages of development are a function of age. Were this implication not present, then the life cycle concept would amount to little more than a statement that if we look at the past we find that all firms had some sort of a beginning, a period of existence and, if now extinct, an end. Even if a careful collection of the relevant facts about groups of firms in like circumstances should establish a statistical pattern in which some affinity in origin, regularity in development and similarity in disappearance could be discerned, it might be interesting history and might enable one to deduce a variety of *ad hoc* theories but it would not be a theory of development without the further generalizations about the principles according to which the life cycle proceeds.

Whatever superficial plausibility such a theory may have had in the days of the 'family firm,'[2] it lost even that when the publicly held corporation became the dominant type of firm. Even Marshall, who was an early exponent in economics of this theory of the growth of the firm, was doubtful about its applicability to the joint-stock company, and I would not spend much time on it now had it not been recently adopted and put forward with vigor by one of America's foremost economists. Kenneth Boulding has virtually called for a 'life cycle' theory of the firm[3] and has categorically insisted that there is an 'inexorable and irreversible movement towards the equilibrium of death. Individual, family, firm, nation, and civilization all follow the same grim law, and the history of any organism is strikingly reminiscent of the rise and fall of populations on the road to extinction. . . .'[4]

The purposes a life cycle theory of the firm would serve are obvious, yet the theory as a bare undeveloped hypothesis has existed for a long time and nothing has been done to construct from it a consistent theoretical system with sufficient content to enable it to be used for any purpose whatsoever.[5] The basic hypothesis is not one from which significant logical consequences can be deduced, such as can be deduced,[6] for example, from the proposition that firms attempt to maximize profits. Supplementary hypotheses about the kind of organism the firm is and the nature of its life cycle are required. Although we have a respectable collection of information about firms, it has not stimulated economists even to suggest the further hypotheses necessary to the development of a life cycle theory of the firm. This, I think, is primarily because the available evidence does not support the theory that firms have a life cycle characterized by a consistent transition through recognizable stages of development similar to those of living organisms. Indeed, just the opposite conclusion must be drawn: the development of firms does not proceed according to the same 'grim' laws as does that of living organisms. In the face of the evidence one is led to wonder why the analogy persists and why there is still a demand for a life cycle theory of the firm.

The purpose of analogical reasoning in which we consciously and systematically apply the explanation of one series of events to another very different series of events is to help us better to understand the nature of the latter, which presumably is less well understood than the former. If the analogy has really helpful explanatory value, there must be some reason for believing that the two series of events have enough in common for the explanation of one, *mutatis mutandis*, to provide at least a partial explanation of the other. This type of analogy must be distinguished from the purely metaphorical analogy in which the resemblances between two phenomena are used to add a picturesque note to an otherwise dull analysis and to help a reader to see more clearly the outlines of a process being described by enabling him to draw on what he knows in order to imagine the unknown. Analogies of this sort are not only useful but almost indispensable to human thought.

The biological analogies of the firm are not of this metaphorical type or there would be no call to push them into service to help *explain* the development of firms. They are clearly related to the whole family of analogies between biological organisms and social institutions that flourished in profusion during the 19th century[7] but which are, for the most part, no longer popular among social scientists, although curiously enough they are apparently still popular among some biologists.[8] In the notion that a firm is an organism akin to biological organisms, there is an implication that, since all such organisms have something in common, we can use our knowledge of biological organisms to gain more insight into the firm. It is not an easy task even for the biologist to state unambiguously what is meant by an organism[9] or what distinguishes the biological organism from non-living matter. But in principle it is characteristic of biological organisms that they reproduce and have an identifiable pattern of development that can be explained by the genetic nature of their constitution.[10] Furthermore, the particular pattern of development that is supposed to characterize firms—birth, youth, maturity, old age, death—is characteristic only of biological organisms that reproduce sexually. Organisms whose reproductive processes are primarily asexual have in general a very different pattern of development in which *death* plays no part,[11] and certainly the development of firms shows no pattern similar even to that of organisms that reproduce asexually. Clearly the one thing a firm does not have in common with biological organisms is a genetic constitution, and yet this is the one factor that determines the life cycle of biological organisms.

The characteristic use of biological analogies in economics is to suggest explanations of events that do not depend upon the conscious willed decisions of human beings. This is not, of course, characteristic of biology as such, for some branches of biology are concerned with learning processes and decision making, with purposive motivation and conscious choice in men as well as animals. In this, biology overlaps sociology and psychology and, in a sense, even economics. Information drawn from these branches of biology can be useful in helping us to understand the behavior of men and consequently of the institutions men create and operate. In using such information, however, we are not dealing with analogies at all, but with essentially the same problems on a more complex scale. But, paradoxically, where explicit biological analogies crop up in economics they are drawn exclusively from that aspect of biology which deals with the non-motivated behavior of organisms or in which motivation does not make any difference.

So it is with the life cycle analogy. We have no reason whatsoever for thinking that the growth pattern of a biological organism is *willed* by the organism itself. On the other hand, we have every reason for thinking that the growth of a firm is willed by those who make the decisions of the firm and are themselves part of the firm, and the proof of this lies in the fact that no one can describe the development of any given firm or explain how it came to be the size it is except in terms of decisions taken by individual men. Such decisions, to be sure, are constrained by the environment and by the capacity of the men who make them, but we know of no *general* 'laws' predetermining men's choices, nor have we as yet any established basis for suspecting the existence of such laws. By contrast no one would seriously attempt to explain the transition from infancy to manhood or the normal processes of aging in terms of such decisions, for we have every reason for thinking that these matters are predetermined by the nature of the living organism.

There can be no doubt, I think, that to liken a firm to an organism and then attempt to explain its growth by reference to the laws of growth of biological organisms is an ill-founded procedure. If it were no more than this, one could still question whether one should take the trouble of seriously analysing the analogy. But, besides being ill-founded, this type of reasoning about the firm obscures, if it does not implicitly deny, the fact that firms are institutions created by men to serve the purposes of men. It can be admitted that to some extent firms operate automatically in accordance with the principles governing the mechanism constructed,[12] but to abandon their development to the laws of nature diverts attention from the importance of human decisions and motives, and from problems of ethics and public policy, and surrounds the whole question of the growth of the firm with an aura of 'naturalness' and even inevitability.[13]

'Viability' Analysis

The second type of biological analogy I wish to discuss claims to have drawn on the principles of biological evolution and natural selection which were first put forth in a comprehensive form by Darwin. The discussion of the processes and progress of human society in terms of natural selection and evolution followed close on the introduction of these concepts into biology.[14] The analogy I am concerned with here avoids the crudities and attempts to avoid the value judgments that characterized the 19th century doctrines of Spencer and his followers in their application of these principles to society. It is very modern in its emphasis on uncertainty and statistical probabilities. Nevertheless, it is open to the same basic objections that in my opinion adhere to all such biological analogies.

The purpose of the theory is to get around a logical difficulty alleged to be inherent in the assumption that firms attempt to maximize profits in a world characterized by uncertainty about the future. If uncertainty exists, firms cannot know in advance the results of their actions. There is always a variety of possible outcomes, each of which is more or less probable. Hence the expected outcome of any action by a firm can only be viewed as a distribution of possible outcomes, and it is argued that while a firm can select those courses of action that have an optimum distribution of outcomes from its point of view, it makes no sense to say that the firm *maximizes* anything, since it is impossible to maximize a distribution. Hence profit maximization as a criterion for action is regarded as meaningless. According to the 'viability analysis,' however, this is not a serious difficulty for the economist if he draws on the principle of natural selection and considers the adaptation required of firms by their environment.

The argument, orginally set forth by Armen A. Alchian,[15] is as follows: To survive firms must make positive profits. Hence positive profits can be treated as the criterion of natural selection—the firms that make profits are selected or 'adopted' by the environment, others are rejected and disappear. This holds whether firms consciously try to make profits or not; even if the actions of firms were completely random and determined only by chance, the firms surviving, *i.e.*, adopted by an environment, would be those that happened to act appropriately and thus made profits. Hence 'individual motivation and foresight, while sufficient, are not necessary,'[16] since the economist with his knowledge of the conditions of survival can, like the biologist, predict 'the effects of environmental changes on the surviving class of living organisms.'[17]

Alchian argues that the introduction of the supplementary and realistic assumption of purposive behavior by firms merely 'expands' the model and also makes it useful in explaining the nature of purposive behavior under conditions of uncertainty.[18] If firms do try to make profits even though (because of uncertainty) they don't know how to do so, then clearly they will have a motive for imitating what appears to be successful action by other firms. This explains conventional rules of behavior (traditional markups, etc.) which can be looked on as 'codified imitations of observed success.'[19] This is the evolutionary aspect of the theory: successful innovations—regarded by analogy as 'mutations'—are transmitted by imitation to other firms. Venturesome innovation and trial and error adaptation are also purposive acts which, if successful, are 'adopted' by the environment. Thus 'most conventional economic tools and concepts are still useful, although in a vastly different analytical framework—one which is closely akin to the theory of biological evolution. The economic counterparts of genetic heredity, mutations, and natural selection are imitation, innovation, and positive profits.'[20]

In accepting and enlarging upon Alchian's argument Stephen Enke has argued that, if competition were so intense that zero profits would result in the long run, economists could make 'aggregate predictions' *as if* every firm knew how to secure maximum long-run profits. For with intense competition only firms that succeeded in maximizing profits would survive.[21] But under these circumstances, Enke notes, the economist can use the traditional marginal analysis and his predictions will be the same as they would be if he employed the 'viability analysis.' Which of the two he uses however is not 'immaterial,' since 'the language of the former method seems pedagogically and scientifically inferior because it attributes a quite unreasonable degree of omniscience and prescience to entrepreneurs.'[22]

There is much to be said about this revival of an old approach to human affairs and about its relation to the traditional marginalist approach in economics, in particular as to whether the two approaches really answer the same types of questions about the effect of 'environmental' changes on price and output. In this paper I am not so much concerned to present an analytical critique of the theory as to discuss the applicability of the biological analogy and the implications involved in its use. Again we find that the characteristic of the analogy employed is to provide an explanation of human affairs that does not depend on human motives. The alleged superiority of 'viability' over marginal analysis lies in the claim that it is valid even if men do not know what they are doing. No matter what men's motives are, the outcome is determined not by the individual participants, but by an environment beyond their control. Natural selection is substituted for purposive profit-maximizing behavior just as in biology natural selection replaced the concept of special creation of species.

In biology the theory of natural selection requires the postulate that competition—a struggle for existence—prevails, but it is a postulate that rests firmly on observed facts. Darwin deduced the struggle for existence from two empirical propositions: all organisms tend to increase in a geometrical ratio, and the numbers of any species remain more or less constant.[23] From this it follows that a struggle for existence must take place. Translated into economic terminology, the explanation of competition in nature is found in the rate of entry. The 'excessive entry' is due to the nature of biological reproduction. But how shall we explain competition in economic affairs where there is no biological reproduction? The psychological assumption of the traditional economic theory that businessmen like to make money and strive to make as much as is practicable performs a function in economic analysis similar to that of the physiological assumption in the biological theory of natural selection that the reproduction of organisms is of a geometric type—it provides the explanation of competition (and in economics, incidentally, also of monopoly). To be sure, the two assumptions rest on vastly different factual foundations and should not be treated as analogous. We can only say that there is some evidence that such a psychological motivation is widely prevalent and that we have found we can obtain useful results by assuming it. If we abandon this assumption, and particularly if we assume that men act randomly, we cannot explain competition, for there is nothing in the reproductive processes of firms that would ensure that more firms would constantly be created than can survive; and certainly from observations of the real world we can hardly assume that competition is so intense that zero profits will result in the long run or that only the best adapted firms can survive.

Although insisting it is not necessary, Alchian is prepared to assume that firms do strive for positive profits.[24] But I cannot see that even this is sufficient to explain the existence of competition sufficiently intense to enable the economist to assume that only the 'appropriately adapted' firms will survive. Even with this modification firms would still be affected by environmental changes only when these changes cause losses. When changes in the environment opened up new opportunities without acting adversely on the old, then, on the assumptions of this analysis, firms would not respond at all to the new conditions since profits would already be positive and firms are assumed to be uninterested in increasing their profits.[25]

Once motivation is introduced the usefulness of the model becomes even more questionable. Great emphasis is laid on the predictive power of the viability theory. Therefore the essence of the theory cannot be that those firms best adapted to the economic environment will survive; this could easily become a circular argument. Rather it is that the economist can know what the conditions of survival are and therefore can know the characteristics of firms that will be required by these conditions of survival. Now, apart from the pardonable notion that economists have a special knowledge denied to firms[26]—which is quite appropriate if firms (but not economists) are treated as non-motivated organisms—this would seem reasonable provided either that environmental conditions are identifiable and are independent of the actions of the firm or, if they are dependent on the actions of firms, that the economist can know how firms will, by their actions, change the environment.

Once human will and human motivation are recognized as important constituents of the situation, there is no *a priori* justification for assuming that firms, in their struggle for profits, will not attempt as much consciously to adapt the environment to their own purposes as to adapt themselves to the environment. After all, one of the chief characteristics of man that distinguishes him from other creatures is the remarkable range of his ability to alter his environment or to become independent of it. Underlying the viability analysis is the assumption that, even if firms can and do make more or less intelligent choices, they can do nothing in unpredictable ways to 'force' the environment to 'adopt,' and thus make successful, the results of their action.

The concept of the environment of firms on which the economist using 'viability' analysis bases his predictions is by no means clear. There is little doubt that there are parts of the external environment of firms which are identifiable and which for all practical purposes we can safely assume will not be quickly or unpredictably altered by firms—geographical factors, the conditions of transportation, established government policies. There are other aspects of the environment that can be altered within fairly narrow limits and in more or less predictable ways—the amount of natural resources, the state of employment. There are still other aspects of the environment, equally important for survival, which we cannot assume are beyond the influence of firms and which can be unpredictably altered by them in a large number of ways—the state of technology, the tastes of consumers.

It is these unpredictable possibilities of altering the environment by man that create difficulties in comparing the economist to the biologist observing the processes of natural selection and studying the nature of adaptation. Animals, too, alter their environment, but in a rather unconscious fashion without much deliberation about different probable outcomes of their actions. The possibilities open to animals of affecting their environment in a given period of time are so much more restricted than those open to men that the biologist has a very much easier task, for the relative consistency of animal behavior and the relatively narrow limits within which animals can act give him a more secure basis for prediction.[27] If firms can deliberate, if they can weigh the relative profitability of assaulting the environment itself and if they can act in ways unknown to the economist, what are the 'realized requisites of survival' that can give the economist confidence in his predictions? Alchian has treated innovations as analogous to biological mutations. But mutations are 'alterations in the substance of the hereditary constitution' of an organism,[28] while innovations, though they may consist of changes in the constitution of firms, more often than not are direct attempts by firms to alter their environment. In other words, innovations are directly related to the environment of firms whereas the biologists tell us that genetic mutations are apparently completely unrelated either to the environment or to the agent inducing the mutation. The biologist cannot explain why mutations take the course they do while the economist, if he can assume with some justification that the activity of firms is induced by a desire for profits, has a plausible partial explanation of innovation.

It is not possible to go very far with this aspect of the matter because the authors of the viability approach have given us no hint of what they mean by the environment. It is vaguely referred to as an 'adoptive mechanism'[29] but in view of the enormous complexity of the interrelationships in the economy, a prediction of the types of organisms that will survive a given change in the environment involves the prediction of a new general equilibrium and does not seem to me to be an 'intellectually more modest and realistic approach'[30] than any other. After all, even the most ardent proponent of the marginal analysis never claimed that his tools enabled him to make such sweeping predictions as are implied here. By its very nature a prediction of the kinds of firms that will survive in the long run must take account of all the reactions and interactions that a given change in the environment will induce. With our present knowledge this is impossible, and the assertion that 'the economist, using the present analytical tools developed in the analysis of the firm under certainty, can predict the more adoptable or viable types of economic interrelationships that will be induced by environmental change even if individuals themselves are unable to ascertain them'[31] places the wrong interpretation on the kind of thing the economist can do. If he can predict the consequences of environmental changes, it is not because certain types of interrelationships are more 'viable' in a long-run sense, but because he has an idea of how people will behave. He knows little about long-run viability since he knows very little about all of the secondary and tertiary reactions that will in the end determine the 'conditions of survival'—at least he has as yet given little convincing evidence of such knowledge.

Alchian's central objective of exploring the 'precise role and nature of purposive behavior in the presence of uncertainty and incomplete information' is important[32] but the biological framework in which he has cast his model has led him to underestimate the significance of the very thing he claims to be exploring. After all, one of the more powerful effects of uncertainty is to stimulate firms to take steps to reduce it by operating directly on the environmental conditions that cause it and men have a greater power consciously to change their environment than has any other organism. A direct approach, stripped of biological trappings, to the problem of what happens when men try to reach an objective but don't know the 'best' route, would not lead to underemphasis on the significance of purposive activity on the part of men. It is by no means 'straightforward' to assume non-motivation,[33] for without motivation economic competition, leading to the elimination of all but the best adapted within a community, cannot be assumed. Hence, if the operation of natural selection through competition is made the guiding principle of the analytical technique, then an assumption equivalent to profit maximization must be made and the professed *raison d'être* of the viability approach disappears.

## Homeostasis of the Firm

A third biological concept that has appeared in one form or another in economic literature is the concept of homeostasis.[34] Organisms are so constructed that there is a certain 'equilibrium' internal condition which their bodies are organized to maintain. Any disturbance of the equilibrium sets forces in motion that will restore it. Kenneth Boulding considers that 'The simplest theory of the firm is to assume that there is a "homeostasis of the balance sheet"—that there is some desired quantity of all the various items in the balance sheet, and that any disturbance of this structure immediately sets in motion forces which will restore the status quo.'[35]

Once again we find the characteristic of the biological analogy—action taking place in human affairs without the intervention of human decisions based on deliberation and choice. But here it is applied to describe a characteristic of organized activity and a possible method by which men may achieve certain objectives. The notion of homeostasis, treated simply as a principle of organization, does not obscure the importance of purposive behavior in human affairs but rather emphasizes its significance and illuminates its rôle in a complex social framework. This analogy is of the helpful descriptive sort: it is not claimed that the principles of physiology can explain the working of a firm.

Indeed, one could legitimately object to the appropriateness of including the 'homeostasis theory' of the firm among the biological theories. Homeostasis is a word drawn from physiology, but it describes a characteristic of any activity that takes place within a framework so constructed that certain types of action are automatically induced without any interference from whatever agency is responsible for the construction. This notion can be extended from the physio-chemical reaction which take place within a living organism in order to maintain a constant internal environment, to include the operation of a thermostatically controlled heating or air conditioning system[36] and even the conduct of a game of tag according to predetermined rules.

Thus the managers of a firm may lay down rules for the operation of the firm which are determined with reference to some 'ideal' interrelationship of the parts of the firm. The desired ratio of assets to liabilities, of inventories to sales, of liquid assets to fixed assets, etc., may be determined in advance and an organization so constructed that any disturbance of the desired ratio automatically sets in motion a process to restore it. There can be little doubt that the more complex an organization becomes, the more necessary it is to establish areas of quasi-automatic operation. The importance of routine as a means of taking care of some aspects of life in order that others may be given more attention has frequently been stressed.[37] The fact that many business decisions are not 'genuine decisions,' but are quasi-automatic and made routinely in response to accepted signals without a consideration of alternative choices has misled many into attacking the assumption that firms try to make as much money as they can—particularly where it can be shown that the rules governing the routine actions are not fully consistent with profit maximization.

This whole area of the behavior of the firm is still not adequately explained. Imitation of apparent success may, as Alchian suggests, account for some habitual and apparently irrational behavior. The persistence of routine action after the conditions for which it was appropriate have passed may also account for some of it; other partial explanations have been suggested.[38] The theory of homeostasis provides a formal framework of explanation into which many routine responses can be fitted, but it throws no light at all—nor does it claim to—on why and how the 'ideal' relationships between the relevant variables which the firm is now attempting to maintain were originally established or on the conditions under which decisions may be made to alter them.[39] Strictly speaking, the basic principle is not a biological one at all in spite of the name given it. It is a general principle of organization, examples of which may be found in biology, in mechanics and in social organization, and if one chooses to introduce into economics another mysterious word borrowed from another science—well, that is a matter of taste.

\* \* \*

The desire to draw biological concepts into the explanation of social affairs is hard to understand since for the most part they add to rather than subtract from the difficulties of understanding social institutions. The observed regularities and the postulated explanations of non-motivated biological behavior are related to chemical processes, thermal reactions and the like; they are unrelated to conscious deliberation by the organism itself. The appeal of such biological analogies to the social scientist plainly springs from a persistent yearning to discover 'laws' that determine the outcome of human actions, probably because the discovery of such laws would rid the social sciences of the uncertainties and complexities that arise from the apparent 'free will' of man and would endow them with that more reliable power of prediction which for some is the essence of 'science.' It should be noted that the distinction to be made is not that between human and non-human beings but between actions that are in some degree bound up with and determined by a reasoning and choosing process, no matter how rudimentary, and actions that are, as it were, 'built into' the organism, or into the relationship between the organism and its environment, and cannot be altered by conscious decision of the organism itself.

Our knowledge of why men do what they do is very imperfect, but there is considerable evidence that consciously formulated human values do affect men's actions, that many decisions are reached after a conscious consideration of alternatives, and that men have a wide range of genuine choices. The information that we possess about the behavior of firms, small as it is, does furnish us with some plausible explanations of what firms are trying to do and why.[40] Biological explanations reduce, if they do not destroy, the value of this information and put nothing in its place.[41]

To treat the growth of the firm as the unfolding of its genetic nature is downright obscurantism.[42] To treat innovations as chance mutations not only obscures their significance but leaves them essentially unexplained, while to treat them directly as purposive attempts of men to *do* something makes them far more understandable. To draw an analogy between genetic heredity and the purposive imitation of success is to imply that in biology the characteristics acquired by one generation in adapting to its environment will be transmitted to future generations. This is precisely what does *not* happen in biological evolution. Even as a metaphor it is badly chosen although in principle metaphorical illustrations are legitimate and useful.[43] But in seeking the fundamental explanations of economic and social phenomena in human affairs the economist, and the social scientist in general, would be well advised to attack his problems directly and in their own terms rather than indirectly by imposing sweeping biological models upon them.

## Comment by Armen A. Alchian*

Edith Penrose's 'Biological Analogies in the Theory of the Firm' appearing in the December, 1952, issue of this *Review* criticizes an article of mine appearing in the *Journal of Political Economy* on the ground that it rests on the theory of biological evolution.[44] A brief reply may serve to bring out some points of scientific value.

The presently relevant aspects of my original article can be summarized briefly. Economics predicts the observable effects of change in exogenous and endogenous factors impinging on the operation of the economic system. It analyses the economic effects of these various factors upon the optimal conditions of firms and other basic units. From these deduced changes in optimal conditions it predicts that the constellation of firms found in a new environment will have characteristics closer to the new optimal conditions than to the old. It does not (or should not) assert that any or all of the firms in the original circumstances will adjust or modify themselves to achieve the conditions which are optimal for the new conditions. What it does (or should) say is that in the new environment the observed characteristics of the population of firms will be found to have changed toward the new optimal conditions. And this will have happened whatever the wisdom, perspicacity, or motivation of the individual firms. Those who like to think that firms are able to make the required adjustments are free to do so; others, among whom the author is to be counted, can be less restrictive in their axioms and still get similar predicted observable circumstances.

These less restrictive axioms do not assert that businessmen try to maximize profits, since, with uncertainty, no definite meaning can be attached to that prescription of behavior. It is true that there is some situation which, if achieved, would, *ex post*, have yielded a larger profit than any other would have. But this situation is unknowable; hence the lack of prescriptive content. But the economist can, from certain generalized production functions and demand functions, infer the directions of changes in the optimal values of the variables of these functions if these values are now to approach the conditions of the

* Published in *American Economic Review*, Vol. XLIII, No. 4, September, 1953.

new rather than the old optimum. The economist can do this not because he has greater knowledge than the individual firm but because he is analysing changes in the optimum conditions of generalized functions. The businessman requires much more than this; he needs to know his particular values, not merely the directions of changes between two different optima derived from generalized functions.

The significant point is that the new optimum is approached even in the absence of foresighted appropriate adaptive behavior of individual economic units. It can be induced by differential growth, viability, or profit rates in a competitive regime in which (1) realization of profits is a necessary condition for survival, and in which (2) there is a diversity of adjustments manifested in the variety of factor-service input ratios or consumption patterns.

Economic analysis is therefore not merely a theory of the behavior of individuals; it is a theory of the operation of an economic system, and it yields predictions about the effects of certain changes in both endogenous and exogenous factors affecting the economy. To regard it as a theory of individual behavior is fatal.[45]

With this prologue I now turn to Mrs. Penrose who says she is 'not so much concerned to present an analytical critique of the theory as to discuss the applicability of the biological analogy and the implications involved in its use.'[46] This is a bit puzzling. The theory I presented stands independently of the biological analogy. Criticisms of the latter are irrelevant to the theory. Mrs. Penrose seems, at the same time, not to have noted another distinction—that between (1) the foundations and development of a theory and (2) the methods of exposition and presentation of it. In my original article every reference to the biological analogy was merely expository, designed to clarify the ideas in the theory.[47]

Having said this much, I could stop if Mrs. Penrose had criticized only the analogy, for then her criticisms would have been irrelevant. But some of her criticisms are directed at the theory, and they are incorrect.

Some of her criticisms rest on logical errors.[48] Most of her criticisms rest on a misconception of what I wrote. In an extremely revealing footnote, she misconstrued the logic of my position, which she restated as follows: 'Economists can know the conditions of survival. . . . Therefore economists can know what firms must do to make zero or positive profits. Therefore economists can know how maximum profits can be obtained.'[49] Not only do the second and third sentences represent a *non sequitur*, but in addition they exactly reverse my position. Let me explain this by a little analogy(!) A football coach knows that the condition of winning is making more points than his opponent. Does knowing this imply that the coach can know what his team must do in order to win? Does the coach know how this can be done? Defining a desired condition is not the same thing as knowing how to achieve that condition. The confusion between desired conditions and the methods of achieving those conditions is a confusion which I attempted to expose in my original article. Profit maximization purports to be a definition of a situation, not a statement of a method of achieving that condition. That is what Enke meant when he said it was a description, not a prescription.[50] That distinction is fundamental. Ability to prescribe behavior is not necessary—however helpful it is—for the economist to perform as an economist. I started my presentation in the original paper with an extreme model of 'random' behavior in order to emphasize that such special knowledge is not necessary. Subsequently, motivated purposive behavior was introduced—without implying profit maximizing because this could not be defined. It was then stated that even with varied motivations the economist had a method for predicting the types of new situations or firms which would have higher probability of survival and thus tend to become the dominant surviving type. It was denied that the economist could predict which particular firms would survive and what adjustments each particular firm ought to make. Thus all of Mrs. Penrose's criticisms on pages 813–15 miss the point.

Finally, she asserts that 'the biological framework in which he cast his model has led him to underestimate the significance' of the precise rôle and nature of purposive behavior in the presence of uncertainty and incomplete information.[51] Whether I am right or wrong in my implicit estimation of the significance of certain undefined types of purposive behavior cannot be judged by examining the axioms from which theorems are derived. Only by testing its predictive value by empirical investigation can the theory and its implications about the significance of a particular type of purposive behavior be evaluated properly.

Surely some of her criticisms must hit their target: but this target is one of her own creation—the utilization of strictly analogous reasoning in which the concepts, conditions and interpretations of a theory in one discipline are blindly picked up and applied in another discipline. Neither Enke nor I did that. And there is a grave danger in shooting so many arrows toward this straw target. Economics may gain much, as it already has, from the concepts and methods of analysis of other disciplines.

ARMEN A. ALCHIAN

## Comment by Stephen Enke

In the interests of brevity, I shall make only one important point by considering Mrs. Penrose's footnote 26. She wonders 'why it is reasonable (on grounds other than professional pride) to endow the economist' employing 'viability analysis' with the ability to predict the characteristics of surviving firms adopted by the environment, while not crediting the entrepreneur with equal knowledge. Let us consider gas stations, all operating on a 5 cent a gallon retail margin; there is strong price leadership and enforcement by suppliers, so that this margin is not infringed. If this margin were cut to, say 2 cents, the economist, or any other informed person, can predict that in the long run (1) there will be fewer gas stations, (2) gallonages per surviving station will rise, (3) services will be cut, (4) more emphasis will be placed on lubrications, etc., and (5) costs will fall towards 2 cents a gallon. However, no economist, and no one else, can predict which operator on which location with which brand and which employees might, in fact, survive. It is within the wit of man to describe in aggregate terms some of the qualitative characteristics of surviving firms—but not to prescribe quantitatively the measures that will maximize the profits of an individual firm. Professional pride has nothing to do with it.

STEPHEN ENKE

## Rejoinder

In his original article Alchian wrote: 'The suggested approach embodies the principles of biological evolution and natural selection. . . .'[52] If this and similar passages were not intended to imply that these 'principles' were in fact part of the 'foundation,' of this theory, I can only plead a (perhaps pardonable) misunderstanding of them. I am happy to accept his assurance that the biological analogy was used merely as an expository device. The criticism nevertheless remains that the biological analogy places the whole problem in a misleading frame of reference. That I say this indicates that the differences between Alchian and myself, as he pointed out, go much deeper than is revealed by my objections to his analogical reasoning.

My original criticism of the 'viability' analysis was simply that it gives an inadequate and inconsistent account of the significance of human motivation in economic affairs. (This is precisely the reason, incidentally, why the biological model fitted it so well.) In trying to avoid the difficulties inherent in predicting human behavior in an uncertain world, Alchian has given us a model in economics which, it seems to me, (1) explains very little—certainly much less than the traditional economic theory—and relies on a grossly misleading and unjustifiable assumption about competition;[53] (2) is mistaken as to the appropriate use of the model of the profit-maximizing individual firm in economic analysis;[54] and (3) is inconsistent in its treatment of knowledge as between economists and businessmen.[55]

I. *The Misleading Assumption of Competition*

Alchian claims as a chief merit of his theory that it does not rely on 'predictable individual behavior';[56] the results yielded by it 'will have happened whatever the wisdom, perspicacity, or motivation of individual firms.'[57] He rejects the 'restrictive axiom' that businessmen try to maximize profits, and substitutes the even more restrictive 'axiom' that there exists competition so intense that firms must conform to 'optimum' conditions in order to survive. Enke has pointed out that 'If there is no competition, a great many policies—all "good" but only one "best"—will permit an isolated monopoly to survive.'[58] Clearly this is true whenever there is a large amount of monopoly in any competitive system. Hence it is necessary to assume intense competition in order to conclude, as Alchian does, that 'Among all competitors, those whose particular conditions happen to be the most appropriate of those offered to the economic system for testing and adoption will be "selected" as survivors.'[59]

Now I should not object to this aspect of his model if Alchian could show one of the following to be true: (a) that intense competition could reasonably be expected to occur whatever the conscious motivation of individuals; (b) that sufficiently intense competition exists in reality; or (c) that the assumption leads to more useful results than the assumption it displaces. Both (a) and (b) are satisfied by the biological theory of natural selection, but my original criticism was that neither (a) nor (b) is satisfied by Alchian's model patterned after it. Alchian himself barred (c) as a criterion for determining the superiority of his theory over the traditional theory when he pointed out that both his and the traditional approach yield the same results.[60] If I am correct in asserting that none of these three conditions is met by Alchian's model, then the assumption of intense competition becomes a kind of *deus ex machina* introduced to make the model work, having no justification except to serve the purpose of replacing the human motivation so summarily dismissed from its former key position.

Competition cannot reasonably be expected to exist if men are presumed to act randomly. Under such circumstances its occurrence would depend on the constant and rapid creation of new firms, and without motivation it is hard to see why new firms should be created at all. Even if motivation is introduced in the modified form of a 'desire to make profits,' there would still be no reason why firms should appear in sufficient numbers in exactly the required industries or why there should be any response at all to many types of change, for example, to a general price rise due to government spending.[61] This type of response can occur only if firms seize on opportunities to make a little *more* profit—and this comes to the same thing, for all theoretical purposes, as the assumption that maximum profits are desired.[62] The existence of competition cannot be explained by Alchian's model, whereas it is explained in the traditional model with its quasi-empirical assumption that businessmen desire maximum profits.

There would be no need, however, for Alchian to *explain* the existence of intense competition providing that it were found to represent the facts of the real world with reasonable accuracy. But from the evidence at hand one cannot well conclude that a notable characteristic of our economic life is the prevalence of the kind of intense competition required to ensure that firms conforming to the 'optimum conditions' of a competitive model have a higher probability of survival than large diversified firms in protected monopolistic positions.[63]

In other words, although useful results are sometimes obtained by assuming the existence of intense competition, the assumption itself is no more 'realistic' nor 'intellectually more modest'[64] than the assumption that individual firms try to maximize their profits. Both assumptions are highly artificial analytical devices, though very useful when appropriately used. The latter is applicable to a wider variety of of problems and explains considerably more than the former. It takes account, albeit in an unrealistically pure and rigorous form, of the undoubted fact that businessmen are moved by a desire to make money, and it thus explicitly acknowledges human purpose. This is straightforward and has many advantages. The argument in favor of rejecting it on *a priori* grounds has to be powerful indeed.

II. *The Inappropriate Use of 'Profit Maximization'*

The assumption that firms try to maximize their profits is rejected by Alchian, not for empirical reasons arising from a study of how firms actually do behave, but for logical reasons arising from the existence of uncertainty. He holds that in the presence of uncertainty no unique maximum profit position exists; that it is therefore impossible to give any meaning to the proposition that firms try to maximize profits; and that, consequently, a mere desire for maximum profits provides no guide for action. On the other hand, he is prepared to admit 'motivated purposive behavior' and to recognize that firms are in business to make a profit, although apparently he feels that even this modified motivation is going a bit far.[65]

Once it is allowed that firms do try to make profits, it is not difficult to go a bit further and assume that in general they tend to try to make, if they think they can, a bit more profit than they are making. The question, then, is what difference does it make whether any individual firm knows the *best* way of going about its business—or indeed whether there is any 'best' way before the event. This would make a great deal of difference if the economist were attempting to predict the actions of any particular firm. But the economist does not attempt such predictions—nor could he succeed if he tried, as Alchian rightly emphasizes. The economist uses the model of the profit-maximizing individual firm, not to predict the actual conduct of any firm, but merely as an analytical technique to assist him in understanding the effect of change on prices, production, employment, etc.[66] For this purpose it makes no difference whether the conduct of any particular firm can be predicted, whether any firm at all can actually succeed in maximizing profits or even whether uncertainty makes it impossible to say that any particular profit is a maximum.[67]

The existence of uncertainty and the fact that different businessmen do evaluate risk differently is one of the reasons why firms do not all rush at the same speed in the same direction. In most cases differences in attitude toward uncertainty will affect the *rate* of movement in the economy, rather than the *direction* of movement. If firms respond at all to any of the disturbances usually analysed by the economist— taxes, changes in demand or cost, etc.—they will move in the direction predicted by the economist using the traditional theory and *for the reasons assumed in that theory*.

On this level all that the traditional analysis does is to provide the economist with some insight into the type of action appropriate if losses are to be avoided or more profits made when a disturbance occurs. All he need assume is that most businessmen are intelligent enough not to act perversely, that there will always be some firms who prefer to make more money rather than less under otherwise equal conditions and thus will try to make such adjustments as they think appropriate to this end. If the *kind* of response which actually occurs in the economy when a change takes place corresponds to the *kind* of response appropriate in the model of the profit-maximizing individual firm, then the model has shown its worth for all the purposes for which it is normally used.

III. *The Treatment of Knowledge*

Alchian's model is inconsistent in its treatment of what economists and businessmen can know. In a footnote to my original article I presented a 'summary' of the logic of his argument, which read in part: 'Economists can know the conditions of survival. *Therefore economists can know what type of firm can escape negative profits.* Therefore economists can know what firms must do to make zero or positive profits. Therefore economists can know how maximum profits can be obtained.'[68] In quoting this in his reply, Alchian omitted the sentence here italicized, presumably because he did not think it important. He then charged me with a *non sequitur* and with reversing his position.[69] But, although the words of the omitted sentence are mine, *the logical sequence is his*,[70] and it is precisely at this point that his entire argument, which starts out with the proposition that it is impossible to know how maximum profits can be obtained reverses itself.

Alchian's insistence that economists can know 'what types of firms or behavior . . . will be more viable' makes all the difference in the world to the argument. Consider his example of the football coach. If the condition of winning is merely the definition of winning, then of course we get nowhere. But if the football coach knows *what type of team or type of behavior can win*, can one seriously argue that he can have no idea as to how the required type of team or action could be achieved (even though he himself might not be able to achieve it)? If the omniscience of the economist extends to the type of firm or of behavior than can survive, it is mere quibbling to insist that he cannot advise as to how that type may be achieved.

Now my point is simply that economists do not and cannot have that knowledge of the long-run conditions of survival in the real world which is necessary to give Alchian's model the predictive power he claims for it. Given uncertainty and, in particular, purposive activity on the part of firms, the knowledge of both economists and businessmen about the type of behavior or of firm that will survive in the long run in the real world is equally limited and for much the same reasons. I readily agree that economists deal with, and do know, certain things businessmen do not deal with and consequently do not know; but Alchian is concerned, not with what people in different occupations *do* know, but with what people *can* know in an uncertain world. The proposition that uncertainty makes it impossible for businessmen to form reasonably accurate ideas as to how profits might be made, yet on the other hand does not interfere with the economist's ability to know what type of firm or activity can make profits, seems to me to be simply inconsistent. Uncertainty does not make the businessman's knowledge useless to him in his search for profits; and if the economist can predict anything about the real world of firms it is only because he has an idea about how firms are likely to respond to certain kinds of change relevant to their operations.

The weakness of the traditional analysis for predictive purposes in either the short or the long run is associated with the pound of *ceteris paribus*. It is clear, even from Enke's example, that viability analysis must use the same pound to exclude all unpredictable actions of firms in response to environmental changes which themselves change the conditions of survival. This is the chief import of the criticisms on pages 813–15 of my paper which Alchian believes miss the point. I did *not* do Alchian the injustice of implying that he thought that the economist could, or should be able to, predict the fate of any *particular* firm. On this point we are in complete agreement.

Attempts radically to alter the framework of existing theory are always likely to meet resistance, but progress comes from the interaction between innovation and resistance to innovation. I do not want my criticisms to be interpreted as an attack on the purpose of Alchian's original, provocative and able article—to approach some of the problems connected with uncertainty from a new direction. I am inclined to agree—if I may accept for once a biological metaphor of Alchian's—that 'the marriage of the theory of stochastic processes and economics'[71] may be very fruitful; but let it be a marriage by mutual consent with a community property agreement, and not a violent seduction.

Edith T. Penrose

NOTES

1 This paper is a by-product of my work on the theory of the growth of the firm in connection with a project on firm growth directed by G. H. Evans, Jr., and Fritz Machlup, and financed by the Merrill Foundation for the Advancement of Financial Knowledge. I am particularly indebted to Professor Machlup for his careful criticism of the manuscript and for many valuable suggestions, and to Professor Bentley Glass for safeguarding my ventures into biology.

2 In a paper published on the sizes of businesses in the textile industries in parts of England from 1884 to 1911, S. J. Chapman and T. S. Ashton came out rather wholeheartedly in favor of a life cycle interpretation of the development of firms: 'Indeed the growth of a business and the volume and form which it ultimately assumes are apparently determined in somewhat the same fashion as the development of an organism in the animal or vegetable world. As there is a normal size and form for a man, so but less markedly, are there normal sizes and forms for businesses.' 'The Sizes of Businesses, Mainly in the Textile Industries,' *Jour. Royal Stat. Soc.*, Apr. 1914, LXXVII, 512. In this article the analogy between the firm and the biological organism is carried very far, but in a 'belated appendix' Professor Ashton, in his characteristically cautious way, very much qualifies the analogy: 'The picture of the growth of an industry outlined here recalls a well-known passage in which Dr. Marshall compared business undertakings with the trees of the forest; and other biological analogies spring so readily to mind that it may be more useful to point out the differences, rather than the similarities, between the life-history of businesses and that of plants, or animals, or men. Businesses are by no means always small at birth; many are born of complete or almost complete stature. In their growth they obey no one law. A few apparently undergo a steady expansion. . . . With others, increase in size takes place by a sudden leap. . . .' 'The Growth of Textile Businesses in the Oldham District, 1884–1924,' *Jour. Royal Stat. Soc.*, May 1926, LXXXIX, 572. Professor Ashton attributes some of the differences between the development of firms in the earlier (1884–1914) and later (1918–1924) periods to the development of the joint-stock company.

3 '... we must go on further to discuss the problem of what determines the "optimum" or equilibrium balance sheet itself, as this is also to some extent under the control of the firm. This should bring us directly into "life-cycle" theory, and indeed one would have expected Marshall's famous analogy of the trees of the forest again to have led economists to a discussion of the forces which determine the birth, growth, decline, and death of a firm. In fact the theory of the firm, and of the economic organism in general, has not developed along these lines . . . much of the static theory of the firm can be salvaged . . . nevertheless, even when this has been done we still do not have a life-cycle theory. . . .' Kenneth E. Boulding, *A Reconstruction of Economics* (New York: Wiley & Sons, 1950), p. 34.

4 *Ibid.*, p. 38.

5 The idea that a firm's vigor declines with age, which follows naturally from the notion that firms have life cycles, did, however, enable Marshall to maintain the possibility of competitive equilibrium even when firms operated under increasing returns to scale. Growth takes time, and Marshall argued that before a business man got big enough to obtain a monopolistic position, his 'progress is likely to be arrested by the decay, if not of his faculties, yet of his liking for energetic work.' And if conditions in an industry were such that new firms could quickly master the economies of scale, then it would be likely that the established firms would be 'supplanted quickly by still younger firms with yet newer methods.' See Alfred Marshall, *Principles of Economics* (London: Macmillan, 1920, 8th ed.) pp. 286–87; also p. 808, footnote 2. The importance of this decline in a firm's luck or skill for the Marshallian use of the concept of the representative firm is clearly brought out by G. F. Shove in the symposium on 'Increasing Returns and the Representative Firm,' *Econ. Jour.*, Mar. 1930, XL, especially 109.

6 Theoretical models of competition between 'populations' or of the conditions of population equilibria do not require the assumption that individuals *develop* in accordance with life cycle patterns but merely that there exist 'birth' and 'death' rates.

7 These analogies are, as a matter of fact, very old and are found in classical literature. It is not even clear whether their first use was to help in explaining the nature of biological organisms by analogy with social institutions or in explaining the nature of social institutions by analogy with biological organisms. See Oswei Temkin, 'Metaphors of Human Biology,' in *Science and Civilization*, Robert C. Stauffer, editor (Madison: University of Wisconsin Press, 1949).

8 See, for example, a series of papers published under the general title 'Levels of Integration in Biological and Social Systems,' *Biological Symposia*, 1942, VIII, in the introduction to which the editor stated: 'What these papers seem to be saying, in most general terms, is this: The organism and the society are not merely analogues; they are varieties of something more general . . .' (p. 5).

9 See J. H. Woodger, 'The "Concept of the Organism" and the Relation between Embryology and Genetics,' *Quart. Rev. Biol.*, May 1930, V, 6 ff. It should be noted that the concept of 'organism' as used in philosophy, notably by Alfred Whitehead, has no biological connotation.

10 Moreover, biological organisms have a form in a sense in which societies (and firms) do not. This was one of the objections to the use of the economic analogy to explain biological facts: 'The economic metaphors . . . do not account for the biological phenomenon of form.' O. Temkin, *op. cit.*, p. 184.

11 And yet Boulding points out that the chief difference between biological and social organisms is the absence of sexual reproduction and argues that the

'genetic processes in the social system are perhaps somewhat more akin to asexual reproduction...', *op. cit.*, p. 7.

12  See the discussion of homeostasis below. If analogies must be used, there is much to be said for comparing a firm to a machine that operates in accordance with the principles governing its physical organization, but the construction, evolution and uses of which are determined by a mechanic. However, neither type of analogical reasoning has much explanatory value.

13  Not the least of the effects of this kind of reasoning is to bring 'natural law' to the defense of the *status quo*. See the discussion in Richard Hofstadter, *Social Darwinism in American Thought* (Philadelphia: University of Pennsylvania Press, 1944) pp. 30 ff. and the quotation (p. 31) he gives from John D. Rockefeller: 'The growth of a large business is merely a survival of the fittest. . . . The American Beauty rose can be produced in the splendor and fragrance which bring cheer to its beholder only by sacrificing the early buds which grow up around it. This is not an evil tendency in business. It is merely the working out of a law of nature and a law of God.'

14  This subject was widely debated throughout the Western world and the literature is far too extensive to cite. For a useful, though limited, bibliography, see Hofstadter, *op. cit*. The idea of the survival of the fittest, however, was first suggested to Darwin by a work in the social sciences—Malthus on population.

15  Armen A. Alchian, 'Uncertainty, Evolution, and Economic Theory,' *Jour. Pol. Econ.*, June, 1950, LVIII.

16  *Ibid.*, p. 217.

17  *Ibid.*, p. 220.

18  'It is not argued that there is no purposive, foresighted behavior present in reality. In adding this realistic element—adaptation by individuals with foresight and purposive motivation—we are expanding the preceding extreme model.' *Ibid.*, p. 217.

19  *Ibid.*, p. 218.

20  *Ibid.*, p. 219. It should be noted that the treatment of imitation as analogous to genetic heredity is essential to give the principle of natural selection any evolutionary significance. Natural selection has two meanings: 'In a broad sense it covers all cases of differential survival: but from the evolutionary point of view it covers only the differential transmission of inheritable variations.' See Julian Huxley, *Evolution, the Modern Synthesis* (London: Allen & Unwin, 1942), p. 16.

21  Stephen Enke, 'On Maximizing Profits: A Distinction between Chamberlin and Robinson.' *Am. Econ. Rev.*, Sept., 1951, XLI. The assumption of intense competition is essential for the results claimed by the authors of this approach, as we shall see below.

22  *Ibid.*, p. 573.

23  See Julian Huxley, *op. cit.*, p. 14.

24  'The pursuit of profits, and not some hypothetical undefinable perfect situation, is the relevant objective whose fulfilment is rewarded with survival. Unfortunately, even this proximate objective is too high.' *Op. cit.*, p. 218

25  Once we admit that when opportunities for making money arise, some firms will prefer to take a chance on making more money rather than to rest content with less, we might just as well assume that firms act *as if* they were attempting to maximize profits, since, for the purpose of detecting the direction of change, we get more useful results from using this assumption than from any other that has yet been devised. It should be

obvious that for this purpose marginal movements are the significant ones, yet the viability approach leaves us no way of predicting marginal movements except under special conditions and leaves us completely helpless if there is a pronounced lag between the introduction of a given environmental change and the effect of the change on the birth and death rates of firms, for 'these long-run forces of adjustment operate in the main through the effect of altered conditions of survival and the births and deaths of firms.' Enke, *op. cit.*, p. 572.

26  For the life of me I can't see why it is reasonable (on grounds other than professional pride) to endow the economist with this 'unreasonable degree of omniscience and prescience' and not entrepreneurs. Although this is incidental to our discussion, it seems to me that the logic of the argument runs somewhat as follows: It is impossible to know in advance what actions will yield maximum profits. Therefore firms cannot know in advance what they should do to maximize their profits. If there is intense competition, zero profits will be maximum profits and firms making negative profits will fail. Economists can know the conditions of survival. Therefore economists can know what type of firm will escape negative profits. Therefore economists can know what firms must do to make zero or positive profits. Therefore economists can know how maximum profits can be obtained. Therefore it is *not* impossible to know in advance what actions will yield maximum profits.

One can only suggest that firms should hire economists!

27  As a matter of fact, it is doubtful if many biologists would agree that their powers of prediction are as sweeping as are implied here, particularly if man is included among the organisms with respect to whom the effects of environmental change are predicted.

28  Huxley, *op. cit.*, p. 18.

29  'The suggested approach embodies the principles of biological evolution and natural selection by interpreting the economic system as an adoptive mechanism which chooses among exploratory actions generated by the adaptive pursuit of "success" or "profits."' Alchian, *op. cit.*, p. 211.

30  *Ibid.*, p. 221.

31  *Ibid.*, p. 220.

32  *Ibid.*, p. 221.

33  'It is straightforward, if not heuristic, to start with complete uncertainty and non-motivation and then to add elements of foresight and motivation in the process of building an analytical model. The opposite approach, which starts with certainty and unique motivation, must abandon its basic principles as soon as uncertainty and mixed motivations are recognized.' *Ibid.*, p. 221.

34  C. Reinold Noyes, for example, who believes that 'economics is fundamentally a biological science' uses the physiological concept of the homeostasis of the body in his discussion of consumers' wants. See *Economic Man* (New York: Columbia University Press, 1948), pp. 29 ff.

35  *Op. cit.*, p. 27.

36  See Kenneth Boulding, 'Implications for General Economics of More Realistic Theories of the Firm,' *Am. Econ. Rev.*, (Papers and Proceedings), May, 1952, XLII, 37 ff.

37  For a well-balanced discussion of the rôle of habitual behavior and routine decisions in business activity see George Katona, *Psychological Analysis of Economic Behavior* (New York: McGraw-Hill, 1951), especially pp. 229 ff.

38  *Ibid.*, p. 230.

39  The homeostasis principle '. . . says nothing about what determines the equilibrium state itself. In biology this can generally be assumed to be

given by the genetic constitution: in social organisms, the equilibrium position of the organism itself is to a considerable degree under the control of the organism's director,' Boulding, *Reconstruction of Economics*, pp. 33–34.

40  As Jacob Marschak observed, 'It would be a pity if we should not avail ourselves of that type of hypothesis provided by our insight—however imperfect or ambiguous—in the behavior of our fellow-men. This is our only advantage against those who study genes or electrons: they are not themselves genes or electrons.' 'A Discussion on Methods in Economics,' *Jour. Pol. Econ.*, June 1942, XLIX, 445.

41  If one attempts to apply the biological evolutionary principle to human activity, one must first show that human activity does not differ in kind from that of other organisms, and the argument of one noted biologist must be shown to be invalid: 'Man differs from any previous dominant type in that he can consciously formulate values. And the realization of these in relation to the priority determined by whatever scale of values is adopted, must accordingly be added to the criteria of biological progress, once advance has reached the human level.' Huxley, *op. cit.*, p. 575.

42  Boulding, for example, finds 'mysterious' the 'problem of death and decay' of firms and asserts that 'the question as to whether death is inherent in the structure of organization itself, or whether it is an accident is one that must remain unanswered, especially in regard to the social organization.' 'Implications for General Economics of More Realistic Theories of the Firm,' p. 40.

It is surely unwarranted to confine the explanation of the disappearance of firms to some obscure thing, 'inherent in the structure of organization itself' or to 'accident,' and it would not have been so confined if the very nature of the problem had not been prejudged and limited by the biological approach. I am not sure that there is any precise meaning at all in Boulding's statement and I suspect that this is the reason his question 'must remain unanswered.'

43  But one should be discriminating in using them. The varieties of biological phenomena are so numerous that a parallel may be found somewhere for every conceivable type of social situation. There is even apparently a type of symbiotic growth among algae and fungi which combine to form characteristic lichens that can be compared to the growth of a firm by merger. Very curious 'parallels' are sometimes drawn. For example, one biologist finds that 'There is an interesting parallel in the need for salt by the organism and the epiorganism [*i.e.*, society]. The commodity is so important for the social group that it was one of the earliest and most prized objects traded . . .'! R. W. Gerard, *Biological Symposia, op. cit.*, pp.77–78.

44  Edith T. Penrose, 'Biological Analogies in the Theory of the Firm,' *Am. Econ. Rev.*, Dec. 1952, XLII, 804–19.

Armen A. Alchian, 'Uncertainty, Evolution and Economic Theory,' *Jour. Pol. Econ.*, June 1950, LVII, 211–21.

45  For example, see the prolonged exchange of views on profit maximization and marginalism beginning with R. A. Lester, 'Shortcomings of Marginal Analysis for Wage Employment Problems' (Mar. 1946) and F. Machlup, 'Marginal Analysis and Empirical Research' (Sept. 1946) and continuing for three years in this journal. Machlup's defense of profit maximization and marginalism against those who were trying to test axioms rather than theorems would have been airtight if he had (a) defended profit maximization analysis as based on a set of axioms postulating accurate foresight and from which

theorems about the operation of the economic system are derived rather than as a theory of individual entrepreneurial behavior, (b) pointed out the difference between testing axioms and testing theorems, and (c) not defended marginalism or profit maximization as a basis for describing individual behavior in the presence of uncertainty.

46  Penrose, *op. cit.*, p. 811.
47  Readers of an earlier draft, containing no references to the biological similarity, urged that the analogy be included as helpful to an understanding of the basic approach. My conviction that they were right has been strengthened by Mrs. Penrose, for, paradoxically, she has revealed that the analogy is even better than I had suspected.
48  For example, she confuses necessary and sufficient conditions in saying that if we 'abandon the assumption' that 'businessmen . . . strive to make as much [money] as is practicable' and that 'if we assume men act randomly, we cannot explain competition, for there is nothing in the reproductive processes of firms that would ensure that more firms would constantly be created than can survive.' *Op. cit.*, p. 812. Except for her insisting on the analytical use of the biological analogy such inferences on her part would be unjustified. Conditions for competition in the two areas, biology and economics, need not be the same; and in any event desire to make a profit is not profit maximization and furthermore random behavior was not assumed; I repeatedly stated that it was used as a starting point for the complete exposition. See also p. 815 where 'long run' is interpreted as an actually realized situation.
49  Penrose, *op cit.*, p. 813, footnote 26.
50  Stephen Enke, 'On Maximizing Profits: A Distinction Between Chamberlin and Robinson,' *Am. Econ. Rev.*, Sept. 1951, XLI, 566–78.
51  Penrose, *op cit.*, p. 816.
52  Armen A. Alchian, 'Uncertainty, Evolution and Economic Theory,' *Jour. Pol. Econ.*, June 1950, LVII, 211 (hereafter cited as Alchian, *JPE*).
53  Edith T. Penrose, 'Biological Analogies in the Theory of the Firm,' *Am. Econ. Rev.*, Dec. 1952, XLII, 812.
54  *Ibid.*, p. 813.
55  *Ibid.*, pp. 813–15.
56  Alchian, *JPE*, p. 211.
57  Alchian, above, p. 600 (hereafter cited as Alchian, *AER*).
58  Stephen Enke, 'On Maximizing Profits: A Distinction between Chamberlin and Robinson,' *Am. Econ. Rev.*, Sept. 1951, XLI, 571.
59  Alchian, *JPE*, p. 213. It should be noted that 'appropriate' in this quotation can only refer to those 'optimal' conditions that can be known in advance by the economist from his knowledge of 'generalized production and demand functions.' Were it otherwise, Alchian's model would be reduced to the circular and empty proposition that only the most appropriate survive because those that survive are the most appropriate.
60  Alchian states that with his 'less restricted' axioms he can get 'similar predicted observable circumstances.' Enke explicitly notes that predictions are the same on the basis of either theory and the economist can predict 'as if each and every firm knew how to secure maximum long-run profits' (*op. cit.*, p. 567). On the other hand, Alchian insists that 'only by testing its predictive value by empirical investigations can the theory . . . be evaluated properly' (*AER*, p. 602). But if we get the same results with two theories, how will *empirical* research help us to choose the superior of them? The conclusions Enke draws about his gas stations certainly do not require the roundabout approach of viability analysis. I should be surprised, incidentally, if both theories would yield the same result when applied to the real world of

# BIOLOGICAL ANALOGIES IN THE GROWTH OF THE FIRM

monopolistic competition; but this is beside the point since Alchian and Enke insist that they do.

61 *Cf.* Penrose, *op. cit.*, p. 813. It is my contention that an explanation of intense competition cannot be satisfactorily given unless an assumption equivalent to profit-maximizing behavior is made. The demonstration of this, however, falls outside the bounds of this rejoinder.

62 When Alchian states that 'desire to make a profit is not profit maximization' he obviously believes that the profit-maximizing assumption implies that individual firms must be able to ascertain what maximum profits are and actually to achieve them—a contention rejected here. For an excellent discussion of this question see Fritz Machlup, *The Economics of Sellers' Competition* (Baltimore, 1953), pp. 53–56.

63 This especially applies to monopolistic positions resulting from the established reputation of the firm as a whole, and not primarily to special positions regarding particular products.

64 'This approach suggested here is intellectually more modest and realistic . . . ,' Alchian, *JPE*, p. 211.

65 The pursuit of profits . . . is the relevant objective. . . . Unfortunately, even this proximate objective is too high. Neither knowledge of the past nor complete awareness of the current state of the arts gives sufficient foresight to indicate profitable action.' Alchian, *JPE*, p. 218.

66 At no point has Alchian demonstrated that his undeniably valid objection to the use of the profit maximization assumptions for the purpose of predicting individual firm behavior is an equally valid objection to its use in the analysis of the economy as a whole. Indeed, he suggests the contrary when he concludes that 'most conventional tools and concepts are still useful, although in a vastly different analytical framework . . .' (Alchian, *JPE*, pp. 219–20). If the new framework were built on a more acceptable 'axiom' than the old, it would be desirable; if it is not, as I am trying to show, then it seems to me that its creation becomes an unnecessary *tour de force*.

67 For this reason it is misleading to call it the 'theory of the firm' at all—but that is another story. It should also be noted that uncertainty is only one factor destroying the applicability of the theory for predicting the actions of any particular firm.

68 Penrose, *op. cit.*, p. 813.

69 Alchian, *AER*, p. 602.

70 'With a knowledge of the economy's realized requisites for survival and by a comparison of alternative conditions, he [the economist] can state *what types of firms or behavior* relative to other possible types will be more viable' (Alchian. *JPE*, p. 216, italics mine).

71 Alchian, *JPE*, p. 221.

II

## LIMITS TO THE SIZE AND GROWTH OF FIRMS*

*Contains the first presentation of the essential argument of the theory of the growth of the firm which is more fully set forth in a book published in 1959.*

THERE are two approaches to the question of the size of firms. The traditional approach attempts to explain size in terms of the balance of advantages and disadvantages of being a particular size. Another approach emphasizes the process of growth and treats size as a more or less incidental result of a continuous on-going or 'unfolding' process. The first—which I may call the size approach—has found its exponents among those who stress the economies and diseconomies of large-scale production and among those who stress the monopolistic advantages and economic power of bigness. The second—the growth approach—has so far been expounded in any systematic form only by the 'biological economists'—by those who view firms as organisms and conclude that they grow like organisms. That variant of the growth approach leaves no room for human motivation and conscious human decision and I think should be rejected on that ground. (See 'Biological Analogies in the Theory of the Firm,' *American Economic Review*, December, 1952, pages 804–819.)

I want here to suggest an alternative growth approach which, in common with the biological variant, insists that a predisposition to grow is inherent in the very nature of firms, but which, in contrast, makes growth depend on human motivation—in the usual case on the businessman's search for profits.

The analysis presented in this paper is most applicable to business enterprise in the form of corporations. It will be concerned only with industrial firms and only with internal growth. Growth by merger will not be considered. While it is true that merger in the broadest sense has been of great importance in the growth of firms—and of greater real significance, perhaps, than statistics of the percentage growth by merger for different firms would indicate—merger by no means accounts (at least in recent times) for even the major part of the growth of most large firms, as the studies of Weston, Schroeder, and

* I am particularly indebted to Carl Christ, Evsey Domar, and Fritz Machlup for helpful criticism of this paper. Presented at the Annual Meeting of the American Economic Association in December, 1954 and published in the *American Economic Review*, Vol. XLV, No. 2, May, 1955.

others have shown.[1] Furthermore, I think it likely that growth can proceed just as far with or without merger, though not at the same rate. To show this, I want to consider growth without merger.

## II. *Causes of and Limits to the Growth of Firms*

First, I shall be concerned with the causes of and limits to the growth of firms. It is convenient to divide the relevant considerations into two categories: those that are external to the firm and those that arise from the nature of the firm itself. For example, difficulties in obtaining capital or the existence of unfavourable demand conditions have been used to explain why some firms fail to grow or disappear entirely. Conversely, a growing economy or changes in technology increasing the so-called 'optimum size' of firms have often been given as explanations of continuous expansion. Such things do indeed have important effects on the rate of growth of firms. They are external factors which influence the speed and direction of growth but the importance of which cannot be fully understood without an examination of the nature of the firm itself. The more interesting question, to me, is whether there is anything in the nature of the economic institution we call a firm that induces growth, makes it possible, yet limits the rate of growth. This is the problem I want to discuss here: the internal incentives to and limits on growth—a theory of the growth of firms that does not relate to fortuitous external events.

*The Problem of Planning.* To begin with, let me stress an obvious fact, but a fact of central importance for the growth of firms: Successful expansion must, in the usual case, be preceded by planning on the part of the firm. Firms do not just grow automatically, but in response to human decisions. And if firms act on the basis of plans, it follows that they have some degree of confidence in these plans. The question therefore arises how a firm obtains the required degree of confidence. Unless it is assumed that knowledge is perfect and that uncertainty is absent—assumptions that are useless and inappropriate in this context—it is clear that a body of knowledge sufficient to sustain rational plans for action must be developed within the firm. What may broadly be characterized as managerial research will be necessary for the purpose. Consequently, some part of the managerial and entrepreneurial services of the firm must be available to work on the requisite plans whenever expansion is considered.

The fact that expansion must be preceded by research and planning, on the one hand limits the amount of expansion that can be undertaken at any given time and on the other hand permits continuous expansion through time. I propose to show that expansion itself tends to create opportunities for further expansion—opportunities that did not exist before the expansion was undertaken—and this for two reasons. First, the execution of any plan for expansion will tend to cause a firm to acquire resources which cannot be fully used at the level of production contemplated by the plan, and such unused services will remain available to the firm after the expansion is completed. Second, planning itself and the execution of a plan will tend to absorb managerial services which will not be required in the operation of the enlarged concern and which therefore will be released upon completion of the expansion program, while at the same time the services that the firm's management is capable of rendering will tend to increase between the time when the plan is made and the time when its execution is completed.

*The Continuing Availability of Unused Productive Services.* Let us begin with the problem of using fully all services acquired. This raises two separate questions: Why does a firm acquire services it does not need? And why, if it acquires them, does it not use (or sell) them?

The answer to the first question involves a familiar analysis with an extended application. As long ago as 1832, Babbage set forth the principle that the most efficient output for any establishment will be that which uses a 'direct multiple' of the 'number of processes into which it is advantageous to divide' the production of an article. This is essentially what Austin Robinson a hundred years later called the 'balance of processes'[2] and what Sargent Florence calls the 'principle of multiples.'[3] If a collection of indivisible productive resources is to be fully used, the minimum level of output at which the firm must produce must correspond to the least common multiple of the various maximum outputs obtainable from the smallest unit in which each type of resource can be acquired. The principle has usually, if not always, been applied only to machines, and even in this case Robinson has suggested that it may be necessary to plan production on a very large scale in order to use all machines at their most efficient level of operation. When, however, the full range of resources used in any firm is considered, including management, engineering, and even research personnel, as well as the minimum resources needed to sell the product, it is clear that this least common multiple may involve an enormously large output. This output will tend to be greater the larger the variety of resources and the more diverse the units in which they come. A firm would have to produce on a vast scale if it were to use fully the services of all the resources required for much smaller levels of output.

Obviously the output that will utilize most efficiently all of the firm's resources can only be reached if there is no limit short of this output on the supply of any one of them or on the market for the product. If there is any fixed factor in the productive operations of a firm, it will only be by chance that the firm will be able so to organize its resources that all of them will be fully used. The firm will, to be sure, use the services for which its resources were acquired to the fullest extent consistent with the lowest-cost combination of these resources for the amount and kinds of product it has decided to produce, but there is no reason to expect that it would be possible to make them come out even, so to speak. If the limiting factor were removed, additional services from some of the resources already acquired by the firm in the course of its operations would be available at no extra cost to the firm.

In addition, of course, most productive resources, including labor and managerial personnel, are capable of being used in many different ways and for many different purposes. Hence a firm in acquiring resources for particular purposes—to render particular services—also acquires a range of potential productive services, most of which will remain unused. This multiple serviceability of resources often gives firms a flexibility in a changing and uncertain environment which may become of great importance in determining the direction of growth.

*Limits on the Supply of Managerial Services.* There is one type of productive service which, by its very nature, is available to a firm in only limited amounts. This is the service of personnel, in particular of management, with experience within the firm. Even if all other resources, that is, factors of production as available in the market, including the best managerial personnel, were in perfectly elastic supply, executives with experience within any given firm can only be found within that firm. The question, therefore, arises whether what I might call internally experienced managerial services are a necessary input in the process of expansion. It would seem so. Both general reasoning and discussions with businessmen confirm the hypothesis that a firm would never have a high degree of confidence in any extensive plan for expansion drawn up and executed exclusively by men with no experience within the firm itself.

There is little doubt that experience in a given environment does increase the ability of individuals to deal effectively with that environment, to anticipate and provide for circumstances they might otherwise have overlooked, and, in particular, to use other men to better advantage. This does not mean that indefinitely large plans for expansion cannot be made by any group, no matter how small or inexperienced, but only that such plans are apt to be less workable or reliable according as the scope of the plans increases in relation to the number of people with the requisite experience. To the degree that internally experienced personnel are not available, the firm (that is, the existing executive group) is likely to have less confidence in its ability successfully to execute an expansion program. Management consultants, industrial engineering consultants, and similar advisory groups undoubtedly can render very important services. They reduce the extent to which a firm's own managerial resources must be left free for expansion purposes and widen the range of problems the firm can successfully handle. Nevertheless, not only the execution of plans for expansion but also the examination and approval of recommendations (after all, it takes time even to read reports let alone to pass on them) require a great deal of time and effort of the firm's own management.

Thus the making of expansion plans in which a firm has the requisite degree of confidence requires services which can only be produced within the firm. The production of these services requires time, and this limits the scope of a firm's expansion plans at any given time, but permits continuous extension of these plans through time. There are, of course, a number of restrictions on what can be achieved within any given period of time. Operations must often proceed in a certain sequence and many services required in expansion must be internally produced. These restrictions must be taken into consideration when the firm makes its expansion plans including its estimates of the time it will take before they are fully executed. But even to discover the size and composition of an optimum output under given circumstances requires an input of managerial and entrepreneurial services, some of which can be supplied only by individuals who have had experience within the firm. It is this restriction on the horizon of the firm, so to speak, that is significant from the point of view of its growth.

Let us imagine a case, for example, in which a firm has a fixed amount of experienced entrepreneurial services available for planning expansion; that the use of these in conjunction with other services will enable the firm to create a plan in which it has absolute confidence as to the outcome; and that for any more extensive expansion it will be necessary to fall back on the use of less experienced services, with the consequence that the firm will have less confidence in the outcome of actions undertaken in accordance with the plans. This lack of confidence can be looked upon as causing increasing costs—costs that are neither short run or long run in the traditional sense. They are not short run, because plant and equipment are considered variable. They are not long run, because the considerations causing increasing costs are associated only with a temporary, though unavoidable, limit on the availability of services required for planning purposes. In both this and the traditional approach, managerial and entrepreneurial services are a fixed factor but costs are increasing, not because of some indefinable complexity due to size beyond the power of human beings to resolve even when organizational techniques are fully adjusted, but because of the limited supply of experienced personnel at the time the firm plans its expansion. So long as executive resources can be added to a firm's staff, the supply of internally experienced management can be increased in the course of time, but the number of new executives that can be absorbed by a given firm in a given period and placed in situations from which they can gain the requisite experience is also limited by existing managerial resources and by the existing scale of operations of the firm. The rate at which the latter can be increased is precisely the problem that we are exploring.

In view of this limit on a firm's ability to extend indefinitely its plans for future operations, it is almost inconceivable that these plans can always, or even usually, be of such a scope that all of the firm's resources will be fully utilized. In general, there will always be services capable of being used in the same or in different lines of production but which are not so used because the firm could not plan extensively enough to use them. Such services are wasted for the time being. The type of resources for which this waste of service is most significant will depend on the nature of each particular firm. For some firms it may be machinery; for other firms the waste of administrative ability, executive talent, or the specialized connections of the sales force will be more important; for still others the significant waste may be of know-how, research-acquired knowledge, and engineering ability. That these last two are not an unimportant type of available, though unused, service is readily seen in the history of the expansion of individual firms in the modern world. The particular services that remain unused need not be of the same type as those that are actually used in current production; unused services are sometimes those of products and materials, skills and ideas, which are by-products of the primary productive activities of the firm.

Unused services are in themselves a challenge and an incentive to the ambitious entrepreneur. They are the excess services of resources which have been acquired and are needed in current operations; they are available for further production at no extra cost to the firm. An increase in production of the existing products of the firm which uses these services would, other things being equal, enable the firm to produce at a lower average cost. Of course, if a firm overestimates this free element in costs and underestimates the costs of other services that will be required, particularly increases in overhead costs due to additional production, a mistake will be made. The danger of 'creep'—imprudent expansion on the basis of existing unused services—has frequently been expounded by accountants and others. But this does not affect the validity of the argument presented here. (See, for example, Robert L. Dixon, 'Creep.' *Journal of Accountancy*, July, 1953, pages 48-55.) If the services were used in the production of products new to the firm, a competitive advantage would obtain over firms not having this free or joint element in costs. In many cases, of course, the ability of a firm to use such services depends on increases in knowledge and improvements of technique. As a matter of fact, many inventions and technological advances can be directly traced to the desire of a firm to take advantage of already possessed services not used in current production. Sometimes the profitability of devoting available resources to the primary lines of production has been so great that exploitation of minor lines has for long periods been neglected. That all profitable opportunities for productive activity known and open to a firm are not exploited together need not be explained by limited supplies of capital or other resources. As the investigations of Heller have shown, limited managerial services are often stated by management itself to be important restriction on expansion. (See Walter W. Heller, 'The Anatomy of Business Decisions,' *Harvard Business Review*, March, 1951, pages 95–103.)

*The Release and Growth of Managerial Services*. If the preceding argument be accepted and it be agreed that the supply of the services of experienced personnel will, other things being equal, limit the amount of expansion a firm can plan at any time, then it is evident that further expansion becomes possible if, after the expansion plans are executed, such services become available to an extent greater than is necessary to operate the expanded concern.

Any substantial expansion involves not only acquisition of new personnel but promotion and redistribution of the old. Not infrequently a new subdivision of managerial organization is affected and a further decentralization of managerial functions takes place. Such reorganization may be done all at once or gradually as the progressive execution of the expansion plans calls for it. But the point is that both the new men brought in and the existing personnel of the firm gain further experience. Insofar as the increase in the experience of the men concerned is necessary to the execution of the firm's plans and was taken into consideration when the plans were made, the newly available services will be absorbed in the process of creating and operating the expanded concern. But it is impossible that all such to-be-created services be anticipated and their utilization provided for, not only because they include the as yet unknown experience of the planners themselves, but also because the variety of the potential services is unpredictable and their creation extends into the indefinite future.

Whether in carrying out an expansion program all of the firm's personnel are fully used depends partly on the policy of the firm. Firms whose general policy is to devote some part of their executive personnel to the planning of expansion will have this personnel available for new planning as soon as any given plan is completed. At the other extreme a firm's position may be such that those who make the plans also execute them and are therefore unable to devote time to additional planning until current plans are fully in operation. This policy is likely to be adopted of necessity by small and new firms whose executive staff is not very large. Nevertheless, once expansion has been completed and the operation of the firm fully geared to the new level of activity, additional productive services will be both created and released to provide the basis for further expansion.

They will be created because all personnel of the firm will gain additional experience as time passes. They will be released because as the range of services the managerial staff can perform increases, the new services will not be fully absorbed in the existing managerial tasks. Not only is there likely to be a generalized improvement in skill and efficiency but also the development of new and specialized services. When men have become used to working in a particular firm or with a particular group of other men in a firm, they become individually and as a group more valuable to the firm because the range of services they can render is enhanced by their knowledge of their fellow workers, of the methods of the firm, of the best way of doing things in the particular set of circumstances in which they are working. Individuals taking over executive functions new to them—either because the firm has added new functions or because the men are in new jobs—will run into difficulties and make mistakes merely because of the relative unfamiliarity of their work. When these individuals become more familiar with their work and succeed in integrating themselves into the organization under their control, not only will the effort required of them be reduced, but the range of services they are capable of rendering will be increased through experience. Both processes leave a residue of certain types of unused services in the wake of an expansion program.

The unused services thus created do not ordinarily exist in the visible form of idle man-hours but in the concealed form of unused abilities. The more complete use of the services of any given individual is supposed to be made possible by the process of promotion. Just as machinery after a point becomes less valuable and is 'down-graded,' so managerial resources may become more valuable and be 'promoted.' That promotion does not take care of all the increase in services that becomes available is the common experience of many firms in periods when growth is slow. Pressure from younger executives for advancement sometimes even creates for the firm a problem of maintaining the morale of personnel and cannot be entirely explained by the hypothesis that individuals tend to overestimate their own ability.

So far, managerial services have been treated as though they formed a pool of services differentiated only by the experience of the people rendering them. We have begged the question of the type of managerial services required to enable a firm to plan effectively and operate efficiently on an ever increasing scale. If the problems of adapting to rapidly changing external conditions are ignored, the simplest function of management is that of co-ordinating everyday operating activity within a given administrative framework, coping with minor problems, and improving existing ways of doing the same things. Growth requires initiative and imagination and thus a type of service different from that of the routine manager. The simplest type of initiating service is called for when expansion takes the form of increasing production of the products already being produced. Here management need only use on a wider scale the specialized knowledge it has already acquired. The managerial services available may be highly specific to the firm's existing products, but this makes it all the easier to expand as unused services develop, provided that additional goods can be sold at profitable prices. The analysis of this type of expansion falls within the framework of the conventional theory of the firm. The firm will expand the output of its products up to the point at which the increment in total costs equals the increment in total revenue, which, in the absence of diseconomies of scale, will be set by the well-known limitations on demand. This is familiar analysis and need not be elaborated.

*The Utilization of Productive Services and Limits on Demand.* The emphasis on demand as a restriction on the output of a firm arises because in the traditional theory of the firm, the firm is usually defined with reference to a given product. The theory of the firm serves primarily as a foundation for a general theory of price and output and hence is concerned with products rather than with the analysis of the firm as a many-sided economic institution. Once it is recognized that, in principle, the market of a firm in the wider sense is restricted only if the kinds of product it can produce are for some reasons limited, then it becomes clear that the flexibility and versatility of its own resources are the important factors governing the possibilities of its expansion. So long as there are profitable production opportunities open anywhere in the economy, a firm can take advantage of them if its resources are versatile—in particular, if its management is imaginative, flexible, and ambitious.

There are many examples of firms with vigorous and creative management who have substantially altered their range of products, sometimes completely abandoning their original lines and expanding their total output in spite of unfavourable demand conditions for their original products. There are, of course, many examples of other firms which were not able to make the required adjustments. In such cases, failure to grow is often incorrectly attributed to demand conditions rather than to the limited nature of entrepreneurial resources.

There is not time here to examine this problem in more detail. Insofar as a firm is organized to produce only a given range of products, any limitations on the market for those products will prevent the firm from expanding the production of them. This is likely to mean that the firm will be unable to use fully its productive resources. A versatile type of executive service is needed if expansion requires major efforts on the part of the firm to develop new markets or entails branching out into new lines of production. Here the imaginative effort—the sense of timing, the instinctive recognition of what will catch on or how to make it catch on—becomes of overwhelming importance. These services are not likely to be equally available to all firms. For those that have them, however, a wider range of investment opportunities lies open than to firms with a less versatile type of management.

The most effective restriction on the versatility of executive services is that which stems from a lack of interest in experimenting or a lack of confidence in dealing with new and alien lines of activity. I say that this is the most effective restriction because mere specialization of managerial knowledge and ability is not itself a serious bar to a firm's branching out into new lines of activity if existing executives are sufficiently interested and confident to bring into the firm people possessing other relevant knowledge and ability. But it often happens that the horizon of management is extremely limited, particularly in smaller firms. Content with doing a good job in his own field, the small entrepreneur may never even consider the wider possibilities that would lie within his reach if only he raised his head to see them. If occasionally he gets a glimpse of them, he may lack the daring or the ambition to reach for them, although he may be an ambitious, efficient, and successful producer in his chosen field. Specificity of managerial resources means that some of the managerial services most essential for expansion are not available to the firm even though other managerial services essential for efficient operation in a particular field are available.

It is not helpful to dismiss this lack of vision, of confidence, or of experimenting ambition as an example of a failure to attempt to maximize profits, for they are intimately connected with the nature of managerial ability itself and must be reckoned as a limitation on the supply of specific types of productive services. After all, even a preliminary investigation into the possible profitability of any particular type of expansion presupposes a prior decision as to the allocation of managerial services between normal operations and the extension of these operations—a decision which, by its very nature, cannot in the first instance be the result of a careful analysis of anticipated costs and revenues, since before such an analysis is undertaken the decision that it should be undertaken must be made. Whether such a decision is made at all is often a matter of temperament.

## III. Are There Limits to the Size of Firms?

The managerial limits to growth discussed above are rather different from the type of limit on the supply of managerial services that is implicit in the traditional theory of the firm. There it is assumed both that the factors of production are in perfectly elastic supply and that the long-run average costs will at some point start rising—a situation which fundamentally can only be due to the quality of managerial services. The chief difference lies in the fact that imagination and ideas are not confined by time and space while the administrative activities of a single individual must be so confined. There is no known limit to the former. But there is to the latter, and from this it has been deduced that there must be a size of firm beyond which the services required for its efficient operation simply cannot be supplied by mortal men. This deduction is frequently stated as if it were an empirically established fact. The complexity of structure and the scope of activity are believed to become such that even the minimum decisions required of the chief executives in order to ensure the requisite degree of coordination are so difficult or so numerous that the firm must suffer in efficiency. Costs of production would therefore be greater in very large firms than in smaller firms—the so-called 'diseconomies of scale' would overshadow all economies.

The planning of appropriate changes in the administrative organization of a firm is part of the planning of an expansion program. To some extent the kind of services necessary to effect this type of reorganization can be hired. Just as techniques of production appropriate to large-scale operations are known and can be adopted from the outside, as it were, so known techniques of administration can be learned from the outside. Changing organization as a firm grows is very largely a progressive decentralization of authority. By its very nature, decentralization of this type cannot precede to any significant extent the expansion of the firm but must develop more or less simultaneously with it. Some firms may be incapable of effecting the appropriate administrative reorganization—for example, the existing executives may be unwilling to relinquish any of their personal control. These firms will find themselves unable to operate efficiently on a larger scale and will run into trouble as they attempt to expand. While some executives may be insufficiently used, others will be overworked, presenting a truly fixed factor, and diminishing returns will proceed with a vengeance.

We do not know how effective the decentralization of authority can be as a means of keeping costs per unit of output from rising as a firm expands. Reliable empirical evidence does not exist and all studies of the matter are inconclusive, but there is no evidence that a large decentralized concern requires supermen to run it. It does require more men who are competent to make high-level business decisions and who have had experience with the particular kinds of problem with which the firm has to deal. Neither is there significant evidence that the ability to fill the higher administrative positions is excessively rare or that the demands on the men occupying these positions exceed their ability to cope with them effectively.

As to the problem of co-ordination, a strong case can be made for the proposition that as a firm increases in size and its various parts obtain a greater and greater degree of autonomy, the real issue is not whether the whole thing becomes unwieldy but whether it should properly be called a single firm in any economic sense. For many analytical purposes, we must treat some of the parts of the firm as having an independent existence, while for other purposes we must recognize the nature of their interdependence in the administrative structure. As administrative co-ordination becomes more tenuous with increasing size, efficiency becomes less a matter of expert co-ordination and more a matter of the efficient operation of virtually independent parts. It is common knowledge that one department of a single firm may be efficient while other departments are inefficient. Some department heads are good and some are bad. I suppose that if an inadequate manager is placed in charge of a department, this is, by definition, poor co-ordination on the part of some central authority. But it need not be looked on in this way. There are plenty of poor managers in the business world—in responsible positions, too—and the large firms undoubtedly get some of them.

Co-ordination is relevant to efficiency only insofar as co-ordination is required and is attempted. That central control can be dispensed with over wide areas has surely been amply demonstrated by many large firms in the present-day economy, and that the end is not yet in sight is generally admitted. Neither the conception of a fixed factor nor the analysis of the diseconomies of scale is relevant in these circumstances. In the process of growth the 'organism' radically changes its form; the firm becomes less 'organic' and, indeed, less of a 'firm' in the pristine economic sense.

This suggests an interesting paradox. The growth of firms may be consistent with the most efficient use of society's resources; the result of a past growth—the size attained at any time—may have no corresponding advantages. Each successive increment of growth may be profitable to the firm and, if it uses otherwise underutilized resources, advantageous to society. *Ex ante*, the economist interested in resource allocation should approve. But once any increment of expansion is completed, the original justification for the expansion may fade into insignificance as new opportunities for growth develop and are acted upon. In this case, it would not follow that the large firm as a whole was any more efficient than its several parts would be if they were operating (and growing) quite independently. *Ex post*, the economist might disapprove.

## NOTES

1 Fred J. Weston, *The Role of Mergers in the Growth of Large Firms* (University of California Press, 1953); Gertrude G. Schroeder, *The Growth of Major Steel Companies, 1900–1950* (Johns Hopkins Press, 1953).
2 E. A. G. Robinson, *The Structure of Competitive Industry* (Harcourt, Brace, 1932), pp. 31–33.
3 P. Sargent Florence, *The Logic of Industrial Organization* (London: Kegan Paul, 1933), pp. 18–20.

# III

# THE GROWTH OF THE FIRM*

### A CASE STUDY: THE HERCULES POWDER COMPANY

*Growth is governed by a creative and dynamic interaction between a firm's productive resources and its market opportunities. Available resources limit expansion; unused resources (including technological and entrepreneurial) stimulate and largely determine the direction of expansion. While product demand may exert a predominant short-term influence, over the long term any distinction between 'supply' and 'demand' determinants of growth becomes arbitrary.*

THE following analysis of the growth of the Hercules Powder Company was originally intended for inclusion in my *Theory of the Growth of the Firm*, but was omitted in order to keep down the size of the book. The Hercules case was designed to illustrate the argument of that study; the interpretation of important factors in the growth of Hercules is shaped by the case histories of other firms studied. Consequently I shall begin with a brief summary of some of the relevant conclusions presented in my larger work. In doing this I necessarily risk appearing either dogmatic, since oversimplification and absence of supporting argument are unavoidable, or trite, since demonstration of the theoretical and empirical significance of the conclusions is impracticable here.[1]

A firm is both an administrative organization and a pool of productive resources. In planning expansion it considers two groups of resources; its own previously acquired or 'inherited' resources, and those it must obtain from the market in order to carry out its program. All expansion must draw on some services of the firm's existing management and consequently the services available from such management set a fundamental limit to the amount of expansion that can be either planned or executed even if all other resources are obtainable in the market. This is as true for expansion through acquisition as it is for internal expansion, although acquisition permits a faster rate of growth and often facilitates diversification. A firm is not confined to

---

* Published in *The Business History Review*, Vol. XXXIV, No. 1, Spring, 1960. Awarded the Newcomen Prize for the best article to appear in the *Review* during the year 1960. Copyright © 1960. President and Fellows of Harvard College.

'given' products, but the kind of activity it moves into is usually related in some way to its existing resources, for there is a close relationship between the various kinds of resources with which a firm works and the development of the ideas, experience, and knowledge of its managers and entrepreneurs. Furthermore, changing experience and knowledge of management affect not only the productive services available from resources, but also the 'demand' which the firm considers relevant for its activities.

At all times there exist within every firm, pools of unused productive services and these, together with the changing knowledge of management, create a productive opportunity which is unique for each firm. Unused productive services are, for the enterprising firm, at the same time a challenge to innovate, an incentive to expand, and a source of competitive advantage. It is largely because such unused services are related to existing resources and partly because of the pressures of competition that firms tend to specialize in broad technological or marketing areas, which I have called technological or market 'bases.' In a sense, the final products being produced by a firm at any given time merely represent one of several ways in which the firm could be using its resources, an incident in the development of its basic potentialities. Over the years the products change: there are numerous firms today that produce few or none of the products on which their early reputation and success were based. Their basic strength has been developed above or below the end-product level as it were—in technology of specialized kinds and in market positions. Within the limits set by the rate at which the administrative structure of the firm can be adapted and adjusted to larger and larger scales of operation, there is nothing inherent in the nature of the firm or of its economic function to prevent the indefinite expansion of its activities as time passes.

Entrepreneurial services are as much productive services as are the services of management, labor, or even machines. Entrepreneurial incompetence, or general cautiousness, including a conservative attitude toward financing, should be looked on not as a failure to 'maximize' profits, whatever that may mean, but as a limitation on the supply of productive services to the firm.

In the explanation of the course of expansion of a particular firm and of the limits on its rate of expansion, it is illuminating to put the chief emphasis on the firm's 'inherited' resources and productive services, including its accumulated experience and knowledge, for a firm's productive opportunity is shaped and limited by its ability to use what it already has. Not only is the actual expansion of a firm related to its resources, experience, and knowledge, but also, and most important, the kinds of opportunity it investigates when it considers expansion. Moreover, once a firm has made its choice and has embarked on an expansion program, its expectations may not be confirmed by events. The reactions of the firm to disappointment—the alteration it makes in its plans and activities and the way in which it adapts (or fails to adapt)—are again to be explained with reference to its resources.

These relationships are portrayed in the chronology of the changing productive opportunity of the Hercules Powder Company. The history of this company illustrates the nature and significance of the areas of specialization of a firm—its technological and market bases—as well as some of the difficulties encountered when an attempt is made to move to new bases markedly different from the old. The outlines of the company's diversification are presented in Chart I. The following story elaborates, explains, and discusses the significance of the movements implied therein.

In 1912 a large United States firm, E. I. Dupont de Nemours, then looked upon as dangerously close to monopoly in the explosives business, was broken into three parts by action of the federal courts as a result of an antitrust suit initiated by the federal government in 1907. One of the two 'new' firms thus created was the Hercules Powder Company. At the time of its formal organization in 1913 Hercules had a thousand employees and nine plants; it produced explosives only: black powder and dynamites.

During the next forty-odd years this amputated piece of DuPont, like a cutting from a plant, continued to grow.[2] It, like DuPont, has over the years branched out in numerous directions in response to external opportunities and internal developments. The parent and its involuntary offspring have not grown in the same directions, and in only a few fields are they in direct competition with each other. Hercules is not only completely independent of DuPont, but has acquired its own personality and its own position in the industrial world quite unrelated to DuPont's position. By 1956 it had 11,365 employees, 22 domestic plants, and total assests of nearly $170 million, making it the 165th largest industrial company in the United States measured by total assets.[3]

The company's rate of growth has been modest (something over 5% per year in terms of fixed assets) but fairly steady. Its financing has been conservative, virtually all of its growth having been financed with internally generated funds. It has engaged in little acquisition, only eight small companies with total assets at the time of acquisition of less than 10% of the company's present net worth having been acquired in its entire lifetime. Its 'entrepreneurship' has been what I have called 'product-minded,' reasonably venturesome and imaginative, but concentrating on 'workmanship' and product development rather than on expansion for its own sake or for quick profits.

#### DEVELOPMENT OF THE TECHNOLOGICAL AREAS OF SPECIALIZATION

The original technological base of the Hercules firm was explosives and for the first few years of its existence it was kept busy with the expansion of this field. Two new plants were acquired and improvements were made in existing plants and in the processes of production of dynamite, smokeless powder, and cordite. One of the innovations—the production of acetone (a solvent used in the manufacture of cordite) and other products from the giant kelp found on the Pacific Coast—involved an extension of the firm's knowledge and experience in a type of organic chemistry which was to become significant in its subsequent diversification.

The manufacture of explosives is still of considerable importance for Hercules (accounting for 18% of sales in 1951) and at times has been its most profitable operation, providing funds for the extension of activities in other directions. Substantial innovations have been made in the field of semigelatin explosives, smokeless powder, packaging of explosives, and explosive supplies. Some diversification into the production of chemicals used in explosives production, notably nitric acid, anhydrous ammonia, and other nitrogenous compounds, has been made, and this development has contributed in recent years to Hercules' position in the agricultural chemicals industry.

In spite of the innovations and enlarged activity, however, the explosives business was not one to permit extensive growth and development of the firm. In particular, it provided little opportunity for the use of the experience in the field of organic chemistry that had been developed by Hercules men in the course of the firm's operations. Furthermore, at the end of the First World War the plant, organization, and accumulated funds of the firm were much greater than could be used in explosives in view of the drastic decline in demand after the war. In the immediate postwar period numerous opportunities for profitable investment were open on all sides in the expanding, changing economy. But which of them would furnish opportunities for the growth of a still relatively small and specialized explosives company?

### Nitrocellulose and New Areas of Specialization

Nitrocellulose is one of the most important basic raw materials in the production of explosives. In 1915 Hercules had bought the Union Powder Company, which had a plant for the nitration of cotton linters (the 'fuzz' on cotton seeds and a by-product of cotton production) into nitrocellulose, then used primarily for smokeless powder,

## CHART I

### DIRECTION OF EXPANSION

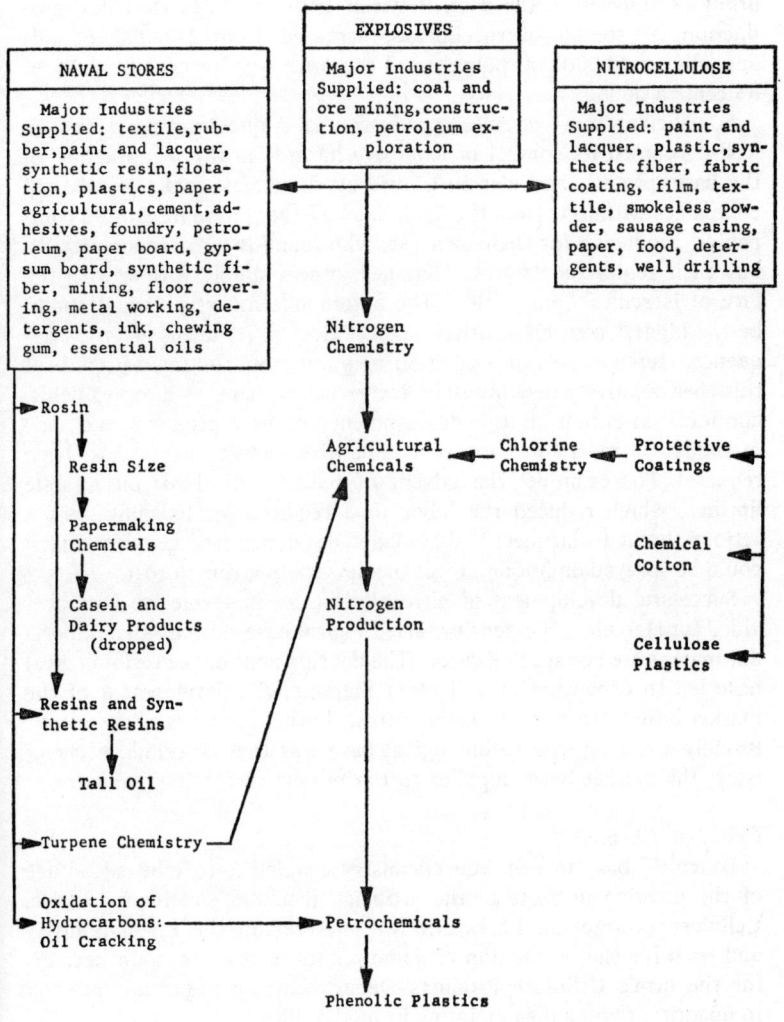

but also for celluloid and collodion ('new skin'). Already by 1917 the company was experimenting with the production of nitrocellulose for industries other than explosives, for if it could produce a suitable soluble nitrocellulose it felt sure of a large market supplying the needs of the lacquer, film, and protective coatings industries generally. It succeeded in developing an appropriate product, and by 1923 was firmly established in the field. Between 1918 and 1944 Hercules' production of soluble nitrocellulose increased from 100,000 pounds annually to 28,000,000 pounds and the price was lowered from 75 to 33 cents a pound.

So efficient was Hercules' production and quality control and so well-developed its control of explosive hazards in the manufacture of the basic product and also in its use by customers, that a number of companies withdrew from the field. Some of these were integrated companies, producing for their own use, who found it more economical to buy their requirements from Hercules; others simply withdrew in the face of Hercules' competition. The automobile industry turned out to be the biggest consumer, using nitrocellulose in its lacquers. In consequence, Hercules was in a position to profit from the rapid growth of this then relatively new industry. Nevertheless, here, as in other fields, continual attention to the developments of new products and new methods to meet or surpass competitive developments has been required. For example, the advent of baked enamel for automobile finishes, which reduced the labor time required for finishing, was a serious threat to lacquer; and Hercules developed new lacquers which could be sprayed on hot and meet the new competition in cost.

Successful development of nitrocellulose for nonexplosive uses provided for Hercules an extensive technological base as well as an important market area of specialization. The development of the technological base led to expansion in still other markets; the development of the market base furthered expansion into still other branches of chemistry. Broadly speaking, the technological base was that of cellulose chemistry; the market base, supplier to the protective coatings industry.

*Cellulose Chemistry*

Hercules' base in cellulose chemistry enabled it to take advantage of the growing markets in the artificial fiber and plastics industries. Cellulose acetate, an important raw material in the rayon industry and used for the production of some grades of plastics, soon became, for the firm's Cellulose Products Department, an important product in quantity, though disappointing in profitability.

The cellulose acetate market is highly competitive, and, in this as well as in most of its other products, one of the firm's biggest competitive problems arises from the ever-present possibility that its customers will integrate vertically and start producing their own requirements. In the long run Hercules can prevent this only by producing a high-quality product and selling it at a price that makes integration unprofitable for customers. Hence a relatively low sales margin is earned and continual research and experimentation are carried on. (Hercules has even experimented with the spinning of fibers in order to acquire knowledge which might be of use to its customers. As we shall see, 'technical service' is one of the 'utilities' Hercules sells with all of its products in order to maintain its market position.)

With the development of synthetic rubber during the Second World War came a new petrochemical base for cheap plastics (polystyrene) which soon began to displace cellulose acetate in molding powders, the basic material from which molded plastics can be made. Petrochemicals, however, involved a branch of chemistry in which Hercules had only limited experience at the time. Many of the companies producing the new plastic material had developed extensive experience during the war which gave them a new 'base' in petrochemicals. Hercules' wartime activities were in very different areas. The firm's lack of an adequate technological base was sufficient to prevent it from taking up the production of polystyrene and similar petrochemical products. Consequently the company attempted to reach new markets with its own cellulose acetate by taking up the production of molding powders.

The extensive knowledge of cellulose chemistry possessed by Hercules has provided a continuous inducement to the firm to search for new ways of using it. For example, during the war Hercules, in an attempt to replace a lubricant no longer available, took up the production of an extraordinary versatile cellulose gum—sodium carboxymethy-cellulose (CMC). The firm was much impressed with the properties of this chemical composition, but was not sure to what use American industry could put it. Perhaps CMC could be used in the sizing of textiles (Hercules already produced some types of fabric coating). No one knew; nevertheless, advertisements were placed in trade papers describing the qualities of the product and inquiring 'What do you see in CMC?'

The product caught on. Here, surely, is an almost perfect example of the creation of consumer demand as a consequence of entrepreneurial desire to find a use for available productive resources. The biggest uses for CMC, initially, turned out to be as a stabilizer in foods, ice cream, lotions, drugs, and cosmetics. CMC also proved to have an industrial application in oil-well drilling mud—an outlet the firm had not anticipated. It is now also used in textile sizes, finishes, and printing pastes; in ointment bases, in thickening rubber latex; in can-sealing compounds and grease-proof paper coatings; in tooth paste; in emulsion paints and lacquers; in leather pasting; in ceramic glazes; and as a binder for crayons and lead pencils. Innovations in use and in the product continue. In 1955 a new type of CMC was introduced which was expected still further to expand the market and the variety of uses.

There are other cellulose products and specialties that have been developed by Hercules which we shall not take the time to discuss here, but our account cannot leave out the firm's early diversification into the production of its own requirements for chemical cotton, the raw material for cellulose products of all kinds. In 1926 Hercules purchased the Virginia Cellulose Company at Hopewell, Virginia, in order to produce one of its own basic raw materials and, in particular, to control its quality. The purchasing of cotton linters—of which the second cut is used for chemical cotton—is a highly specialized business. Hercules buys around 40% of that part of the nation's production of cotton linters destined for chemical uses, and supplies not only all of its own requirements but sells outside as well, for production must be on a large scale to be efficient. When Hercules went into production of chemical cotton its chief use was in paper making and in nitrocellulose products. The expanding rayon industry provided a new and growing outlet, and later another important use was found in the manufacture of high tenacity viscose rayon for tire cord.

*Developing a New Base—Naval Stores*

Broadly speaking, as can be seen in Chart I, the operations of the Hercules Powder Company, apart from explosives, can be divided into two large chemical branches, with a third becoming clearly evident in recent years. They all overlap in the markets they serve, and each leads in its later stages into new areas of chemistry which may well provide new technological bases for further diversification. The movement into new aspects of cellulose chemistry, just described, was an obvious entrepreneurial response to the postwar decline of nitrocellulose markets in the explosives field. The subsequent branching out of the company was the logical (though not inevitable) effect of its continually increasing knowledge of cellulose chemistry as well as of its developing position in its various market areas. Later we shall discuss the interaction between the technological and market bases of the firm; for the present we are concerned primarily with the technological aspects of its diversification, although clearly technological developments are of use only if profitable markets can be found.

Important as the opportunities were in the field of cellulose chemistry, however, they did not appear to the firm to promise sufficient scope for the entrepreneurial, managerial, labor, and technical services available to it at the end of the First World War. In 1919 the company had created an industrial research department for the express purpose of investigating products Hercules could profitably produce. This department decided that the firm could go into the production of wood naval stores (rosin, turpentine, and pine oil) obtained from the stumps of the long-leaf southern pine—like linters, a waste product of another industry.

Naval stores production was not as obvious an opportunity for Hercules as was cellulose chemistry, but, again it was expected to provide openings for the use of the existing resources of the firm. Hercules believed that it could use its knowledge of organic chemistry to produce a purified wood rosin good enough to compete with the gum rosin when gum prices were high; incidentally, the naval stores operation would also provide a use for dynamite in the blasting out of the stumps. Extensive lumbering operations during the First World War had resulted in a large-scale cutting of suitable trees, and it was widely believed that the consequent reduction in the supply of gum rosin would lead to an extended period of high prices which would make profitable the production of wood rosin.

Consequently, in 1920 Hercules built a plant in Mississippi for the steam distillation of rosin from pine stumps; it also bought another company owning a deteriorated plant but with a large supply of stumps. The mining of the pine stumps produced three joint products: rosin, turpentine, and pine oil, the main product for Hercules being rosin. Hercules did succeed in developing a purified wood rosin suitable as a substitute for gum in many uses, and it improved the productivity of the old plant it had bought, increasing rosin output by one third in three years and liquids production by 100%. But the firm's original expectations of demand for wood naval stores turned out to have been too optimistic. Although wartime lumbering did sharply deplete the supply of standing pine available for the production of gum naval stores, the second growth of trees came in, output rose, and gum prices fell drastically. Wood rosin could not compete with gum when gum prices were low, and the naval stores business of Hercules went into the red for many years. Notwithstanding its heavy investment in rosin chemistry research, Hercules came close to withdrawing from the business.

*Rosin and Terpene Chemistry*

But research paid off; unable to sell rosin in its existing forms in competition with gum, Hercules learned how to modify the product by hydrogenation, disproportionation, and polymerization and thus to convert it into various kinds of rosins for which many new uses could be found. Rosin is essentially abietic acid; when esterified with various polyols it makes hard resins valuable in the manufacture of paints and varnishes, and Hercules already had a position in the protective coatings field. Customers could be found for hard resins and rosin esters, and these, together with a variety of specialty resins as well as straight esters, were developed into an important outlet for rosin production. The naval stores operation became the equal of cellulose chemistry as a central technological base of the firm and in 1928 was organized as a separate department. In 1936 still another department was created, charged with the task of developing new uses and new outlets for resin-based products. As we shall see, the knowledge generated in this department, together with its market opportunities, soon led it outside the field of rosin chemistry and into new areas.

Further description of the range of products produced by Hercules and based in rosin chemistry would involve us in too much detail (and, incidentally, in too much chemistry), although we shall return to some of the more interesting developments in our discussion of the interaction between the technological and market bases of the firm. It is fairly easy for small firms and 'in-and-out' producers to take up the production of resins; profit margins are consequently low, and the profitability of the industry for Hercules depends on large volume. To sustain its position, the firm has to rely on technological knowledge, low production costs, service facilities to customers, and continual improvements in production, in quality, and in variety of its products. Hardly a year passes without the introduction of several new products or improved varieties of old products, developed under the stimulus of actual or potential competition, the pressures of technical men with ideas to put across, and the hope of profit from innovations in which the firm has special advantages because of its accumulated experience.

But rosin is only one of the three joint products of the pine stump, and markets had to be found both for turpentine and for pine oil, a relatively new industrial product in the 1920's, derived only from the wood operation. Fractional distillation methods were perfected which permitted the production of higher grades of turpentine and pine oil. In 1929 pine oil outlets were not developing fast enough to keep pace with production, and Hercules intensified its research into pine oil chemistry looking for derivative products. Thanite was developed, a terpene thiocyanoacetate providing a toxicant for insecticides, and later toxaphene, a chlorinated camphene. These products put Hercules firmly in the field of agricultural insecticides which in turn stimulated research into agricultural chemicals generally. At times the demand for pinene has exceeded the company's output and it has had to buy crude products from pulp mills for refining. As was the case in cellulose chemistry, a large variety of chemical products and processes has been developed in the field of terpene chemistry. One of

THE GROWTH OF THE FIRM. A CASE STUDY

the latest processes bids fair to give Hercules a more established base in petrochemicals, a field which, as was noted above, had up to recently been outside Hercules' major fields of specialization, thereby handicapping the firm in its ability to meet competitive developments.

*Petrochemicals—A New Base*

In several of its manufacturing processes Hercules has always been involved in petrochemical operations. Although the manufacture of explosives is not in itself a chemical process, the production of the essential ammonia is. Nitric acid used for making explosives is obtained from ammonia, and the process used by Hercules to produce ammonia involves the cracking of natural gas. Furthermore, some of the processes in rosin and terpene operations of its naval stores activities are similar in nature to the cracking of oil. Indeed, some of them can be and are used in oil cracking. Finally, in experimenting with the chemistry of terpenes and with the oxidation of the hydrocarbon by-products of naval stores, Hercules developed a reaction that utilized benzol and propolene and that resulted in a new process for making phenol. These developments opened up two new branches of chemistry for the firm: air oxidation processes and petrochemicals; new plants have been built for operation in both areas. The phenol plant, established near an oil refinery, uses a by-product of the refinery. Among the important uses of phenol is the manufacture of synthetic resins for phenolic plastics; it is also used in the manufacture of varnishes, enamels, herbicides, and pharmaceuticals. Lack of any raw material 'base' in petrochemicals had prevented Hercules from participating fully in the rising markets for rubber-base paints and for phenolic plastics. One of the primary hopes of management in establishing the phenol plant was to open the way for the acquisition of further knowledge in order to provide a base for expansion in this wide field of chemistry, as well as to put the company in a position to keep up with competitive developments arising in petrochemistry and affecting the market for some of its major products.

Finally, Hercules in 1955 took up the production of polyethylene for plastics. The technology was based on the work of German scientists who had discovered in experimenting with new types of catalysts that ethylene could be polymerized at low pressures to give a new type of high molecular weight polyethylene. This not only further extends Hercules' activities in plastics, but also takes it further into catalytic chemistry, which may, in time, lead into still further technological areas.

#### INTERACTION BETWEEN TECHNOLOGICAL AND MARKET BASES

Hercules is a producer of chemical products for other industries; it does not manufacture final products for the nonindustrial consumer.

To obtain knowledge of the 'demand' for its products, one of its principal tasks it to watch industrial developments in all relevant sectors of the economy in order to discover where its products might be made to supply the requirements of industrial consumers as well as or better than existing products. It is a conscious policy of the firm systematically to review its resources with an eye on external developments, asking the question, 'What have we got to offer?'

Because of the nature of its market, Hercules stresses 'technical service' to customers; salesmen are for the most part technically trained men. In selling their products the salesmen are expected to take an active interest in the production and market problems of their customers. This permits them to acquire an intimate knowledge of the customers' businesses and not only to demonstrate the uses of their own products and to suggest to customers new ways of doing things, but also to adapt their products to customers' requirements and learn what kinds of new products can be used. It is standard practice in the development of new products to get customers to try them out on a 'pilot plant' basis and thus to assist Hercules in the necessary research and experimentation.

Obviously, it is in those areas where Hercules' personnel have the greatest experience and the most extensive relationship with customers that opportunities for the sale of existing products and for the promotion of new products will be widest. Hence, in spite of the enormous variety of possible end uses of Hercules' chemical products, the firm nevertheless remains in a relatively few broad 'areas of specialization.' Approximately 40% of the total value of sales are accounted for by three industry groups: protective coatings, paper, and mining and quarrying, and an additional 40% by six others: synthetic fibers, plastics, agricultural chemicals, petroleum, rubber, and identifiable military uses (the last including fees obtained from the operation of government owned ordnance plants).

The interaction between the market opportunities of the firm and the productive services available from its own resources can be seen in the development of almost any field we examine. A few examples will illustrate.

*Paper-making Chemicals*

The biggest customer of rosin is the paper-making industry which uses rosin largely in the form of rosin size, a sodium soap of rosin. As a result of the close association with the paper industry consequent upon its entry into naval stores production, Hercules in 1931 acquired the Paper Makers Chemical Corporation, a diversified, loosely organized company producing a variety of industrial chemicals. On acquiring the corporation, Hercules reorganized its productive activities, consolidating production in the more efficient plants and getting rid of others; it eliminated alum production and the jobbing activities of the old company. Eventually a separate department, called the Paper Makers Chemical Department, was created to take over the remaining collection of activities.

Although the basic reason for the acquisition of the old PMC was the outlet it provided for rosin and the possibilities for growth that Hercules saw in the rosin-size business, the activities of the new department in Hercules rapidly extended not only to many other chemicals useful in the paper-making industry but also to other industries using the same or similar chemicals. Thus, with the advent of synthetic rubber production, Hercules looked into the possibilities of using rosin soap as an emulsifier in the production of synthetic rubber, and now sells a very large proportion of its rosin soap to the synthetic rubber industry.

This in turn stimulated interest in the general field of synthetic rubber production, now one of the more important areas of Hercules' research. Hercules' interest in the paper industry, arising from rosin sizes, has in recent years been substantially reinforced by the growing uses of chemical cotton in paper making. Much research has gone into the characteristics imparted to paper when chemical cotton is substituted for other raw materials. As a result, Hercules has been able to establish its raw material for many uses in paper making.

Among the activities of the old Paper Makers Chemical Corporation when it was acquired by Hercules was the production and sale of casein, a milk product used in the paper industry. Hercules retained this business for some twenty years and attempted to develop the field. For a while the operation was profitable, but owing to rising support prices of dairy products, imported casein became so much cheaper than the domestic product that it was no longer profitable to produce it. On the other hand, since the firm had an organization and a sales staff that it wanted to use, attempts were made to develop a chemical to displace casein in paper manufacturing. These attempts continue, but the casein operation itself was finally discontinued in 1953, after many years of unsatisfactory performance.

*Protective Coatings*

Protective coatings is a broad term including paints, lacquers, and other forms of providing a 'coating' to protect wood, metal, cement, textiles, and other materials. Hercules' market position in this field goes back to its early production of soluble nitrocellulose for the lacquer industry; it was subsequently extended as the firm developed rosin products, also valuable in the paint and lacquer industry. The interest in the general market area of protective coatings imparted by these important uses of its basic raw materials led to developments within the firm which took it into the production of other products from other raw materials, but products that served the same types of customers and involved similar types of technological processes.

One of the early successful innovations in the field was the development of Parlon, a chlorinated rubber, valuable as an ingredient in paints for chemical plants and in other places where resistance to alkalies and acids is important. This product is produced in the large cellulose products plant of Hercules but is not related to cellulose through either raw materials or production processes. It was introduced to broaden the firm's base in the market for protective coatings. During the Second World War, rubber was in short supply and the firm, in order to use its plants, produced Clorafin, a chlorinated paraffin used as a plasticizer in synthetic rosins and as an ingredient in compounds for imparting flame, water, and mildew resistance to textile materials. After the war, the production of this product was continued and the production of Parlon resumed.

Development of the general field of protective coatings and of plasticizers also led the Synthetics Department beyond its original specialty of finding outlets for rosin in various forms, into research with chemical materials, unrelated to rosin, for the manufacture of new ingredients for protective coatings, new types of plasticizers, polyols used in rosins, and raw materials for synthetic fibers. By 1951, substantially more than 50% of the sales of this department were of nonrosin-based products.

*Agricultural Chemicals*

All three of the major technological fields of Hercules have combined to give it an interest in the field of agricultural chemicals. The fact that nitrogen chemistry, in particular ammonia, is important in the manufacture of explosives and also one of the major bases of commercial fertilizers early gave Hercules a connection with agriculture. With the progressive development of chlorine and terpene chemistry and the introduction of the new insecticides, Thanite and toxaphene, mentioned above, this interest broadened. Although the original stimulus to the entry of Hercules into agricultural chemicals stemmed directly from the types of resources it possessed, once the firm had entered the field in a major way and created a technical and sales force to serve this market, the market possibilities became the primary stimulus. Extensive research activities were undertaken to develop further the firm's position in the field. A new laboratory for research into agricultural chemicals was opened in 1952, and in 1954 Hercules, together with the Alabama By-Products Corporation, set up the Ketona Chemical Corporation to produce anhydrous ammonia using by-product coke-oven gas as a raw material, the first ammonia plant to use this process in the United States. The plant produces for both agricultural and industrial nitrogen users in southeast United States.

*Plastics*

Celluloid, which is virtually nothing but nitrocellulose and camphor, was the forerunner of modern plastic materials (and, incidentally, is still important in many uses). This product was produced by Hercules from the very beginning; the development of cellulose acetate further committed the firm to the plastics industry. The various kinds of chemical plastics, which in a broad sense can often be regarded as the same 'product,' are made by substantially different chemical processes. Hence, the widening of Hercules' position in the plastics field stems from different types of chemical technology, the development of which has itself been stimulated by the firm's attempt to maintain and improve its position as a supplier to manufacturers of plastic products. Thus much research effort has been directed specifically toward the development of plastics, not only based on the firm's primary raw materials and on chemical processes used in its several other operations, but also going far afield into new processes and new raw materials. Hercules' research is broad and many different areas of activity are being explored, but which of the possible products are finally selected for 'basic' expansion depends on the firm's estimate not only of the new markets they may create for the firm but also of how they fit in and can be developed along with existing resources and market areas. Many of the technological developments discussed above, such as the development of phenolic chemistry, were to a large extent stimulated by a desire to take full advantage of the growing opportunities in plastics.

*Oil Additives*

The story of Abalyn, tall oil, and Metalyn, is a minor one in the history of Hercules but is interesting from our point of view as an illustration not only of an interaction between technological and market bases, but also of the way in which new raw material sources can be developed in order to maintain an existing market position.

We have noted that rosin is one of the primary raw material bases of the firm. Under the pressure of the 1947 recession, Hercules was eagerly looking for new outlets for rosin. As we have seen, one of the primary measures adopted by the firm lay in the conversion of rosin to other products for which markets could be found. Among these products was Abalyn, a methyl ester of rosin, useful as an oil additive in high pressure greases because of its ability to hold grit and other foreign matter in suspension. The important competitive substitutes were lard and sperm oil, which were expensive compared to rosin when Abalyn was introduced. These, however, fell in price and Abalyn became relatively expensive. To keep its markets, Hercules decided to buy tall oil, a by-product of paper mills, from which rosin could be obtained more cheaply than from the naval stores operations. This by-product, esterified, yielded a substance which used the same equipment as Abalyn for its production, did the same or better job as an oil additive, and was substantially cheaper. The product, called Metalyn, became an important product of the Synthetics Department, but at the same time lost for the firm an outlet for its own rosin. On the other hand, tall oil became a new and significant raw material for Hercules, and in 1954 the firm announced plans to build a plant for the processing of crude tall oil and the manufacture of rosin and fatty acids from it, thus establishing itself in another new field also based on a by-product of another industry. This development may be of especial importance in the future in view of the fact that the naval stores production is essentially a mining operation (the supply of existing stumps is being steadily depleted and no new stumps are being 'produced'). Hence this source of wood rosin will eventually run out and substitutes will be required.

*Food Industries*

Finally, the latest venture of Hercules again illustrates the constantly changing and cumulative process involved in the interaction between the resources and markets of a firm. In 1965 Hercules acquired the Huron Milling Company, a small firm processing wheat flour to produce amino acids, food supplements, and wheat-based food flavoring, including monosodium glutamate. At first sight this acquisition looked rather far afield, although Hercules did have earlier connections with the food business through its CMC, discussed above, as well as chewing gum (a rosin derivative) and anti-oxidants for food products. Hence, although food chemistry and food markets had not been of primary concern to the firm, they were not completely alien to its experience. Nevertheless, the primary incentive for this particular acquisition and for the choice of this specific direction of expansion was somewhat different.

It will be recalled that Hercules produces its own chemical cotton from cotton linters. Production is carried on in the Virginia cellulose plant and the scale of activity depends not only on the demand for the product but on the supply of linters, which is a function of the size of the cotton crop. The supply of linters has not been sufficient in recent years to employ fully the services of the personnel connected with the Virginia cellulose operation, and the firm has been looking for some suitable activity to absorb these 'unused services.' The Huron

THE GROWTH OF THE FIRM. A CASE STUDY    59

Milling Company was on the market. It was a family firm whose owners wanted to get out and retire from business and also to put their assets in a different form. (Estate tax considerations may well have had something to do with their desire to sell.) At the same time, the firm's activities were of such a nature that Hercules saw an opportunity to extend its knowledge in the food field, especially in the chemistry of amino acids, and to use the personnel of the Virginia cellulose plant. The Huron Milling Company was accordingly purchased with an exchange of shares and is now operated as the Huron Milling Division of the Virginia Cellulose Department. Whether a new base will develop for Hercules remains to be seen, but a start has been made which, if it fits in well with the general nature of Hercules' activities, may not only mean new markets for the firm, but new technology as well.

THE CHANGING PRODUCTIVE OPPORTUNITY OF THE FIRM

The diversification of the Hercules Powder Company, while unique in its details, is by no means unique in its general pattern and will be found repeated in greater or less degree in the story of any number of long-established successful firms. The company's history illustrates the impossibility of separating 'demand' and 'supply' as independent factors explaining the growth and diversification of a firm. The Hercules story illustrates the crucial role of changing knowledge about its own resources in the determination of a firm's course of expansion; at the same time it illustrates the restraining influence of a firm's existing areas of specialization, in particular its technological bases. Whether or not the appearance of new industries, of new 'demand,' in the economy as a whole will provide profitable opportunities for the expansion of a particular firm depends largely on whether that firm has, or can obtain, an adequate 'base' in the relevant field.

Although no single group of industries served by Hercules accounts for more than around 16% of Hercules' sales, two of its primary technological bases, cellulose chemistry and rosin and terpene chemistry, have until recently accounted for over three quarters of its business, with nitrogen chemistry a third important base. Within these bases new products and new markets are continually being created; at the same time petrochemicals have become a leading activity and the emergence of new bases for future operations can already be discerned. By 1926, a bare thirteen years after the firm's creation, new product lines accounted for 35% of total sales; by 1952, 40% of its sales consisted of products that had originated from the firm's research activities after 1930.

The market-creating activities of Hercules are of two kinds; we have discussed one, its extensive reliance on 'technical service' to its customers. The other lies in extensive promotion activities related to its customers' products and only indirectly to its own products. For example, Hercules does not manufacture hot lacquers, but it devotes considerable effort to developing the market for these lacquers; only if the end product is extensively used will the demand for the components made by Hercules be high. The firm even goes as far as to promote the sale of aerosol lacquers (lacquers packaged in aerosol cans under pressure), although it produces neither the lacquers nor the cans.

Because of the nature of its market, Hercules is peculiarly sensitive to business fluctuations. When the demand for final products falls off in the economy, the decline in sales affects the intermediate products and raw materials produced by Hercules in magnified form. The question of the desirability of vertical integration therefore arises, for a producer of intermediate products can usually reduce the sensitivity of its total activities to fluctuations in demand by itself undertaking to produce products destined for the final consumer. This has been a solution adopted by many firms, but it is a solution largely denied Hercules by the nature of its market connections. Forward integration would immediately adversely affect one of the pillars of the sales and market policy of Hercules, for customers would no longer be willing to open their plants, disclose their processes, and discuss their problems with the technical servicemen of Hercules. The technical relationship with customers so carefully cultivated and so important for the creation of new opportunities would be impaired if customers had any reason to fear that Hercules would itself become a competitor.

*The Rate of Growth of the Firm*

The discussion so far has been concerned exclusively with the direction of expansion. What about the rate of growth of the firm? Hercules has not grown so fast as some other firms in related fields of activity, but it has grown faster than industry as a whole. Can one identify a basic factor limiting the firm's rate of growth? Here, of course, we can only speculate, draw inferences from the course of events, and attempt to interpret statements made by the officials of the firm.

Practically all of the growth of Hercules has been financed with internally generated funds. There has been some criticism within the firm of its conservative financing, and the allegation is made by many, particularly by junior executives who feel that their opportunities have been unnecessarily limited on this account, that the firm's growth has been restricted by its preference for internal financing and its insistence on a strong 'cash position.' On the other hand, one of the older executives, long a senior official in the firm, asserted categorically that it was not finance but rather the availability of profitable opportunities for expansion which controlled the firm's rate of expansion. He said that if Hercules found new opportunities for profitable investment exceeding its own financial resources it would borrow the money (or preferably raise it from existing stockholders) to take advantage of them.

The same executive stated that neither was expansion held back by the ability of the firm's personnel. He felt that the war record of the firm showed that if the opportunities were there it could do a great deal more than it was doing. In contrast, another senior executive took a different view: 'Give us the men,' he said, 'and we will do the job.'

These appear to be conflicting explanations of the limits on the rate of expansion of the firm. Although it is obvious that an insistence on financing all expansion from retained earnings would limit the firm's growth, it is unsafe to assume that this has provided the effective limit on expansion merely because little outside capital has in fact been raised. On the other hand, it is undoubtedly true that from a purely managerial point of view the administrative organization of Hercules could have been expanded much more rapidly than it was, In other words, it is probable that the managerial services available from the administrative and technical staff of the firm have rarely been fully used. Under these circumstances we must examine the nature of the firm's 'entrepreneurship.'

Hercules has clearly been imaginative, versatile, and venturesome in the introduction of new products, even at times going into production on a small scale before any market for a particular product was clearly evident; at the same time it has been cautious and conservative in entering new and alien fields of technology. It has been willing to venture extensive funds in speculative research in new fields; it has been unwilling to move into production and invest in plant and equipment in new fields before it had established a research base of its own. And it has been conservative in the methods chosen for entering new fields. For example, it was long after petrochemicals had become an important and growing aspect of the field of industrial chemistry that Hercules decided to enter in a significant way, and then it moved cautiously, relying largely on production processes the firm itself had developed. Another firm, technologically less conservative, might have entered much earlier and through extensive acquisition; Hercules has tended to emphasize the importance of establishing a technological position based on some specialty arising from its own experience. On the other hand, once the firm has become 'basic' in a field, as some of the officials of the firm like to put it, this conservatism largely disappears and the variety and quantity of product is expanded as rapidly as developing technology and markets will permit.

This means, in effect, that the growth of the firm is fundamentally constrained by the knowledge and experience of its existing personnel. Hercules has apparently been loath to go into new fields of activity except through the relatively slow process of building up its internal technical resources. New people are continually being brought into the firm and trained in the processes and methods of the firm; new ideas are eagerly sought from the outside, particularly from abroad, and incorporated into the firm's research program. But new *bases* are not acquired 'ready-made,' so to speak, through extensive and rapid absorption of new people in new fields that are not easily integrated with some existing and internally developed unit in the firm.

The profitability of opportunities for expansion is examined not only in the light of the expected market for certain products or types of products, but largely in the light of how Hercules, with its existing resources and types of operation, could take advantage of and develop them. If the growth of the firm has been restrained by a 'lack' of profitable opportunities for expansion, this merely reflects the lack of entrepreneurial confidence in the profitability *for Hercules* of areas of activity with which the officials of the firm are insufficiently familiar. Since a 'technological base' consists not of buildings, kettles, and tubes, but of the experience and knowledge of personnel, the basic restriction comes down to the services available from existing personnel; the problem of entrepreneurial confidence is fundamentally a problem of building up an experienced managerial and technical team in new fields of activity. Here, again, we can see the nature of the market as a restraining influence on expansion. To the extent that limited opportunities in existing fields force firms to go into new ones, the rate of growth is retarded by the need for developing new bases and by the difficulties of expanding as a co-ordinated unit. The speed with which the firms *try* to move, however, is to a large extent a question of the nature of their 'entrepreneurship.'

The above interpretation of the growth of Hercules is based on a study of past history and of recent attitudes. It is clear that entrepreneurial attitudes, the 'firm's conception of itself,' have had a pervasive influence not only on its direction of growth but also on the method of growth and on the rate of growth. Whether these attitudes will persist depends on the way in which the entrepreneurial resources of the firm change as time goes on. Hercules takes pride in the long service of its people and in the fact that its board of directors is not only a 'working board' but also drawn from men who have spent a great part of their working life within the firm. The first president of the firm served in that capacity for 26 years, was chairman of the finance committee until 1952, and only retired from the board in 1956; of the 15 members of the board in 1950 all but 2 had been with the firm at least 25 years. As the men who built up the firm and carried it through its first few decades retire, it remains to be seen whether the growth of Hercules will be shaped in the future by the same considerations as it has in the past, for in spite of the importance of technological and market considerations, the entrepreneurship of a firm will largely determine how imaginatively and how rapidly it exploits it potentialities.

## NOTES

1 NOTE ON SOURCES: This study of the Hercules Powder Company was made possible by a Fellowship granted me by the Foundation for Economic Education in cooperation with the company, which enabled me to spend six weeks studying the company from within in the summer of 1954 with the full cooperation of all of its personnel. The paper was completed in 1956; when I decided to publish it now [1960] I inquired of the company about subsequent developments, receiving the following reply: 'More recent events, while of great interest within Hercules (and we believe in the industry), are largely a continuation of the types of growth you have shown to be typical and more or less to be expected, except at possibly a somewhat faster rate. Actually, the manuscript can never be quite up to date in an expanding company, nor for your purpose does this seem to be necessary.' I agree with the last statement and for this reason have made no attempt to bring it to the present.
2 The story of Hercules also illustrates the point that the splitting up of large companies will often not have an adverse effect on efficiency if the advantages they have in expansion are economies of growth and not economies of size. For a discussion of these two types of economies and their significance see Penrose, *The Theory of the Growth of the Firm*, Chap. VI.
3 This rank is the one given in the *Fortune Directory* of the 500 largest United States industrial corporations. Supplement to *Fortune* (July, 1957). In addition to the above, Hercules had three plants in wholly owned subsidiaries abroad and employed some 6,000 workers in government owned Hercules-operated ordnance facilities.

## IV

## FOREIGN INVESTMENT AND THE GROWTH OF THE FIRM*[1]

*An analysis of foreign investment which takes place through the reinvestment of the retained earnings of foreign-owned subsidiaries, with special reference to General Motors-Holden's Ltd., in Australia.*

### I

SUCH is the desire for an accelerated rate of capital formation in most of the less-industrialized countries to-day, that fears about the ultimate problem of servicing foreign capital invested in the country are often pushed, rather uneasily, into the background, while inducements are held out to attract such capital, in particular to attract private direct investment. In Australia, however, a recent event has emphasized once again the controversial aspect of foreign investment when the investment takes the form of the successful establishment of a foreign subsidiary enterprise in which the ordinary capital, or common stock, is largely held abroad. Most successful firms grow, and if a firm grows by virtue of its own earnings, that is to say, if the rate of new investment in the firm does not exceed the rate of earnings on its past investment, then the foreign equity in the firm grows without any net import of new foreign exchange—*i.e.* foreign investment in the firm is increasing while there may be a net outflow of foreign funds.

Thus, a relatively small initial dollar investment, for example, may establish a firm whose earnings are sufficient to permit extensive expansion through ploughed-back profits, each increment of expansion increasing the foreign liabilities of the country. It is therefore likely that dividend remittances, when they become an important proportion of profits, will be enormously high in relation to the original dollar investment, although not necessarily high in relation to the total foreign investment in the firm. This paper is concerned with the economic implications of this form of foreign investment for the economic policies of some of the less-industrialized countries to-day.

We shall consider first the particular case of General Motors-Holden's Ltd. in Australia, secondly, the general question whether it makes any difference if foreign investment takes place through the expansion of existing foreign-controlled firms rather than in some other form, and finally what solutions, if any, there are to the problems raised.

* Published in *The Economic Journal*, Vol. LXVI, June 1956.

## General Motors-Holden's in Australia

With the publication of its Annual Report for 1954, General Motors-Holden's Ltd., the Australian wholly-owned subsidiary of General Motors Corporation in the United States, raised a hornet's nest of controversy about its corporate ears, and became more firmly than ever committed to the necessity of convincing the Australian public that 'What's good for General Motors is good for the country.' The storm arose over the revelation that: (a) GMH made a profit A£9,830,000 after taxes in 1953–4, which is variously portrayed as 560% on ordinary capital (the ordinary capital representing in this case the original dollar investment of GMH), 39% on shareholders' funds (net worth), 24% on funds employed or 14% on sales, depending on how the speaker feels about the size of the profit; and (b) that a dividend was declared to the parent company of A£4,550,000, which again is 260% on ordinary capital, 18% on shareholders' funds or 11% on funds employed.[2] The dividend declared is about 8% of the dollar export receipts in the Australian balance of payments for 1954–55.[3]

Newspaper stories and editorials critical of the profit (the highest ever declared by a firm in Australia) and of the dividend payment were followed by an attack by Dr. H. V. Evatt, Leader of the Labor Party (the Government Opposition) on the GMH 'Colossus,' and letters pro and con appeared in the correspondence columns of the Sydney and Melbourne Press. Economists immediately began debating the issues raised for the Australian economy, and H. W. Arndt, Professor of Economics of Canberra University College, stated on the air 'As far as policy towards future overseas investment in this country is concerned, the Holden case may induce some caution in giving indiscriminate encouragement.' He suggested that Australia might do well to 'concentrate less on attracting American capital and more on hiring American technical and managerial know-how.'

The background of the General Motors situation is as follows: The firm began in Australia in 1926 with chassis-assembly plants in various States, and in 1931 bought out the Australian-owned Holden's Motor Body Builders Ltd. with an issue of Preference Shares. Until 1945 the combined firm was engaged in building bodies and assembling imported chassis units, but in that year began planning the manufacture in Australia of the Holden car. This car appeared in 1948, and has steadily gained a larger and larger share of the automobile market in Australia.[4] All Australian produced, it is the cheapest and most popular car in its class (medium weight—by Australian standards—and in the middle price range). GMH has been unable to produce enough cars to meet demand at existing prices, and the waiting list was announced as being equal to twelve months production. In 1953–54 the retail price of the car was reduced by A£84, or 7%. Further reductions in price would only intensify the shortage and put profits in the hands of 'grey-market' dealers. (It is alleged that at present prices many people buy new Holdens and immediately resell them at A£100 profit.)

Along with the release of the Annual Report, expansion plans were announced involving a 50% increase in production of the Holden (from the present 65,000 a year to 100,000), expansion of the household and commercial refrigeration business and entry into the production of central-heating units and other household appliances. The planned expansion is estimated to cost A£21,622,000, bringing the Company's post-war expenditure on new plants and facilities to A£47,750,000 and the funds employed in the business to more than A£61 million. (The total assets of the largest firm in Australia, a firm which comprehends the entire steel industry, were A£76·4 million in 1954.)

GMH has successfully established a completely Australian automobile,[5] thus reducing import requirements for automobiles and promoting efficient industrialization. In other words, it has done 'just what the doctor ordered'; for the promotion of industrial development by 'import replacement' through efficient home production is established national policy. But if the post-war rate of expansion continues for long GMH will not only be far more important in the automobile market than all other producers combined,[6] but will also be reaching into more and more industries. This continued expansion will not involve any further import of dollars; on the contrary, it will be accompanied by an ever-increasing stream of dollar dividends paid out as the equity increases on which the dividends are paid. Taking the ten-year period 1945–54, total profit of GMH was about 10% on sales and 27% on the average capital over the period, but 72·4% of the profit was retained in the business.[7] And there is little reason for believing that either smaller increments of expansion or lower earning rates are going to mark the reasonably foreseeable future.

At first sight it might seem that the problem is basically one of monopolistic profits; the task, accordingly, being to increase the amount of effective competition in the automobile industry. In general, and in spite of the periodic pronouncements of the Labor Party against 'monopolies,' neither the Australian Government nor the Australian public tend to be much concerned about monopoly as such. There is a very high degree of 'concentration' in the economy—the largest seventy-five firms owning nearly 45% of the total fixed assets in manufacturing.[8] The steel industry and the glass industry are each in the hands of a single firm, 70% of the paper industry in the hands of another, 50% of the rubber industry is in the hands of still another and so on for many important industries. For one firm to gain a dominant position in the automobile industry would not be contrary to the general development of the structure of industry in Australia, and would not be expected to raise many eyebrows, especially since the industry is already highly concentrated. In addition, efficiency in meeting market demand seems to be the chief reason for the predominance of GMH—it is hard to trace the failure of competition to reduce its profits to any significant monopolistic practices on the part of GMH itself. Other companies have not produced a car as popular as the Holden (Ford is the only company significantly close to GMH in the market). The tariff on cars not wholly manufactured in Australia is of some importance, but not so much as is generally assumed. To encourage manufacturing in Australia, there is a duty on automobile parts of around A£87 on a car equivalent to the Holden when imported from Britain already assembled, and A£38 on a vehicle to be assembled in Australia. Since automobiles other than the Holden are assembled in Australia partly from imported and partly from domestically manufactured parts, the duty naturally raises their cost, depending on the extent to which parts are imported.[9] Import restrictions imposed for balance-of-payments purposes are much more important and at present are very severe.

One simple and obvious step to take would seem to be a reduction in the tariff and of import restrictions. But this might have two unwanted effects. First, it is very probable that it would hurt other local manufacturers of component parts more than it would GMH. There is a large scope for the reduction of the price of the Holden in the face of competition, and there is strong reason to believe that while GMH would stand up well to increased competition, especially after the new expansion program is well under way, the others would not. In view of the Government's attitude towards the motor-car industry up to the present, it seems unlikely that it would be prepared at this point to risk a discouragement of further local manufacture. Second, and even more compelling from the Government's point of view, is the possible effect on the balance of payments. Any increase in imports is considered highly undesirable under present circumstances. Because of a current deterioration in the balance of payments, import restrictions have recently been severely increased. Since it is not at all certain that increased competition from imports would in fact go far to reduce GMH profits but might involve a partial destruction of the rest of the automobile manufacturing industry, a reduction of the tariff, or of restrictions, might have little effect on GMH and merely cause a further deterioration in the balance of payments.

Nevertheless, the tariff is one of the factors in the situation, and a full scale inquiry into the state of the automobile industry, which has been scheduled by the Tariff Board for some time, is now in process.

It is possible that tariff revision may to some extent increase competition and reduce the monopolistic element in the GMH profit rate. But competition in this sense will not eliminate the fundamental difficulty—the continued profitable expansion of the firm—unless it eliminates the differentials in expected profit rates which are the very basis for international direct investment.[10]

Various other approaches to the GMH problem have been offered. A favorite of some of the financial commentators is the suggestion that GMH should forthwith accept Australian equity capital. Even apart from the fact that this implies a repatriation of capital for which special permission is required from the Government, the proposal is highly impractical if it is expected to be on more than a token scale designed primarily for its psychological effect. To reduce significantly the foreign-exchange payments if profits and dividends near the present scale are continued, the Australian-held proportion of the total equity in the company would have to be substantial. Let us assume that the parent company would have no objection to 49% of the equity being transferred to Australian hands. Given the present net worth of the company, this might involve a diversion of Australian investable funds to GMH of something like A£90 million. The total investment funds raised in Australia on the capital market by Australian companies and semi-government authorities was not over A£200 million in 1954. Clearly it would take many years before a transfer of funds, including the repatriation of the capital, on this scale could be absorbed by the Australian economy without serious complications.

Finally, it is possible for the Government to limit the remittance of dividends either by direct limitation of the amount that can be exported or by very heavy progressive taxation. Some of the problems raised by this are discussed below; it is sufficient at this point to note that in order to attract foreign capital the Government has made it more or less clear that this type of action would not be taken; hence it would inevitably be viewed as a repudiation of previously incurred obligations and would create bad feeling all around. As is suggested below, once a foreign firm is established, it is not, from an economic point of view, desirable to limit its growth or to buy it out at its market value, in whole or in part, although from a political point of view a case can be made for acquiring it.

II

Let us now examine the problem in more general terms. It will be argued that the growth of foreign investment through the reinvestment of retained earnings by firms is subject to different influences from those determining the inflow of foreign investment from other sources.

Once a foreign firm is established its continued growth is an increase in foreign investment, but an increase which is more appropriately analysed in the light of a theory of the growth of firms rather than a theory of foreign investment. The issues raised for the receiving economy will then be discussed.

*The Growth of the Firm*

In the absence of markedly unfavorable environmental conditions, there is a strong tendency for a business enterprise possessing extensive and versatile internal managerial resources continually to expand, not only in its existing fields but also into new products and new markets as opportunity offers. The expansion is not usually a continuous straight-line process; rather it fluctuates with external conditions, and may be retarded by internal difficulties. But for the successful firms over the long haul—and there always are such firms—there seems to be no reason to assume that the process cannot continue indefinitely, or at least for any relevant future. The 'productive opportunity' which invites expansion is not exclusively an external one. It is largely determined by the internal resources of the firm; the products the firm can successfully produce, the new areas in which it can successfully set up plants, the innovations it can successfully launch, the very ideas of its executives and the opportunities they see, depend as much on the kind of experience, managerial ability and technological know-how already existing within the firm as they do upon external opportunities open to all.[11]

Direct foreign investment—and by this I mean the ownership and operation of business organizations in a foreign country[12]—can, in its origin, be of several types. Here we are interested only in direct investment in manufacturing activity which takes the form of the establishment of branches or subsidiaries in foreign countries. In the realm of manufacturing this kind of investment is probably more effective in expanding productivity and promoting industrial efficiency than any other kind of foreign investment.[13] Its advantages derive largely from the fact that behind the new foreign firm are the resources and experience of the parent concern, including not only managerial and technical personnel but also that indefinable advantage in its internal operations which an efficient going concern usually has over a new one. Consequently the receiving country, in addition to foreign capital, foreign technicians and management, also obtains an unlimited drawing account, as it were, on the intangible resources of the investing company. The establishment of foreign subsidiaries or branches is, for the parent company, not essentially different from the establishment of subsidiaries or branches in its own country. To be sure, greater allowance for risk must be made, and greater profits are expected if the venture succeeds according to plan. But the new expansion is still part of the process of growth of the parent company.

Once established, however, a new subsidiary has a life of its own, and its growth will continue in response to the development of its own internal resources and the opportunities presented in its new environment. This means, in the first instance, expansion in the production of the products contemplated when the subsidiary was originally established, an expansion which may well continue for a considerable period if, as is usually the case, the firm is introducing a new product or was established in an industry with the prospect of rapid growth. In time, however, the possibilities of expanding in other fields will appear attractive, either because expansion in the original lines at the same rates as before is no longer profitable, because new market opportunities have appeared, or because the firm has itself developed, or can draw on the parent company for productive services suitable for other types of products (*e.g.*, by-products, innovations related in production or consumption to the original product, a market standing facilitating the sale of other products, etc.). One of the notable characteristics of the growth of large modern corporations is the extent to which they change the range and nature of the product they produce as they grow. They introduce entirely new products, they improve and alter existing products, they enter into a wide range of industries, continually adding to their product lines. In other words, firms are not limited to their original fields; they tend to branch out in many directions. The extent to which this process of diversification can continue depends upon the flexibility of management and upon the resources of the firm—no clear limit is as yet discernible, even in the largest United States corporations.

GMH in Australia illustrates the principle: even if the profitability of automobile production declines as competition develops, GMH can in principle, and very likely will in practice, continue to expand by going into other activities where a higher rate of return is still available. As noted above, the firm is already making provision for expanding production into a wide range of fields. In short, so long as there are openings in industry in which the firm expects a rate of return on investment sufficient to justify entering it, there is nothing in principle to limit its continued expansion. And if foreign firms have any advantage in management, technology, capital or other resources, foreign firms may be expected to grow somewhat faster than domestic firms, even in the absence of any exorbitant degree of monopoly power. This conclusion is reinforced by the fact that foreign firms tend to enter the newer, and therefore faster-growing and more profitable, industries.

On the whole, foreign subsidiaries have, for a variety of reasons, a greater degree of independence of the parent than have domestic subsidiaries. Where they are in a distant country the distance itself tends to restrict the mobility of personnel; where the subsidiaries operate in a radically different political, economic and social environment more weight is often given to the judgment of their executives, who are likely to possess an understanding of conditions that is not easily available to the parent company; where foreign subsidiaries are concerned, the area over which a close co-ordination of policy is considered necessary is often smaller than it is with respect to domestic subsidiaries operating in a more closely connected national market. For these and similar reasons, a foreign subsidiary, once it is established, is, with important exceptions, more appropriately treated in many ways as a separate firm. The most important of these exceptions relates to finance.

Foreign subsidiaries can obtain additional capital from their parent companies or, like local firms, from the local capital market. If funds for expansion are obtained from local equity sources the firm will represent foreign investment to a progressively smaller degree as it grows, and the problems discussed in this paper will not arise.[14] But a preference for expansion through retained earnings is becoming increasingly characteristic of the modern corporation, and of particular interest from the point of view of foreign investment is the situation, especially favored by American firms, in which the parent company holds all, or nearly all, of the equity and permits the subsidiary to expand with its own earnings.

*Internally Financed Foreign Investment*

The preference of American firms for nearly complete ownership of their foreign subsidiaries contrasts sharply with British practice.[15] Several reasons for the contrast have been put forward by businessmen and economists—on the part of the Americans, a desire for freedom from interference from minority stockholders, a desire to avoid shareholder criticism of a ruthless dividend limitation in the early stages of growth even when profits are high, a desire to avoid precise inter-company book-keeping (such as charging part of the overhead of the parent concern to the subsidiary, charging it for managerial and technical advice, service and 'know-how'), a desire for secrecy;[16] and on the part of the British, an appreciation of the political importance of a shared ownership. Whatever the reasons may be, the preference is clear. The result is that virtually the entire growth of most American-owned subsidiaries is properly classed as a growth in foreign investment. And a very large proportion of the total increment in United States investment abroad in recent years has taken the form of re-investment of retained earnings of such subsidiaries.[17]

Because of the potentialities for expansion inherent in the modern corporation with its extensive techniques of decentralized management, its emphasis on developing new markets, on innovations and on research, and its power to generate internal funds for expansion, there are several reasons for believing that the flow of direct foreign investment when it is the result of the growth of foreign firms through retained earnings will proceed at a faster rate, in larger amounts and for a longer period than was characteristic of direct investment of the past. Furthermore, a decision by the parent firm to embark on a new foreign venture is taken under substantially different circumstances from those surrounding a decision to permit an already existing (and profitable) venture to expand.

Although it is true that foreign investment is undertaken because a rate of return on capital that is higher than alternative rates obtainable elsewhere is expected (although not necessarily in the short run), it should not be forgotten that a very considerable input of the managerial and technical resources of the investing firm may be required to ascertain what foreign opportunities exist and how they may best be taken advantage of. Indeed, the original investment of many a foreign firm has been little more than an exploratory venture which can be classed essentially as part of the cost of discovering whether or not investment is desirable. The cost of investigation, together with the cost of planning, organizing and actually establishing the new firm must, other things being equal, obviously limit the number and extent of such ventures undertaken. When, however, a profitable foreign subsidiary presents a request that it be permitted to retain some or all of its profits for further investment, the case is different. The parent firm is presented with a program for expansion which has already been evaluated by the management of the subsidiary. It has to consider this program to be sure, and it may modify it, but the exploratory work has already been done, and been done by men who know local conditions and (presumably) are trusted and responsible officials of the firm. In other words, an interested group in the firm itself presents a case and makes a plea. The degree of uncertainty surrounding such investment, and the cost of making it, are surely much lower than that associated with a new venture, and the investment, therefore, much easier to make.

Secondly, the very operations of the subsidiary in the foreign country, its knowledge of the market and of the conditions of production, the experience gained by its officials and the position it may have established for itself with its own customers and in the country generally all tend to create new opportunities for further investment, opportunities that did not exist at the time the firm was established. These, since they are new opportunities for the growth of the foreign firm, are by definition, new opportunities for the growth of foreign investment.

Finally, there is the fact that it is often easier not to go into an activity than it is to abandon it once it has been firmly established. In the modern corporate world, maintaining an established position often requires net new investment to keep up with the innovation of products, production techniques and marketing methods of competitors. Thus new investment may be required indefinitely to enable the foreign subsidiary even to maintain its position in the foreign market. And here enters, in a sense, a non-economic factor. More is involved in a going concern than an investment of funds. An organization has been created, men and women have vested interests in the concern, a publicly recognized entity exists. It seems highly probable, and the suggestion has been confirmed in various conversations with business-men, that there will be a strong tendency for the parent company to take its 'cut' of the profits and to permit the subsidiary to retain a part long after a comparison of rates of return on capital would attract new investments in the absence of the existing subsidiaries.[18] In other words, even if increasing investment reduces, or even eliminates, the original differential between foreign and domestic rates of return on capital, foreign investment will still increase because foreign-owned firms will continue to grow through their own earnings. There is very little information available to enable us to evaluate the importance of this aspect of the matter. But from what we do know of the general processes of growth of business firms, there seems no good reason for expecting that the growth of the successful foreign firms will cease at any time, and the only important question relating to the growth of foreign investment is whether finance will be obtained from retained earnings or whether the parent company will gradually permit a dilution of the equity with domestically raised capital, or rely on locally raised loans.

In effect, the profitable operation of foreign companies means, in the first place, that there are likely to be more funds available for foreign investment than there would otherwise have been, and in the second place, that the opportunities for foreign investment are enlarged. More funds are likely to be available because of the general bias in favor of permitting a successful subsidiary to retain some of its earnings for expansion, and because of the reduction in uncertainty and risk associated with foreign investment of this kind in comparison with the launching of entirely new ventures. New opportunities are created in the same way in which any growing and aggressive firm creates its own opportunities; it just happens that opportunities for expansion created by foreign firms are at the same time new opportunities for foreign investment.

Such are the reasons for expecting that the flow of foreign investment will increase at a more rapid rate and continue longer when it depends on the expansion of existing firms than it would if it depended on fresh imports of capital for new firms. And I suggest that the increasing importance of direct foreign investment in manufacturing in recent years is partly explained by these factors—some investment that would not otherwise have been made is being made for the reasons outlined above. It is direct investment, but of a much less speculative and risky kind than is usually assumed.

*The Balance-of-payments Problem*

Such investment, though ardently desired by the less-industrialized countries, is not accepted without misgivings. The misgivings relate to two potential difficulties: the political implication of foreign control and the economic problem of paying for the investment. Both are greatest in the case of direct investment; here we are concerned only with the economic problem. There can be little doubt that in order to provide the means of paying for foreign investment, an economy may, in the absence of expanding exports, be faced with the difficult task of inducing an unpalatable reduction in domestic consumption and investment.

This task may be especially difficult when incomes are expanding rapidly in response to a high rate of investment pursuant to government industrial-development and full-employment policies. Under such circumstances, with consumption and investment at high levels, import demand is likely to be very high, not only for consumers' goods but especially for capital goods. The problem of maintaining economic stability—sustained full employment without inflation—becomes crucial, and is particularly difficult in countries whose export proceeds are subject to wide and unpredictable fluctuations from season to season. The goal of governments is to keep inflationary pressures under some sort of control without precipitating deflation, or if deflation threatens, to counteract the recession quickly and thus prevent depression. In pursuance of this goal there is often little hesitation in applying direct exchange and import restrictions or in using Central Bank credit to maintain financial liquidity in order to ensure that unfavorable movements of the foreign balance do not interfere with domestic policy. Yet the foreign balance is likely to be a continual source of trouble because of an excessive demand for imports consequent upon the difficulty of controlling 'excess demand' within the economy without creating unacceptable unemployment. At the same time depreciation as a remedy is looked on with great suspicion, to be used only rarely and *in extremis*.

Such policies have several effects on the problem under discussion. High levels of activity mean high business profits, which not only attract foreign firms but also furnish an easy source of funds for the expansion of already established firms. In a raw-material-exporting economy, where industrial development does not contribute substantially to exports, large amounts of foreign lending may well be necessary to sustain the development program unless demand for the country's raw-material exports is also expanding rapidly. The high levels of activity attract the foreign investment required in the form of direct private investment, and particularly, after foreign firms have become well established, in the form of the growth of foreign firms through retained earnings. We have already shown that direct private investment will tend to grow rapidly and without practical limits as foreign firms grow. So long as these firms retain the greater part of their earnings for re-investment the effect on the balance of payments of servicing the investment is masked, for the re-investment of profits is equivalent to an import of foreign funds. In not exercising their right to export their profits, the firms leave foreign exchange available to the country that would otherwise have gone into dividend remittances.[19] The effect is similiar to the familiar process by which old loans are serviced partly out of the proceeds of new ones. No problem is apparent until the new lending stops. When profits are not paid out it is as if new loans were being made, but when this new 'lending' slows down and dividends are paid on the existing investment the country may find that some difficult readjustments in the economy are required. It has been traditionally held that loans with fixed interest payments are likely to be more of a burden on the economy than equity investment, because in the former case the payments have to be made at the same rate in periods of depressed economic activity, while in the latter dividend payments fall when profits are low. This appraisal of the two forms of investment must be revised if, as seems likely, the more severe depressions can be prevented by appropriate internal government policy.

When the readjustments in the economy required to service foreign investments are difficult to make they create what is called a 'balance-of-payments problem.' And they will be the more difficult to make the greater the pressure of domestic demand on imports and the larger the foreign payments required. Rapidly developing economies attempting to maintain consistently high levels of employment will have continually to struggle to control excess domestic demand. Under the circumstances assumed in this paper dividend remittances from direct private investment will tend to involve larger transfers of foreign funds than would have been the case if only loans had been received. Consequently, a real 'balance-of-payments problem' in this sense may arise. But it is, in essence, a 'technical' problem, a problem of how best to reduce domestic consumption and investment—whether to use import restrictions, tariffs, exchange rates, taxation, monetary policy or some combination of these, and how to mitigate, if desired, the effects on the distribution of income and on the direction of domestic investment. However, in so far as foreign investment goes into industries producing for the domestic market, total dividend payments are likely to move slowly upward without much fluctuation. If export proceeds fluctuate widely from season to season the dividend payments may cause severe strains on the country's reserves from time to time. In so far as foreign investment takes place in export industries, the profits, and hence the dividends, of foreign firms are more likely to fluctuate with export proceeds and less likely to cause temporary strains. Economic authorities have an understandable reluctance to force far-reaching economic adjustment because of temporary difficulties. In the absence of sufficient flexibility in exchange rates the only alternative lies in the creation of adequate international reserves to maintain service on investment in periods when export proceeds are low.

Australia is an excellent illustration. Exports are only a little influenced by its industrial development.[20] Import demand is continually straining the country's foreign resources. Foreign investment, in particular direct dollar investment through the growth of American firms, is increasing rapidly. The balance-of-payments problem that this investment may create is masked so long as substantial proportions of the profits of foreign firms are retained. But as the total profits of foreign firms grow and the firms become more solidly established, the tendency to pay out a goodly portion of their profits to the parent companies will increase, particularly if foreign firms become nervous about the balance-of-payments position. It is highly probable that in time this portion paid out of a growing total will become large enough to make an appreciable contribution to the already existing difficulties of keeping the demand for foreign exchange within bounds, for the profits of the foreign firms, and the firms themselves, will continue to grow so long as the efforts of the Government to maintain domestic incomes at a full-employment level are successful. The adjustments required, together with the foreign payments made, are indeed a 'cost' to the economy, and the basic question is whether the foreign investment is worth the cost. Is there anything in the nature of the type of foreign investment considered here—the growth of foreign firms—that makes the high cost of this investment 'too high'?

## Costs vs. Benefits

In principle, the growth of foreign indebtedness, like the growth of domestic indebtedness, need not be of particular concern in a growing economy, providing that the net income out of which the indebtedness can be serviced also grows correspondingly. If the increments of investment involved in the growth of foreign firms are of such high productivity that domestic real income is raised to an extent equal to, or greater than, the return on the investment to the owners of the capital it will pay the economy to accept them. There may be 'technical' problems associated with the remittance of dividends, but in principle if the increase in the economy's productivity is not less than the cost, it will pay the economy to make the required adjustments, provided that the same increase in productivity could not have been obtained at a cheaper price.

There are two types of benefits realized from foreign investment: additional supplies of capital, on the one hand, and, on the other, new techniques of production and management, entrepreneurial skill, new products, new ideas. We have suggested that loan capital is in general cheaper than equity capital, but that probably less foreign investment would take place if loans alone were relied on. But it is primarily in the second category of benefits that the special advantages of direct investment lie. The benefits of direct foreign investment, when it takes the form of the establishment of new foreign firms, the introduction of new technology and the provision of experienced managerial and technical services can hardly be exaggerated.[21] In addition, the development of subsidiary industries and the improved productivity of other resources which the entry of a progressive foreign firm into a major industry may stimulate is of great importance in increasing the total productivity of the economy. The importance of the opportunities afforded for training not only the domestic labor supply but also domestic managerial and technical personnel is attested to by the lengths to which many countries go in insisting that foreign firms make special efforts in these directions. If the benefits of direct investment can be obtained in some cheaper way it will, of course, pay to do so. The intrinsic difficulties of controlling direct private investment once it gets established, the problem of paying for it, together with the political fear of foreign 'exploitation' and domination, have led many countries to explore ways of hiring foreign management and technical personnel and purchasing access to foreign technology.[22] Such efforts have met with considerable success in some of the Latin American countries, but from an economic point of view it seems doubtful that they would obtain for the country the full set of advantages which are brought in by new foreign firms, although this will vary appreciably between industries. Nevertheless, a great deal can be gained in these ways which benefits the economy. Furthermore, it must be conceded that from a political point of view there is much to be said for this approach.

Even if a great many of the advantages of foreign investment can be obtained by these cheaper methods, it is still true that so long as still more capital and still further 'intangible' benefits can be acquired by permitting the establishment of foreign firms whose contribution to the domestic product is greater than the cost of the investment, it will pay the economy to accept them. And so long as this condition holds true for further increments of expansion of foreign firms, it will pay the economy to permit the indefinite expansion of these firms.[23] If a point is reached where it does not hold true, *i.e.*, where the profits of the foreign firms contain a strong element of private monopoly gain, the situation is changed and the increase in foreign liabilities may be, to the extent of this gain, a net loss to the economy. This brings us up against the moot question whether the monopoly elements almost inevitably associated with the position of the very large firms are almost necessary conditions for further research, innovation and progress, as has been powerfully argued by many, Schumpeter in particular. In some cases this may be true; in others it plainly is not true. In any event, this is the element in the growth of foreign firms that will cause the increments of foreign investment it represents to take place at 'too high' a cost to the economy, and it is this element, consequently, that receiving countries should pay especial attention to. It is obvious that this particular type of foreign investment is more likely than most others to contain an appreciable amount of unproductive monopoly profit.

Now all of this bears on the proposition, so often put forward, that there is some gain to the economy if arrangements are made after a point to buy out the foreign investor. It is probably true that the contributions of direct foreign investment are of greatest importance in the early stages of the growth of the foreign firms and of industrial development. It seems very likely that as the firm grows the point is reached where the specifically foreign contribution to the operation of the firm becomes less important in relation to the contribution of the locally recruited staff who have grown up with and have been trained in the firm. But if the firm is still paying its way—if the contribution to the productivity of the economy is greater than the return to the owners—it will not be to the domestic economy's advantage to divert scarce domestic capital to replace the foreign capital so long as there are opportunities for the profitable use of domestic capital. On the other hand, if the firm is making monopoly profits it would even less pay the economy to buy the firm at its market value, which would inevitably include the capitalization of prospective monopoly profits. If there is no other way of eliminating the monopoly profit, then the purchase of the firm at a price which did not reflect prospective monopoly profits may be the only economic solution to the problem—this is usually referred to as 'expropriation.'

Thus it is clear that if it is economic to accept direct private foreign investment at all and to permit the establishment of foreign firms, that is to say, if the benefits of foreign investment cannot be obtained in cheaper ways (perhaps over a longer period of time), then it is desirable to permit the continued growth of foreign firms so long as they pay their way.[24] Furthermore, once they are established, under no circumstances does it pay to buy them out at their market value if domestic capital can be more profitably used elsewhere.

The fact still remains, however, that if extensive amounts of direct foreign investment are obtained, the balance-of-payments problem of the type discussed earlier may require adjustments which create awkward difficulties for many established government policies. If, therefore, there is not a reasonably large gain to be had from direct foreign investment, the country may well consider that the investment is not worth the cost of the adjustments required to pay for it. Furthermore, the political repercussions of policies designed to restrict domestic expenditure when it is realized that the restrictions are partly for the sake of paying foreign 'capitalists,' may make such adjustments virtually impossible. By calling the balance-of-payments problem a 'technical' one I do not mean to minimize its difficulties. Policies designed to force the rate of industrialization, coupled with the potentialities for almost indefinite growth inherent in the modern corporation, the modern trend for increasing reliance on internal financing of expansion and the preference, particularly of United States firms, for nearly complete ownership of foreign subsidiaries, raises the possibility that much larger amounts of direct foreign investment may take place in a much shorter period of time when political conditions are favorable than was generally the case in the past. When and if the foreign payments become significant, the countries accepting the investment must also be willing to accept the adjustments in domestic consumption and investment required to service it. One suspects that for some countries there may be a basic incompatibility between the economic objectives of fostering very rapid industrial development and at the same time promoting domestic full employment at all times regardless of the state of foreign balance, and the acceptance of an unlimited, unknown and uncontrollable foreign liability.

## NOTES

1 I want to acknowledge here the criticism of the staff of the Economics Department of the Australian National University, and in particular that of Professor T. W. Swan, who saved me from some serious errors in the latter part of the paper. I am also indebted to Professor Fritz Machlup and to Professor E. F. Penrose for much helpful criticism, and to the John Simon Guggenheim Memorial Foundation for granting me a Fellowship for research in Australia.

2 The par value of the ordinary shares is A£1,750,000; there are in addition A£561,600 of Preference Shares owned by Australians. The net worth of the company (or shareholders' funds) is A£25 million; total funds employed are A£41 million.

3   There is a 15% tax on dividends, and hence the actual dollar remittance was less by this amount.
4   From 1953 to 1954, however, the Holden share of the market actually declined from 36·5 to 35·1% as a result of the inability of the company to keep up with the demand in an expanding market.
5   As the saying goes, 'Australian except for the key'—which is made in Wisconsin.
6   The three largest producers are GMH, Ford and British Motors Corporation. Since 1949 the GMH share of the market has risen from 23 to 35½%, that of Ford from 14 to 21%, while that of BMC has fallen from 31 to 20%. Thus the 'Big Three' have 76% of the market, and Ford and GMH together have risen from 37 to 54%. Of these only GMH publishes a balance sheet in Australia; nothing is known about the profits of the other two.
7   Consequently, the actual dividends paid out have not until this year been appreciable, only some 7·4% on the average capital employed in the ten years. But this, of course, means that the equity upon which profits are expected to be earned and eventually realized in signficant amounts is so much the greater. From now on it may be expected that the GMH dividend will remain high.
8   This figure is subject to a fair amount of error because some of the largest companies in Australia are subsidiaries of foreign firms and do not publish balance sheets. Hence we have no information about their fixed assets. The figure was obtained by taking the largest manufacturing firms listed on the stock exchanges and comparing their fixed assets net of depreciation (because the gross figure is not usually available) with the net fixed assets in manufacturing as given by the Commonwealth Statistician.
9   However, several companies are well forward with plans to produce in Australia.
10  See Section II below for discussion of this point.
11  For a fuller, though still incomplete, discussion of the theory of the growth of the firm upon which part of the argument in this paper is based see my 'Limits to the Growth and Size of Firms,' *American Economic Review*, XLV, No. 2 (May 1955), Papers and Proceedings, pp. 531–43. See Essay II above.
12  This begs the question of what constitutes ownership or, more precisely, a controlling interest. For the purposes of this paper we need not worry about the kind of precise definition necessary for statistical analyses.
13  By the emphasis on manufacturing I want to exclude from consideration the development of public utilities, railroads and similar industries basic to manufacturing through inter-governmental loans or through international financial intermediaries. Obviously such investment may be equally effective in promoting development.
14  There are a variety of ways in which foreign concerns can co-operate with local firms and financial interests in establishing and developing manufacturing activity and without creating the kind of problem typified by the GMH case. See, for example, the methods discussed in *Processes and Problems of Industrialization in Under-developed Countries* (New York. United Nations, Department of Economic and Social Affairs, 1955), pp. 85–6.
15  It has been estimated that on the average American companies hold 85% and British companies 40% of the Ordinary Shares of their subsidiaries in Australia.
16  Foreign subsidiaries in Australia at this time did not have to publish balance sheets unless they were 'public companies' (some shares being publicly held); GMH issued Preference Shares in order to acquire the Holden company, and for this reason had to publish its accounts. The controversy discussed here

continued when reports were published in subsequent years. But in 1959 General Motors Australia Pty. Ltd. was formed as a wholly-owned Australian subsidiary of General Motors Corp. (U.S.A.) and purchased the preference shares of GMH, which promptly stopped publishing its reports. In 1961, however, Australia plugged the loop-hole by its new Companies Act, which required all foreign-owned subsidiaries to issue reports on their Australian operations. GMH resumed publication of its Annual Reports in 1962, [I am indebted to Professor Harcourt for this information, which I have added to the original article.]

17 'Between 1946 and 1951, for example, no less than three-quarters of all United States new direct foreign investment in manufacturing industry was the result of the ploughing back of profits earned in foreign branches and subsidiaries.' United Nations, *op. cit.*, p. 84. According to a very interesting paper on the contribution of overseas companies to Australian post-war industrial development presented to the 1955 meeting of the Australian and New Zealand Association for the Advancement of Science by D. M. Hocking, which was based upon figures developed by the Australian Department of National Development, direct investment in secondary industry in Australia from the United States increased by A£80·1 million between 1945 and 1953. The United States Department of Commerce (*Survey of Current Business*, December 1954) estimates that the undistributed profits of Australian subsidiaries of United States firms amounted to A£65·2 million in the period from January 1946 to December 1953.

18 This is particularly true if capital repatriation is not permitted and the foreign firm has no alternative but to continue to invest to protect its existing investment.

19 We are not, of course, taking into account here the effect on the demand for imports (or supply of exports) of the investment which is itself the alternative to the building up of cash balances for the purpose of paying dividends. This is part of the whole question of how industrialization in a primarily raw-material-exporting country affects the demand for imported goods—a question we cannot explore in this paper.

20 However, we should not overlook the fact that industrialization may have *reduced* exports by diverting exportable products to domestic consumption and by attracting factors of production away from the export industries.

21 For a survey of the enormous contributions made to the Australian economy by foreign firms see D. M. Hocking, 'New Products, New Skills Follow Overseas Capital,' *The Australian Financial Review*, September 29, 1955, p. 2. John H. Dunning has discussed a significant aspect of the importance of the contribution of American firms to the British economy in an article 'United States Manufacturing Subsidiaries and Britain's Trade Balance,' *District Bank Review*, September 1955, pp. 20–30.

22 See United Nations, *op. cit.*, pp. 85–7.

23 This is, of course, apart from the political question of whether large and powerful firms—foreign or domestic—should be countenanced.

24 It has been implicitly assumed in this entire argument, of course, that industrialization itself is, in the long run, a profitable use of the country's resources, or rather that the rate of industrial development implied in the investment program is not excessive. If this is not true and if real income would be increased if more resources were devoted to the raw-material-exporting industries and less to domestic industry, then further foreign investment, just as further domestic investment, in industry will be unprofitable to the economy as a whole.

V

## PROBLEMS ASSOCIATED WITH THE GROWTH OF INTERNATIONAL FIRMS*

*Analyses four types of problems: transfer pricing, monopoly, 'self-financing' and equity participation for local investors.*

THE term 'international firm' is used in this paper to describe the business organization that is engaged in production in a number of countries through branches, subsidiaries, or affiliates, which may or may not be separate corporate entities in the several countries in which they operate. The term 'organization' implies that the entire group, including the head office as well as the various types of subsidiary units,[1] is operated within an administrative framework which knits the whole together in such a way that the general policies and administrative and financial procedures of the group are reasonably consistent and coherent throughout the firm. Individual subsidiaries may have considerable autonomy in their own operations, but if they operate entirely independently of the group as a whole they are better treated from the point of view of the growth of the firm as a simple investment by the parent companies and an extension of its financial influence than as an integral part of the industrial organization, or 'firm'.[2] It should be clear, therefore, that we shall not be concerned with financial firms nor with firms that merely import and export; we are concerned with firms that engage in production in a number of countries and whose activities are organized within a coherent administrative framework.

The large international firm has today become one of the most important types of international economic organization, dominating a wide variety of industries ranging from pharmaceuticals and automobiles to the petroleum industry. A very great proportion of these are American, but there are also important European firms, including particularly British, Swiss, German, Dutch and French, as well as Japanese. Although we speak of international firms because of the international spread of their operations, each such firm with few exceptions, has a distinct nationality, which is the nationality of the parent firm.[3] The controlling center, or head office, is usually located in the country in which the parent company is incorporated, the firm having acquired its international status by extending is productive operations

* Published in *Tijdschrift voor Vennootschappen, Vereinigingen en Stichtingen*, No. 9, 1968.

from its home country to other countries. This process, which includes the establishment and subsequent expansion of producing units abroad, is called 'direct foreign investment', and is usually, though not always, undertaken by enterprises that are predominately privately owned. Hence, in discussing problems associated with the growth of international firms we are at the same time discussing problems of direct foreign investment.

Foreign investment of this kind may be undertaken by a firm for a wide variety of reasons: there may be geographical advantages in producing abroad which relate to the nature of markets, the existence of raw materials, or the availability of other factors of production; there may be a variety of government restrictions on trade and payments which make local production economically advantageous or even necessary; or there may be tax advantages associated with production in particular countries. In the space available here it is not possible to examine these considerations nor to discuss the entire range of problems associated with the growth of firms as they spread activities internationally. Although many of these problems involve aspects of organization and co-ordination, and of adaptation to particular circumstances and markets, perhaps the most interesting relate to the peculiar problems a foreign firm may face in making itself acceptable to the peoples and governments of the countries in which it operates. I shall therefore select for discussion four types of problems from the latter category, all of which become more important as a firm grows in size, and especially if it comes to account for an increasing proportion of the output of an important industry in any particular country.

The first of these problems arises from the significance of transfer pricing, since an international firm is, by definition, an integrated firm, and thus there is likely to be considerable trade in goods or services between its subsidiaries across national frontiers. The second relates to the fact that an international firm, especially if it is in the newer technologically progressive industries or has access to large economies of scale or unusually extensive financial resources, is likely to enjoy a fairly strong monopoly position in its host countries, and particularly in the less-developed of these countries. The third problem arises when the firm produces for the local markets of the host countries and attempts to finance local expansion largely from local earnings. Finally, we shall discuss the problems that may arise if the firm desires, or is forced to accept, the participation of local equity capital or local management.

## Transfer Pricing

When a firm establishes manufacturing subsidiaries abroad the new plants will, in general, be producing or distributing products similar to those which the firm is also producing elsewhere, or will be producing raw or semi-finished materials which it uses in its own operations elsewhere. The former is commonly known as 'horizontal' integration and the latter as 'vertical' integration. Thus, oil companies may establish refineries or distribution networks to serve local markets, and import the crude oil or products required, often from their own subsidiaries elsewhere, or they may undertake crude-oil production largely for the purpose of exporting crude-oil to their own refining subsidiaries in other countries. Pharmaceutical companies establish plants for the manufacture of drugs in other countries, but often supply raw materials or semi-finished products to them from other subsidiaries in still other countries. In both cases, services such as managerial and technical services and patent rights are also made available to subsidiaries, often at a substantial fee.

For some such products and services, market prices may exist; others are of such a specialized nature, or subject to such a degree of monopoly in production (e.g. patented products with no close substitutes or, until recently, crude-oil outside the United States), that the price at which they are transferred between subsidiaries cannot be compared with any reasonably free market price. In these circumstances, international firms possess a very large amount of discretion with respect to the prices at which they transfer products and services across national frontiers between their subsidiaries. And governments have a legitimate interest in these prices.

Without entering into the controversy over whether firms actually attempt to 'maximize' profits, it is probably reasonable to assume that most firms like to make as much profit as seems reasonably practicable under the circumstances, and that by and large they are interested in profits *after* taxes on income. An international firm is subject to income taxes in many countries and will thus have an incentive so to price the products (or services) transferred between their subsidiaries as to minimize the total income tax payments for the consolidated firm by allowing as much profit as possible to appear in those countries where tax rates are lowest. Thus, there is, in a sense, a built-in incentive to discriminate in transfer pricing among the several countries in which the firm operates, and such discrimination has often brought international firms into conflict with host governments.

The discrimination may be in favour of the home country or in favour of certain of the other countries in which subsidiaries are located. The Swiss pharmaceutical firms, for example, seem to favour the former, while the inter-affiliate pricing of crude oil by the international petroleum companies has clearly favoured the crude-oil producing countries.[4] In any case, as the firm grows in size, the significance of its transfer prices may also increase with respect both to the balance of payments of the countries importing or exporting the transferred products and to the income tax receipts of their governments (which are, of course, interrelated). Problems may arise which can lead to considerable public criticism, and sometimes to acute political controversy, with the governments in the strongest bargaining positions being able to obtain concessions which cannot be obtained by weaker governments. Different firms adopt very different policies with respect to transfer pricing, including charges to their subsidiaries for managerial services, patent rights, etc., but in the absence of competitive 'arm's length' market prices, there are no unambiguous criteria that firms can use for the determination of what is 'fair' to all countries concerned.

## The Problem of 'Monopoly'

When a firm establishes subsidiaries abroad, it presumably does so because it believes that it has some competitive advantage over other actual or potential producers. It may possess specialized technological, managerial, financial or marketing expertise; it may have been able to acquire some special type of protection or privilege from governments (including patent rights) for its activities or products; it may operate under private market-sharing arrangements; or it may simply have a head start over other producers in a market too small to support additional efficient producers. Such competitive advantages, whether they be attributable to the firm's own superior efficiency, to government policy, or to private monopolistic arrangements, may enable the foreign subsidiary to obtain a dominating position in an important industry, or even in the economy as a whole, of the host country. Such a situation is common in developing countries, but it is also important in some industries in industrial countries, especially in an industry where patent protection plays an important role. Clearly, the expansion of a foreign subsidiary in such circumstances can give rise to difficulties with host governments, particularly in countries that are sensitive to the prospect of foreign 'domination' of their economies.

To the extent that the competitive advantages giving rise to a monopolistic position are the result of the superior productive efficiency of the foreign firm, the economy of the host country gains even if profits are considerably above 'normal', but at the same time the local pricing policy of the firm may come under fire as an example of 'foreign exploitation'—perhaps especially, but not uniquely, in underdeveloped countries. Again, there is no unambiguous economically 'optimum' policy for the firm to adopt, although so long as the return on investment in one country is greater than that to be obtained elsewhere, the firm would presumably expand output to the point where returns are roughly equalized in each of the countries in which it operates. Some international firms apparently attempt to charge approximately the same prices for their products all over the world and thus to maintain a world-wide system of prices; others seem to price according to the elasticity of demand in different markets and thus practice geographical price discrimination. But if the growth of a foreign subsidiary increases its monopoly position in particular markets, the problem of pricing may become acute partly, if not largely, *because* the subsidiary is foreign owned, even if we can assume that much of the abnormal profit is properly classified as an economic rent attributable to its superior factors of production or management rather than to some form of protection or private monopoly. The problem here is often one of market structure which the firm can do little about, and in the setting of prices for the local market the firm may have to take public and government attitudes into consideration more than its own commercial interests. These last two are likely, of course, to be related.

It is common, however, for governments, especially in the developing countries, to insist that some proportion of the profits of foreign firms be reinvested in the country, for it is assumed that if the foreign capital is welcomed in the first instance as a desirable contribution to the growth of the local economy, then any increase in foreign investment that results from the reinvestment of the profits of foreign firms will be equally desirable. This brings us to the question of 'self-financing', and the problems this may raise for the position of the firm as it expands.

*'Self-Financing'*

The term 'self-financing', or financing of investment from retained earnings, describes a situation in which a firm raises on a current basis all or the greater part of the funds it invests from its customers through the prices it charges. The expansion of local subsidiaries can be financed through borrowing from the local market or from abroad, through the issue of equity shares in the local market, through additional capital investment from the parent company, or, if the profits of the subsidiary are high enough, through the reinvestment of these profits. If local equity shares are not issued any expansion of the subsidiary will increase the amount of foreign investment by the amount of the expansion. Finance through an influx of new investment from the parent will increase the foreign exchange receipts of the country concerned over what they would have been in the absence of the subsidiary. Finance through borrowing or through the reinvestment of retained earnings does not increase the foreign exchange receipts of the country over what they would have been in the absence of the firm, but the former permits local lenders to share in the gross return on investment, and the latter reduces the amount of foreign payments that the country would have had to make in the absence of the reinvestment and thus leaves its balance of payments better off, at least in the short run, than it would have been.

There are a large number of intricate aspects of these questions which cannot be dealt with in the space available here. The policies of different firms differ with respect to the weight they give to the desires of the managers of their subsidiaries to retain some of their earnings for expansion; the situation is different according to the type and amount of outside equity capital in any particular subsidiary; and, of great importance, the international firm may not itself be free to make its decisions according to it own commercial interests because of intervention from its home government as well as intervention from the host governments. Moreover, from the point of view of the economic effects on the country concerned, it makes a difference whether the foreign subsidiary is producing for export or for the local market, that is, whether it derives its income primarily or entirely from local sales.

Governments often put pressure on foreign firms to reinvest a substantial proportion of their profits, but the effect is, of course, to increase the foreign liabilities of the country since eventually dividends will have to be paid on the increased amount of foreign investment. This will create few problems if the foreign firm itself earns foreign exchange from exports, but if the firm is producing for the local market, then the servicing of the investment may at some point claim a large proportion of the foreign earnings of the country.[5] If the profits of the firm are due to its efficiency in production, then in principle the increase in the national income of the country due to the firm's activities will leave the country better off even after the payment of dividends abroad. This will not necessarily be the case if there is a significant monopoly element in the profits.

Moreover, in permitting very large amounts of investment in foreign firms producing for the domestic market, a country may find itself very vulnerable to sudden changes in the policies of the capital exporting countries respecting their own balance of payments. Both the United States and the United Kingdom have in recent years required their own 'international' firms to reduce their investment abroad, which in many cases means an acceleration of the repatriation of earnings. This is a means of putting pressure on the balance of payments of other countries in order to improve their own and may occur at a particularly awkward time from the point of view of the countries in which the subsidiaries of the international firms operate. Foreign firms have been especially interested in attempts to create an international 'charter' for foreign investors to protect themselves against arbitrary actions by host governments. They should perhaps also protest against arbitrary actions by their home governments which create a legitimate wariness in receiving countries, and to some extent justify a refusal on their part to permit the free repatriation of profits in the normal course of commercial operations.

It must be remembered, however, that capital may not be the most important contribution that a foreign firm makes to the local economy; the inflow of managerial skills, technological know-how, and marketing efficiency may be much more important and may be the real benefit for which the economy pays in the form of repatriated profits. At the same time, as the foreign subsidiary grows, it often trains local people and takes them into the firm, even into positions of top management. In a developed economy, the point may be reached at which the subsidiary has become 'local' in virtually everything but ownership, and if profits are abnormally high the economy may find itself paying a high price for capital alone. This may also be the result when an international firm expands by taking over a local firm and when the profitability of the take-over rests significantly on monopolistic advantages possessed by the international firm, as may be the case, for example, when an American oil company takes over a European refinery in order to obtain an outlet at non-competitive transfer prices for crude-oil it produces in subsidiaries elsewhere.

Partly for reasons such as those discussed above, some countries have pressed strongly for the acceptance of local equity in the subsidiaries of foreign firms. Others may want and need the foreign managerial and technical skills but be reluctant to accept the relatively high costs of equity capital investment as well. This brings us to our final set of problems, those associated with a fear of foreign domination and control of the local economy.

*Local 'Participation'*

Local participation in ownership and management, or 'partnership', is often proposed as a solution to some of the problems we have touched on, and some international firms have adopted the general policy as they expand abroad of proposing joint ventures with local businessmen. Others have accepted local participation under pressure. Except in countries where local managerial participation may raise political problems, most large international firms today go out of their way to train and accept local people in their managerial hierarchy, although 'top management' of an international firm is still largely confined to the nationals of the home country. As noted above, however, if 'indigenization' reaches the point where local subsidiaries are entirely managed and staffed by local people, the foreign contribution made by the subsidiary to the economy may become reduced to its capital investment and whatever benefits may be derived from the remaining links with the parent company. These benefits may be considerable, but so may also be the costs, especially if problems relating to integration and transfer pricing are important.

The acceptance of local equity raises a different type of problem. In particular, it may destroy the coherent administrative character of the international firm. A wholly-owned subsidiary of an international firm is an integral part of an international organization. Its managers may argue strongly and effectively in the various committees of the organization for their own views and for the interests of their own subsidiaries, but decisions on important matters are made by committees representing the firm as a whole, not by these managers alone. Usually the use to be made of the profits of the subsidiaries is one of these important matters, and the decisions are made in the light of the interests of the firm as a whole, as are decisions about large investment programmes or about sources of funds. Similarly, high managerial appointments are matters of concern to the entire firm. But if there exists a considerable amount of outside equity, all of these decisions will have to take account of the interests and demands of the outside owners, and the problems of administration may become magnified. In consequence, decentralization may go so far that it will no longer be possible to treat the entire group as a coherent administrative organization falling within the definition of an international firm in the sense used here.

Partnerships, or 'joint ventures', are nevertheless often useful means of reducing local antagonisms and thus of facilitating the growth of the international firm. It is sometimes alleged, however, that they bring about a substitution of local for foreign capital to the disadvantage of the receiving countries, for when an established firm admits local equity capital by selling shares to local investors, it is, in effect, repatriating part of its investment unless the new local capital is used for an expansion that would not otherwise have taken place. At the same time, it is likely that the supply of local savings for investment in countries allegedly 'short' of capital is very much a function of existing opportunities for profitable investment, and that much saving which would otherwise flow in non-productive directions might become available for investment if more secure productive openings for local funds were also available. The shares of foreign firms could well provide very attractive opportunities for such savings and thus even increase the total supply of capital instead of merely effecting a substitution of local for foreign capital.

In this short paper I have raised and commented on certain types of problems associated with the growth of large international firms; I have provided no solutions. It is difficult enough to describe both reasonably adequately and also briefly even the nature of the problems, let alone to provide solutions. But unless solutions are found to the types of problems discussed here, it is highly probable that the acceptability of large international firms, already seriously questioned in many countries, will become increasingly undermined as they expand operations and grow in importance in more and more countries. They may then become little more than international investment companies whose activities are closely circumscribed by governments everywhere. And the fact that American firms are becoming increasingly dominant in some areas does not make the problems less difficult.

## NOTES

1 From now on I shall use the term 'subsidiary' to include branches and affiliates as well as true subsidiaries, since the distinctions between them are not important for my purposes.
2 The degree of autonomy that can be granted to subsidiaries without destroying the administrative coherence of the firm is almost impossible to define in the abstract. The boundaries of any given firm can only be determined with reference to the actual organization of the operations of that firm. For further discussion of this point, see my *Theory of the Growth of the Firm* (Oxford, 2nd ed. 1965).
3 There are exceptions, of which perhaps the Royal Dutch/Shell is one of the best-known examples. This 'firm' is not a firm in a formal sense but is a kind of agreement between two international holding companies, the Royal Dutch Petroleum Company and Shell Transport and Trading Company, under which the assets of the firm are held in undivided ownership in the proportions 60% Royal Dutch and 40% Shell Transport, with income being divided in the same way as well as membership in the controlling Board of Directors. This grouping can be called a 'firm' on my definition because its activities are carried on within a co-ordinated administrative framework, but it has no single nationality and consequently no clear national orientation. I have elsewhere suggested that Royal Dutch/Shell, partly because of its lack of a distinct nationality, has a more truly international outlook than that of any other international oil company. See *The Large International Firm in Developing Countries: The International Petroleum Industry*, p. 109.
4 See *ibid.*, Chapter VI, for a discussion of the petroleum industry. For evidence respecting the drug industry see the *Report of the Committee of Enquiry into the Relationship of the Pharmaceutical Industry with the National Health Service, 1965–1967* (Sainsbury Report). London. HMSO. Cmnd. 3410. 1967. Appendix I, paras. 15, 23.
5 An excellent example of this can be found in the experience of General Motors Holden's in Australia. See 'Foreign Investment and The Growth of the Firm', *Economic Journal*, Vol. LXVI, June 1956. [See Essay IV above.]

# VI

# INTERNATIONAL ECONOMIC RELATIONS AND THE LARGE INTERNATIONAL FIRM*

*Deals with the political economy of the operations of international firms: foreign influence and control; exploitation; special problems.*

LARGE international firms are among the most important international organizations in the world today. A small number of them control some of the world's most important industries and dominate the economic life of some countries. The total receipts of some of these firms exceed the national budgets of many countries and even the national income of some. Their administrative controls are supranational, reaching across all frontiers; although each firm is subject to the laws of each of the countries in which it operates, no one country has a general jurisdiction over all of its operations. International firms are expected to be 'good citizens' of all the countries in which they work, and they must satisfy the demands of many increasingly exacting governments whose interests often conflict. Not surprisingly they often fail in this task, sometimes creating considerable international ill will in the process and occasionally bringing serious conflicts between governments.

The purpose of this paper is to analyse, with particular reference to the underdeveloped countries, the nature of some of the political and economic problems that bedevil the relationship between an international firm and the countries in which it operates. We begin with a brief description of the large international firm and of the contribution expected of the foreign investment that it undertakes.

## The Nature of the Large International Firm

A large firm, national or international, is not a single corporation, but a network, often of astonishing intricacy, of limited (or incorporated) companies whose direction and ownership are closely interlocked. At the top there will usually be a holding company, the administrative functions of which will vary substantially from firm to firm, but which in general acts as the parent company, and owns,

---

* Published in *New Orientations: Essays in International Relations*. E. F. Penrose and P. Lyons, eds. (Frank Cass, 1969).

entirely or in part, directly or indirectly, the shares of the subsidiary companies which make up the firm as a whole; some of the larger subsidiaries are also likely in their turn to have subsidiaries of their own, which are thus owned only indirectly by the parent group. The intricacies of the corporate structure of firms arise from the almost unlimited possibilities of ramifying the arrangements for ownership or control—subsidiaries, or subsidiaries of subsidiaries, may even own the shares of their own parents.

From our point of view a group of companies of this sort can be considered a 'firm' when its several subsidiary units all operate within an administrative framework established for the group as a whole by its management. The more important of these units may have a very large degree of autonomy within the scope laid down by the general policies of the firm, but in financial matters, particularly investment, pricing and dividend policies, their actions will generally be subject to approval by their head offices. The management of large firms is conducted through a variety of committees, and the major subsidiaries are likely to be represented on a number of such committees. These are then in a strong position to urge the acceptance of their own views and may obtain compromises in their own favour, but in general the overall interest of the firm as a whole must remain the primary consideration. Clearly, there is scope for very great variation among different firms in the degree of autonomy accorded subsidiaries. In particular, there are differences in the weight given to the views of subsidiaries in foreign countries, whose claim to special status, as, in a sense, 'representatives' of the interest of the countries in which they operate, is sometimes recognized.

Firms are owned by shareholders, who may be individuals, other industrial companies or institutional investors or governments themselves, but the rights of shareholders with respect to the management of firms are very circumscribed, and their ability to influence policy may be very small indeed. By and large, the policies of large firms are determined by their management and not by their shareholders, although managers may themselves own substantial blocks of the firm's shares, and some shareholders, especially institutional shareholders such as banks or insurance companies, may be extremely influential. Nevertheless, management is in a position to take the initiative in proposing policies, and it alone possesses the necessary information. Moreover it is also more often than not in a position to control shareholders' meetings through the proxies it can obtain from shareholders who are unable (or insufficiently interested) to attend annual meetings of shareholders. Large firms with a great number of shareholders are almost always autonomous bureaucracies, generally 'responsible' in law to their shareholders and required by law to publish certain types and amounts of information, but in practice determining their own policies, constrained not by shareholders' wishes as such but by the necessity of maintaining a financial and market position favourable to the long-run survival and continued growth of the firm.

To sum up, large international firms are autonomous international organizations with very widespread influence in international economic relations and often possessing great power. They are *autonomous* in the sense that they are effectively accountable to no outside body for their actions, although they are constrained by the policies of governments, by the actions of competitors, by the demands of the financial markets, and by similar considerations. They are *international* in the sense that they operate in many countries in the world, although such firms do in fact have a 'nationality'—that of the country in which the parent company is incorporated. The several subsidiaries are themselves often (but not always) incorporated in the countries in which they operate; if they are locally incorporated they can be looked on as 'nationals' of these countries, but they cannot independently form their own policies respecting many very important aspects of their operations. Finally, international firms are *organizations*, in the sense that the activities of the international group are carried on within an overall administrative framework and subject to overall policies laid down by their central administrative units.

A large firm necessarily possesses considerable economic power even if it does not have a strong monopolistic market position: by definition it controls a large amount of resources the use of which is decided by its management; it commands a large market over which it can exercise various kinds of influence; it raises and spends very large sums of money; it is a large buyer and a large employer. This implies possession of economic power, for economic power in its widest sense consists of the ability significantly to influence the use of resources, the distribution of goods, the prices of products, the tastes of the consumers, the development of new technology, and the distribution of income.

These international organizations are primarily interested in their own profitability and continued growth, and left to themselves will be concerned with political matters only to the extent that government policies or political conditions affect their economic operations. In principle, they, like other firms, have only economic, that is to say 'commercial', objectives, but since the political attitudes as well as the political conditions in a country may be of decisive importance for economic operations, an international firm may be unavoidably drawn into political affairs in a number of countries.

## The Contribution of Direct Foreign Investment

In setting up or expanding subsidiaries and affiliates (or branches) abroad international firms engage in direct foreign investment, which is by definition the building (or acquisition) of manufacturing plants, distribution facilities, or other productive assets in foreign countries. If the firms choose their projects sensibly their search for profit should lead them to invest where the productivity of their investment, as measured by costs (which reflect the conditions of supply) on the one hand, and selling prices (which reflect the intensity of demand) on the other, is greatest. Thus, the presumption is that investment by foreign firms, whether it be in the production of food, raw materials, services, or manufactured goods, will increase the productiveness with which resources are used and thereby raise incomes all around. In addition to foreign capital, such investment usually, but not always, brings entrepreneurial skills, knowledge (often patented), marketing techniques and marketing connections, which are not available to the country, or are available only in inadequate quantities or only for restricted uses. Moreover, receiving countries may not only gain from the addition to their resources, but often also from the stimulus provided to their own businessmen by the competition from, or the example of, the foreign enterprise.

The foreign investment is itself a liability for the domestic economy, for it gives foreigners a claim to receive income from the economy. Unlike a contractual debt, however, this claim can only be effective if the investment is profitable. If it is not, the foreigner's claim is reduced to what he can get by letting the enterprise run down or selling it for what it will bring. But if profits are made, part at least of the profits will, in due course, be paid abroad as dividends, while some will usually be reinvested in the enterprise. The reinvestment of profits increases the amount of foreign investment in the economy and reduces the flow of dividends abroad in any given period, thus keeping down the impact of the firm's profitability on the country's foreign exchange resources. But even if all profits were repatriated, it would not follow that the country would lose from the foreign investment. The usual presumption is that the investment will be worth its cost to the receiving country if (a) the private profitability of the firm is not offset by social costs (such as soil erosion, pollution of air or water, etc.) which are attributable to the firm's operations and for which it does not pay; (b) if profits are not significantly enhanced by continuing government subsidies or protection; or (c) by artificial restraints on the entry of new competitors or on the effectiveness of competition. In other words, if the profitability of the firm can be attributed primarily to the superior productiveness of its own resources and organization, then it is presumed that the country gains from its operations, since the national income is increased even if all profits are repatriated. The country should therefore gain even more if profits are reinvested.

Nevertheless, there is clearly a widespread suspicion of foreign firms in all countries in which they have an important place in the economy, even in countries where they are welcomed by the government and received reasonably cordially by the financial community. Some of the suspicion is aroused by the presumed political implications of extensive foreign influence in the economic system; some of it arises from a fear and dislike of foreign economic control of the country; and some of it rests on the belief that powerful foreign groups exploit the country in the sense that they make and retain for themselves 'too much' profit at the expense of the local economy. We shall examine each of these sources of suspicion in turn.

*Foreign Political Influence*

It is not always clear exactly what is meant by the statement that a country is foreign 'controlled' or subject to foreign interference. In its simplest form, 'foreign control' would presumably imply that all, or nearly all, of the important private as well as public decisions affecting the economy are taken by foreigners who are also in a position to prevent the national authorities from interfering seriously with their activities. Thus, in addition to the classical examples of foreign control—colonial economies, occupied territories, or complete political satellites—there are those in which not only is the economy of a nominally independent country dominated by foreign groups, but these groups are also able to manipulate the governmental authorities, often through outright corruption. In one way or another an identity of interests is established between the foreign groups and the ruling authorities. Cuba before the Castro revolution seems to provide a modern and reasonably clear-cut example of this type of foreign economic control—in this case primarily by United States sugar and business interests—leading to subservience of venal political authorities to the foreign groups.

In all countries the policies of governments must take account of the interests of the important economic groups in the country, and everywhere 'vested interests' do in practice often succeed in obtaining favourable treatment at the expense of the general economy. In many countries such influence is relatively mild—ranging from success in obtaining government contracts or other privileges on grounds other than merit, to tariff protection or tax subsidies—and poses no very crucial political problem for the nation as a whole. But in some countries, political authorities may come almost entirely under the domination of powerful economic groups, which may be local landowners, financiers or industrialists, or foreign capitalists. This type of situation forms one of the central problems of political development in a number of countries, especially when the economic interests are allied with the army. Important as this kind of thing may be in some countries, however, its general prevalence can be much exaggerated, and is probably rarely associated with direct foreign investment in manufacturing alone, being more likely to arise when there has been a long-standing alliance between foreign plantation and mining interests and local political families or military groups.

Clearly in a politically independent country the national government has the legal power to control or to prohibit foreign operations. Even if the government is plainly not a puppet of foreign groups, there is nevertheless likely to be controversy over the degree of independence it in fact has. There will always be critics who will find evidence of subservience if the authorities allow foreign firms to operate at all, or who complain of 'foreign interference' if the firms make their interests known. Any firm must inevitably be concerned with government policies which affect its operation. Foreign, as well as local firms, may therefore ask for or protest against economic legislation of various kinds, including commercial, financial, or labour legislation and even government policy regarding state enterprises. A country wanting to attract foreign firms or to encourage the expansion of firms already operating will recognize that such firms will not invest unless they can expect reasonably profitable returns on their investment. It would be unreasonable for a government to expect foreign firms to refrain from making representations on matters of this kind which affect their profitability, and hence their willingness to invest. Such representations should not be looked on as a type of 'foreign interference' in the affairs of the country so long as the foreign firms do not call on the political support of their home governments, intrigue in local politics, bribe local politicians, etc. On the other hand, while many large firms actively engage in domestic political controversies in their own countries, and spend considerable sums of money attempting to influence domestic politicians and obtain legislation or other governmental action that is favourable to their own interests, if such firms do similar things abroad, the line between 'foreign interference in local politics' and legitimate representation of the economic interests of the firm, becomes very thin indeed, especially if the firm looms very large in the local economy, as is sometimes the case in a small country.

So far we have been discussing the political influence exercised by foreign firms over the local government which arises from their economic position in the economy. Some influence of this type is unavoidable so long as foreign firms are active and important in the economy. It is not necessarily economically undesirable. But from the point of view of the receiving countries, political problems may also arise from another direction, for foreign firms operating within the economy may on occasion be used by their own governments as instruments for carrying out the international policies of these governments. In principle, international firms are economic organizations and conduct their affairs in the light of economic or commercial criteria, but since the parent is subject to the laws of the country in which it is incorporated and has a legal responsibility for the activities of its subsidiaries, it may require even the subsidiaries operating in foreign countries to implement within those countries the political policies enjoined on it by its home government. This problem has arisen in recent years most acutely with respect to the international policies pursued by the United States as part of its drive to counteract what it sees as communist influence, especially in the developing countries.

The United States imposes restrictions on the commercial and financial relations that its nationals are allowed to have with certain countries. These may range from an outright prohibition of almost any kind of contract, as is the case, for example, respecting Cuba and China, to regulation of the types of goods and transactions permitted, for example, the regulation of sales of 'strategic goods' to some countries. In consequence, if a particular industry in a foreign country is largely in the hands of subsidiaries of American firms, decisions may be made in accordance with the political policies of the United States government which run directly counter to the economic interests (or political policy) of the foreign country. For example, because of the political relations of the United States with Cuba, an oil refinery in Japan owned by an American oil company refused to supply lubricating oil to a British ship in a Japanese port on its way to Siberia because it was carrying Cuban sugar. This is a direct restriction of Japanese exports in the service of the political quarrels of the United States. Similar cases have been very common in recent years when trade with communist countries has been involved, and a strong argument can be made for an international convention outlawing such intervention in the economic life of other countries except in accordance with internationally agreed sanctions. This type of action becomes a source of tension between countries, especially when a smaller country has no effective means of asserting its position vis-à-vis a larger one.

It is, however, worth asking *why* a politically independent country may not have effective means of asserting its position, since it can always nationalize foreign business, refuse foreign investment or foreign aid, and otherwise eliminate any control or even influence exercised by foreigners operating inside the economy. This question raises the problem of foreign economic control since, given political independence and assuming that the government is not an outright puppet in the pockets of foreign interests, nor need fear the cruder forms of physical pressure from a foreign government such as military action or naval blockades, the restraints on its actions vis-à-vis foreign firms will tend to be related to its fear of economic retaliation. So leaving aside for the moment the question of economic control linked with political control, let us examine the notion of 'foreign economic control' taken by itself. It can hardly be seriously maintained, for example, that the government in Canada is 'controlled' by the United States government (although it may on occasion moderate its policies in deference to or for fear of United States economic pressure), but it is widely maintained, and with some plausibility that the Canadian economy is in some sense controlled by United States economic interests because of the extent to which Canadian industry is owned by United States firms.

*Foreign Economic Control*

The charge of foreign control cannot simply be a complaint that the government is not free to determine its policies independently of foreigners, or that the economy is influenced by the activities of foreigners, for the economic links among nations are so extensive that no country, not excluding the largest, can boast that its policies or the state of its economy are completely independent of the actions or policies of other countries, i.e., of 'foreigners'. In the context of this discussion, the foreigners referred to are not outsiders taking decisions which merely affect the economy from the outside, but rather they are foreign groups owning assets and engaged in economic activities *within* the economy.

At the very minimum, a charge that such groups *control* the local economy must imply (a) that their actions have an important effect on the economy as a whole, *and* (b) that they are not themselves subject to the effective control of national authorities. The latter implies that for one reason or another (and excluding direct political subservience) the national authorities are unable to intervene effectively if they do not approve of the foreigners' behaviour. In a country that is politically independent, such inability to intervene can arise because of ignorance or incompetence on the part of the government officials, or because the government feels that the cost would be unacceptably high to it, or to the economy as a whole, of any intervention that would be unacceptable to the foreign firms. To treat this latter reason for a government's refusal to intervene as foreign economic 'control' would be most misleading if it implied that a government could complain of economic 'control' merely because it did not want to accept the cost consequent on losing the foreign investment.[1] The problem is somewhat different, however, if the expected costs are largely due to the fear that the firms will appeal to their own governments who will then impose wider economic or political sanctions to support them.

The possible effectiveness of such wider sanctions may well be related, albeit only indirectly, to the extent of direct investment from the relevant investing country, for the larger is such investment, the greater are the general financial and commercial ties likely to be between the two countries. Hence, wider sanctions, such as the imposition of restrictions on the granting of ordinary commercial credits by the foreign banks, or the intensification of tariff or other trade restrictions, or deliberate pressures on the external value of the currency or the withdrawal of previous support of it, may be more effectively imposed the more extensive are the economic ties that have arisen in the wake of the activities of foreign firms. In this sense, a larger amount of foreign investment in the economy can lead both to an enlargement of the area within which disagreements may arise between countries and to a greater vulnerability of the receiving country to foreign pressure.[2] On the other hand, the greater is the commitment of particular foreign firms, or of a particular investing country, in the economy of another, the greater is the possible loss in the event of conflict. Hence there is always a strong incentive for both sides to resolve their difficulties amicably.

However, there is no escaping the fact that the cost to a large country of a dispute with a small country may be very small indeed compared with the cost to the small country, unless the latter has a virtual monopoly over commodities that are of great importance to the large country. Large countries do in fact often make 'representations' to smaller countries in support of the commercial interests of their own firms. To take another example from the oil industry in Japan: in 1962, the Japanese government was considering measures to deal with the problems of the oil refining industry, some of which related to local refineries that were owned by or tied to foreign oil companies who were pursuing common policies in an attempt to maintain the price of crude oil. It was known that the Japanese government was uneasy about the extent of foreign control over the industry, and the United States and the United Kingdom governments, fearing that some of the measures would discriminate against the affiliates of American and British companies, made representations to the Japanese governments about the proposed measures. The representations involved heavy pressure and were partly successful, since clearly the Japanese economy could ill afford a serious conflict with these countries, especially with the United States.[3]

Control over affiliates of international firms in a foreign country is also used to achieve the economic as well as the political objectives of the parent country. In order to strengthen their own balance of payments, both the United States and the United Kingdom have for some years put pressures on the international firms based in their respective countries to force the firms to discriminate in the conduct of their financial affairs against the other countries in which they operate. In particular, British and American international firms have been asked to reduce their rate of investment in foreign countries by repatriating more of their earnings arising from foreign operations. Since the foreign income of an international firm will have been derived from sales made in foreign countries, the demand from the parent government that it should repatriate its profits instead of reinvesting them abroad in the normal course of its international expansion is a way of putting pressure on the balance of payments of other countries in order to improve that of the parent country. Such pressure is most likely to occur during times of international financial strain and countries may well find themselves subjected to this kind of discriminatory treatment when they can least afford it. Such government policy is apparently not seen to be inconsistent with the heavy pressure often exerted on other countries to 'liberalize' their attitudes toward foreign investment, especially by the United States.

Thus, for a number of political and economic reasons, a government, when accepting large amounts of investment by a foreign firm, will often uneasily fear that it may be inviting a cuckoo into its nest, which at a later stage will create much trouble.

Nevertheless, it is probably true that much of the dislike of foreign investment is a dislike of foreign influence in the economy regardless of whether the investment is believed to be economically or politically damaging. An attitude of this sort cannot be appraised against economic criteria, for it essentially reflects a political preference. It seems reasonable, however, to take for granted that among any group of people there will be a widespread dislike of extensive foreign influence, and a general preference for running their own affairs. In spite of such a general preference, however, the same people may be willing to give up some freedom from foreign influence in return for economic gain. In other words, the economic advantages to be gained from foreign investment must always be offset against the loss of 'psychic welfare' that admitting more of it, and thus accepting more foreign influence, may bring. This is a familiar type of economic problem and an economist can immediately begin constructing curves depicting an assumed 'trade-off' or rate of substitution between dislike of foreign domination on the one hand, and the economic gain that may arise from accepting some amount of it on the other; or, if foreigners already have an important influence in the economy, a trade-off between the gain in 'satisfaction' to be obtained from curbing their power and the risk of economic loss due to the reduced investment that may follow any actions to this end. In general, it seems reasonable to assume that the more foreign investment there is in the economy, and the greater the influence of foreigners, the greater will be the total 'disutility' felt by the community. This may be expressed in an increasing volume of complaints about foreign investment and an increasing pressure on the government to restrict the flow. Hence it is not enough to argue that some gain always accrues to the receiving country, for the gain must be enough at least to offset the disutility of having one's industry dominated by foreigners.

This way of putting the matter makes it immediately clear that one aspect of the problem of foreign investment takes us into the slippery area of the attitudes of 'the people' or of a community as a whole. In general, we have to assume that these attitudes are expressed by governments, in spite of the obvious fallacy of the assumption. Sometimes a government may resist popular pressures to take action against foreign investors because it is much more aware of the benefits they bring (or of the cost of attacking them) than are the critics; and sometimes governments may strike hostile attitudes that are not widely reflected among the people. Hence the position of different countries as expressed by their governments will change depending on the groups in power and on the pressures on them. As a result, there is always an incentive for outsiders who want a change to meddle in domestic politics, and not least among these may be foreign interests.

In the above discussion I have tried to separate the question of an objection 'in principle' to foreign investment because it is held (correctly) to lead inevitably to foreign influence (whether good or bad), which seems to me to be a purely political matter, from the question of the net economic benefit or loss expected from it. Such distinction cannot, however, be carried very far because there is a reciprocal relationship between political dislike of foreign investment and the feeling that the foreigner is economically 'exploiting' the country.

*The Problem of Economic Exploitation*

I once tried, not entirely successfully, to give a clear economic meaning to the term 'exploitation' in this context when I suggested, in connection with the returns received by the oil companies in the

Middle East, that if a foreign company, *because of its monopoly position or political power*, were able to obtain an after-tax profit significantly greater than that required to induce it to undertake the investment, one could argue that the company was 'exploiting' the country concerned.[4] This definition has been criticized by Professor Kindleberger, who argued that the situation of a company bargaining with a government is one of bilateral monopoly in which the two parties attempt to agree over the distribution between them of the total gain from the investment. There may be a wide gap between the maximum 'price' that the government would be willing to pay to get the foreign investment and the minimum 'price' that the company will demand in return for making the investment. One could then equally argue that if the country succeeds in paying the company less than it would be willing to pay, it is the company that is being exploited. In practice, both sides are very likely to get more than their minimum requirements, and can thus be said to exploit each other.[5]

There is force in this argument,[6] and it is probably true that the definition I offered is far too wide, since 'exploitation' is a pejorative word and it is misleading to define it to cover situations where pejorative implications are not helpful. If exploitation means anything at all in this sense it means that the stronger takes advantage of the weaker for his own ends, but this plainly implies that exploitation will occur in any bargaining situation where the parties are not equally balanced, and hence the use of the word in a condemnatory sense in this context is too ambiguous to be useful.

Even if we conclude that 'exploitation' cannot easily be given a clear objective meaning, however, we cannot ignore the fact that the sense of being exploited is one of the powerful reasons for much hostility towards foreign investment. It might, of course, be said that if all the gain goes to the foreign firm, the country obtaining nothing is being exploited,[7] but this is not what is meant by those who complain of exploitation. The notion is rather that the foreign company is obtaining a *disproportionate* gain—a gain greater than he ought in some sense to get when it is compared with the gain accruing to the receiving country. In other words, it is a feeling of 'unfair' treatment, and this is what makes the notion so extremely difficult to analyse from an economic point of view, for it does not depend on any objective measurement of the absolute gain received by either side, but on their *relative* positions in the light of what seems fair to one side only. It is for this reason that the question of exploitation must be treated as one aspect of a political attitude toward foreign investment in general, for the feeling of being exploited increases any existing resentment against foreigners, and the psychological disutility of having to put up with extensive foreign influence is intensified if it is felt that the bargain is not fair or, conversely, is reduced if it is felt that the foreigner is going out of his way to be fair.

In this respect, the introduction by the international oil companies in the Middle East of 'fifty-fifty'—the equal sharing between governments and companies of profits attributed to crude oil production—was a stroke of genius, for equality of sharing has a distinct ring of fairness about it. Because of this, the oil companies enjoyed a relatively long period of reasonably amicable relations with their host countries during the 1950's (an equal sharing of profits did not obtain in Iran at the time of Moussadeq and the nationalization of Anglo-Iranian). Although changes were demanded in other aspects of the oil concession arrangements with the companies, attacks on profit-sharing agreements for a very long time related not to the principle of equal sharing but to various ways in which the companies were alleged to be evading it by falsifying accounts, failing to expense royalties, obtaining unjustified marketing allowances, and finally by making unjustified reductions in prices. The arrangement broke down and gave way to 'tax reference' prices as a basis for determining the producing country's share of the profits largely because the governments of the oil-producing countries refused to accept the right of the companies unilaterally to alter crude-oil prices—a situation they considered to be 'unfair'.

I have argued above that the economic gain to a country of direct foreign investment must offset its psychological disutility to the people. The disutility seems often to be greater with respect to investment from large and powerful countries (or firms) than to investment from smaller ones with no imperial past or pretensions. Thus, in some industries the underdeveloped countries sometimes accept more readily investment from Japan or from the countries of Western Europe than from the United States, and from smaller firms than from larger ones even if they can expect somewhat greater economic gain from the larger. When this happens it is reasonable to presume (and indeed is sometimes asserted) that the smaller country or firm is feared less than the larger one. There is often a generalized uneasiness that the more powerful the foreign investor, the more likely it will be that the country will get an unfair bargain because, even if the actual gain expected for the country is the same, the *distribution* of the gains will unduly favour the foreigner. (This implies, of course, that the total return on the investment is expected to be greater the larger and more powerful the investor.)

The types of attitudes described above are familiar in different contexts to economists. It is, for example, often argued that the 'welfare' of an individual is related not only to his own income but to the relation between his income and that of others, for even if he is comfortably off he may be miserable if neighbours whose equal he feels he is receive more than he does, unless he also feels that the differences are fair and reasonable. Wage differentials are as acute a source of tension in industrial relations as are the absolute levels of wage rates. Again, psychological 'utility' is at the basis of the theory of consumers' choice, for consumers are presumed to attempt to maximize their 'utility'; and in the theory of income distribution it is recognized that workers (and businessmen) may prefer leisure to work, and therefore that a sacrifice of measurable economic receipts may increase satisfaction and in consequence welfare. Similarly, the notion that a whole community may have preferences of this kind frequently appears in economic analysis. Thus, there is nothing 'irrational' in principle if a community prefers to run its own affairs and to demand an economic gain from admitting foreign influences that is commensurate with the disutility of submitting to it; nor is it unusual that felt 'unfairness' should in itself be a 'disutility' which would raise even more the gain demanded to offset it. This, together with the problem of potential political domination, is at the basis of many of the difficulties facing international firms today.

When a foreign firm is contemplating investing in an underdeveloped country it usually enters into negotiations with the government, since special arrangements which are not required in the developed countries are usually necessary. Many developing countries have specific legal provisions governing the activities of foreign firms, and the firms themselves require assurances with regard to the treatment they may expect once they have entered the country. Fear of exploitation may influence a country's attitude towards the acceptance of foreign investment and may thus affect the terms demanded in such negotiations. Similarly, the hostile political attitude that such a fear may evoke will make the foreign firm more hesitant to enter, with the result that the level of expected profit necessary to attract it will be raised. The really difficult problems in international relations, however, do not arise so much over the terms on which investment from new firms is admitted by countries, although the controversies may be lively and the bargaining intense, as over the treatment of foreign firms already established, since a country cannot allege exploitation and take action against a firm unless the firm is already there and its operations are successful.

In the initial bargaining, attempts are made to agree on rules, conventions, or procedures to govern the activities of the firm and the treatment to be accorded it by the government, but after the investment takes place and the firm is operating successfully the terms of the original agreement may appear to be inadequate and a reappraisal may be demanded by the host government. Such a demand may arise for purely political reasons—a hostile government comes to power, conflict over the role of foreign firms in general or of individual firms in particular becomes a more acute political issue, 'socialist' ideology spreads and private enterprise in general comes under fire, etc. It may also arise simply because a government subsequently decides it can obtain a greater share of the profits of the firm, or force it to reduce its overseas payments, after the firm is established and its bargaining position is perhaps somewhat weakened. Hence the government may attempt to increase taxation, manipulate foreign exchange rates in order to force the firm to buy or to sell local currency at unfavourable rates, restrict dividend remittances in order to force it to reinvest more of its proceeds locally, etc.

These are common devices and governments, especially if they do not take very seriously the sanctity of property or contractual rights, are often in a very strong bargaining position vis-à-vis firms with an extensive financial commitment in the country which they are unwilling to risk. Charges of exploitation may be used as a bargaining technique and at times deliberate attempts may be made by the government to arouse popular feeling. Thus, firms may find it necessary to increase their 'payments' to the economy in order to 'buy off' an enhanced political resistance. Much of the conflict in recent years between the international oil companies and the governments of the countries producing crude oil has been the result of little more than a simple attempt on the part of the governments to increase their revenues, although more recently other issues of substance have been involved, in addition to a strong feeling of exploitation.

We shall not deal further with the political questions or with issues that arise simply because governments see an opportunity to get more out of the foreign firms. More interesting from the point of view of the operations of large international firms are those demands for a reappraisal of the position of a foreign firm which are specifically related to actions or policies of the firm that are understandably of concern to the host country and that were not dealt with or even envisaged at the time when the original arrangements were made for the establishment of the firm. Some of the situations resulting from the policies of foreign firms, and which are *reasonably* a cause for concern to host governments, are fairly easily dealt with, but others are more intractable because they raise very wide issues affecting the international policies and operations of the firms. On some issues, the actions of a foreign firm in a particular country are more closely related to its world-wide policies and its position in other countries than to its position in a particular country, and hence cannot easily be modified without consequent modifications of its operations in other countries.

*Problems Related to Specific Policies of Foreign Firms*

Until fairly recently foreign firms tended to employ foreign personnel, especially in skilled and managerial positions, even where they might, with little extra cost, have trained nationals for the job; they tended to import their materials, especially from their parents, from other subsidiaries in their own group or from their own countries, even when they might have procured them locally with little loss in quality and cost.

Since the Second World War there have been marked changes in this respect. For a number of reasons which cannot be examined here all countries, developed and undeveloped alike, have been laying increasingly greater emphasis on economic growth, and governments everywhere have been taking an increasingly active role in promoting development. As a result almost all countries nowadays have some sort of an economic 'plan' designed partly to stabilize, but largely to promote the growth of their economies. From a strictly economic point of view, the attitudes of developing countries toward investment by foreign firms have become shaped by their desire that the firms should contribute to the development of the country and the activities of the firms are increasingly appraised in this light and in the light of the 'plan'.

It is now common practice for the receiving countries to insist that the firms accept a variety of obligations wherever possible, such as the training of skilled labour, the employment of local people in management, the local procurement of supplies, the recognition of trade unions, and occasionally the admission of local equity capital and the establishment of joint ventures with local entrepreneurs. Such demands have often created problems for established firms, and sharp conflicts have on occasion arisen over government attempts to interfere with what many firms considered to be their 'management prerogatives'. But except for the question of equity participation, demands of the type outlined do not often raise really serious questions of *principle* for the foreign firms and hence agreement has usually been reached. At present as much 'indigenization', or 'integration' into the local economy, as possible has become the established (or at least the proclaimed) policy for most of the larger international firms; they have recognized that such a policy is 'good business' even if it entails some increase in local costs, because it may sharply reduce local feeling that the firm operates in an alien enclave from which it exploits the economy. Some firms have also welcomed local equity capital, but others have felt that the resultant interference with their freedom of action was unacceptable. We shall return later to a discussion of this question, about which there is much dispute in principle as well as in practice.

Government efforts to persuade firms to alter their policies in order to increase their contribution to the local economy of the type discussed above are different in kind from those which merely attempt to increase taxation or to force a greater contribution to the country's foreign exchange earnings. The latter actions of government are not a response to any particular policies of the firms concerned. It may be of course that some profits are largely due to local monopolistic practices on the part of the firm. In such cases the government should attack the offending practices directly or, if this proves impossible, it can introduce some form of control over monopolistic pricing. This raises no problems related specifically to foreign firms in principle, for such controls may equally apply to domestic firms.

Although a government may be in a position to control monopolistic practices relating to transactions conducted within the domestic economy, it will have more difficulty if the practices relate to external transactions, and in particular to imports. A subsidiary of a foreign firm will usually import both goods and services – raw materials, semi-finished or finished goods, patent rights, technical 'know-how', managerial advice, etc. – from its parent or from subsidiaries of the same group. The prices charged for these goods and services become costs for the local subsidiary but not costs for the international firm as a whole since no payments are made outside the firm. Such payments by one subsidiary become receipts of the parent or another subsidiary and hence cancel out in the consolidated accounts of the firm. The consolidated net receipts of a firm are the difference between the payments made outside the firm (including wages and salaries but excluding dividends) and payments received from outsiders, primarily from customers, by units of the firm.

Thus, if we ignore income taxes, the policy of an international firm with respect to the internal prices charged different subsidiaries will affect the accounting distribution of total profit among the subsidiaries, but not the total profit of the consolidated firm. However, since taxes paid by subsidiaries to the government of the countries in which they operate are among the payments made by a firm to outsiders, the group as a whole has a strong interest in minimizing the amount of taxes for any given volume of receipts in order to increase its profits after tax. Clearly, if effective tax rates are different in different countries, the firm will have an incentive to adjust its internal transfer prices wherever possible in such a way that the cost of intra-firm transactions are highest to those subsidiaries in countries where tax rates are highest, for the higher the costs the lower will be the profit shown in the accounts of the subsidiaries and therefore the less will be their income tax liability.[8]

This is roughly the principle of the 'tax haven'. Firms may set up special subsidiaries in countries where tax rates are very low and arrange the routing of paper titles to goods so that a 'purchase' and 'sale' takes place through such subsidiaries at relative prices that permit most of the profit on the international transaction to appear in their accounts in the 'haven' country. It may be that such extreme practices are not commonly adopted by the more important international firms nowadays, but in less obvious forms internal prices are widely used to shift taxable income in order to minimize the tax liability of firms. They may also be used as a deliberate means of augmenting the value of exports from the parent country, and some international firms deliberately conduct their operations in the interests of their home countries.[9]

Some of the problems that the major firms in the international petroleum industry have had in the last ten years with the governments of countries in which they operate have been the direct result of their attempts to maintain a pricing system which had great tax advantages but which discriminated heavily in favour of countries producing crude oil and against those importing either crude oil or products. In consequence, serious conflicts arose with a number of importing countries, and especially with some developing countries whose refineries or distribution facilities were almost entirely owned by subsidiaries of international groups which imported the crude oil for their refineries, or the products for their distribution networks, from other subsidiaries of the same group.

The international petroleum firms are vertically integrated; each operates at all levels of the industry, producing and transporting crude oil, refining it and distributing the products. An overwhelming proportion of the international industry at all levels is in the hands of seven of these firms (the 'majors'), and international trade in crude oil consists largely of transactions among the subsidiaries of this group of companies. After the Second World War until the end of the 1950's there was little price competition either from outside or among the majors themselves, and the prices of crude oil and products entering into international trade were those established by the companies to govern their internal transactions as well as their outside sales. So long as there was no source of supply other than from companies pursuing roughly the same price policies, importing countries whose refineries or distribution facilities were owned by the international firms were not prejudiced by this ownership because, even if their industry had been domestically owned, they would have had no alternative than to import from those companies and pay their prices.

If, however, a new competitor appears in such circumstances and offers supplies at lower prices, an importing country in such position will be immediately placed at a disadvantage if the international companies refuse to lower their transfer prices and insist on continuing to import supplies from their own affiliates. The government of the country will not be able to import the cheaper crude oil because the crude is of no use without refining facilities; nor can it distribute the cheaper products unless it has independent distribution facilities. In a very real sense this aspect of the country's industry is 'controlled' by foreign firms, and although these firms may bring advantages to the country in other respects, their policies in respect of pricing may be distinctly disadvantageous. If the government attempts to do something about it, however, it may have to embark on a very expensive programme of investment. If it attempts to tax non-existent income or enforce price controls it risks serious conflict not only with the foreign firms but with foreign governments as well. Illustrations of the type of problem posed for a government in this position can be found in the experience of both India and Ceylon in the early 1960s.

*Difficulties with Oil Companies in India*

In 1958 the only important refineries in India were subsidiaries of major international oil companies, who had a few years earlier established these refineries at the request of the Indian government under special refinery agreements with the government. These agreements gave the companies the right to import crude oil from any source providing it was imported at 'world market prices'. (India had little domestic crude production.) Since the international companies that constructed the refineries were also producers of crude oil for which they wanted markets, they naturally imported from their own crude-oil producing subsidiaries in other countries.

At this time, the Indian government was becoming increasingly dissatisfied with the monopoly position of the foreign companies, and in addition the belief that the refining industry should be in the 'public sector' was becoming the accepted policy of the government. Moreover, in international markets the price of crude oil was under pressure for a variety of reasons, including Russian competition, and discounts off the 'world market prices' that had been established by the large companies were being granted in one form or another to some importing countries in Europe and to Japan, even by some of the large oil companies themselves. But the price of crude oil to India was unaffected by these changes and criticism of the pricing policies of the foreign companies was growing stronger. The matter came to a head in 1960 when Russia offered the country crude oil on very much better terms than those at which the companies were importing. A government Committee to investigate oil prices was appointed in that year and its report in the following year was not favourable to the companies.

The government brought great pressure on the oil companies and finally used its power to allocate foreign exchange as a means of forcing reductions in import prices. The companies eventually agreed to grant discounts off their 'posted' or published prices for imports. But by this time the damage had been done, and the government refused to allow the privately owned refineries to expand their capacity unless they agreed to give up some of the privileges granted them under the refinery agreements of 1951 and 1953 and also to import crude at prices comparable to those offered by the Russians. It also refused to let them participate in the building of new refineries unless they agreed to a majority shareholding for the government. The companies declined to do any of these and the government went as far as it could consistent with the refinery agreements to deny the established majors any further share in the Indian market. It pushed the development of a state-owned company and the building of government refineries and extended its efforts to find domestic crude oil. As a result of all these policies—the restrictions placed on the expansion of the activities of the private companies, the completion of the government refineries (which brought over half of the country's refining capacity under government control), the preferences granted to the state-owned Indian Oil Company in the importation and distribution of products, and the vigorous efforts to expand the local production of crude oil—the Indian government had by 1966 effectively destroyed the monopoly position of the international companies and was on the way to building up its own domestically controlled industry.

It cannot be pretended that the government's policies were based entirely on cold economic calculations, but the attempt of the international companies to maintain their transfer prices of crude oil to India while giving way elsewhere offered considerable political provocation and strengthened the position of those opposing foreign (as well as private) enterprise in oil. The companies were struggling to maintain a world price structure in the face of competitive pressures from outsiders and they tried wherever possible to hold the line. Until outside competition arose they had always met the demand of governments of importing countries for lower prices with the assertion that their prices were 'competitive'; to prove it, they simply invited the governments to find cheaper sources of oil. Since this small group of seven or eight major companies produced around three-quarters of the world's crude oil outside the United States and the Communist countries, and very little of the remainder entered into international trade in 'arm's length' transactions, the governments had to recognize that the supplies their countries needed were not available at lower prices.

With the advent of Russian competition the companies were desperately concerned to establish the argument that Russian prices were 'political'. Price-cutting by others was dismissed with the argument that it related to 'distress' sales. In 'captive markets' such as the Indian market was when the majors controlled its refinery capacity, and where in consequence there was no room for competition among the suppliers of crude oil, only pressure from the government could be expected to force reductions in the transfer prices of the companies. In its power to allocate foreign exchange the government had an effective weapon which it could use without violating its agreements with the companies so long as the 'world market prices', at which they were required to buy under the agreements, were indeterminate.

Although the Indian government did gain national control of the industry, it clearly did so at considerable cost, for scarce Indian resources had to be diverted to an industry that foreign investors were willing and able to build up at a price. Whether the government would have taken the measures it took regardless of the policies of the companies cannot be discussed here; but that it needed to take some measures to establish a degree of independence in the import policies of its refineries can hardly be questioned, for the foreign firms clearly used their control over the country's refining capacity in the period to the disadvantage of a country that could ill-afford unnecessary increases in its import bill.

*Crisis of Ceylon*

Before 1963 the internal distribution of petroleum in Ceylon was exclusively in the hands of three international companies who owned all of the distribution facilities in the country. There was no local refinery, and refined products were imported by the companies from their subsidiaries in the Persian Gulf at the 'posted' prices set by the companies. In 1960, when oil products began to become available at lower c.i.f prices in world markets, the government requested the companies to reduce their import prices. This was refused, and in 1961 the Ceylon Petroleum Corporation, a government-owned enterprise, was established and given the power to take over some of the properties of the international companies. Arrangements were made by CPC to import products from Russia, and the government imposed maximum c.i.f. prices on imports by the international companies which were only slightly above the prices offered by the Russians. The companies refused to accept these prices, arguing, as they had in India, that the prices were 'political' and 'unrealistic' and would force them to abandon 'normal commercial pricing'. They also pointed out that 'prices charged in Ceylon are the prices prevailing in the market area of which Ceylon is only a minor part.'[10]

As might be expected, the whole issue was highly charged politically, and there is no need here to recount the details of the subsequent controversy. In the beginning, the CPC took over only selected parts of the companies' properties, arguing that it would be wasteful of the country's economic resources to construct new (and redundant) facilities on the Island to enable the state company to distribute the petroleum products that the government was importing from Russia. The foreign companies and their governments protested against the expropriation and demanded compensation at 'fair market value'. No agreement was reached on compensation and in February 1963 the United States government suspended aid to Ceylon in accordance with the 'Hickenlooper Amendment' in the U.S. aid legislation (two of the three companies were American). The companies continued to refuse to accept the government's maximum prices for their imports, and in 1964 the CPC took over the rest of the companies' facilities with the exception of bunkering and aviation refuelling, which served international customers.

The companies had much less incentive to come to an agreement on prices with Ceylon than they had with India, since the Ceylonese market was much less important and the investment at stake much less. At the same time the government was less astute in handling the problem than the Indian government had been. But the issue was essentially the same: an insistent attempt on the part of the international companies to maintain their international price structure in those markets where they had a complete monopoly over the facilities required to handle imports, while giving way elsewhere. The cost to the Ceylonese government of expropriating the companies at the 'fair market price' demanded by the companies would have been very high; at the same time the country was being forced by the companies to pay higher prices for imports than those at which alternative supplies on a continuing basis were available, simply because the distribution facilities were in the hands of foreign-owned subsidiaries whose policy was to maintain prices in order to protect the consolidated profits of the international firms to which they belonged, rather than to import into Ceylon at the lowest available prices.[11]

It is not necessary to condemn or approve the way in which the Ceylonese government handled the affair to recognize the difficulty of the underlying issue. To be sure, the companies were in an awkward position in view of their international interests, and of course they had a strong case in international law over the issue of compensation. But they were treading on very sensitive toes and they never seem to have removed their heavy boots, refusing to see the broader implication of their position from the country's point of view. For example, they refused to lend, lease or sell tankage or other facilities to the government in order to enable the CPC to get a start. They were in fact unwilling to 'meet competition' on prices. Ceylon is a poor country and the only ways she had of legally protecting herself against this type of policy on the part of the companies were expensive: to build new—and redundant—facilities to displace those of the companies or compulsorily to take over the existing ones with compensation which the companies would accept. She chose expropriation, but no agreement could be obtained on compensation, and in any case the government gave no assurance as to when any would be paid.[12]

The United States government had no choice under its aid legislation, for the 'Hickenlooper Amendment' of 1962 requires the President to suspend aid under the Foreign Assistance Act to any country nationalizing property owned by United States nationals and failing to make prompt, adequate and effective compensation in convertible foreign exchange. The United States has, of course, the right to lay down the conditions under which it will give economic assistance to foreign countries, and it has always laid great stress in its aid legislation on the importance of promoting private enterprise. AID is specifically enjoined to take account of the 'progress' being made by recipient countries toward recognition of this importance. But unconditional support of U.S. private firms abroad clearly strengthens the hands of these firms in negotiations with the countries, even with respect to such commercial policies as were adopted in Ceylon, and may pose a very difficult dilemma for any government whose economy is heavily supported by both U.S. private investment and U.S. aid.

*Conclusion*
I began this paper with the assertion that direct private foreign investment can make a great contribution to the economy of the receiving country; because of this contribution most countries welcome it (with reservations) in spite of the problems of the type we have discussed and of which they are aware. Some of the sources of conflict that may arise between governments and companies are fairly easily removed; others are inherent in the institutional structure and legal position of international firms. Such firms are subject to the political and economic pressures of many governments and must give in to the stronger. On occasion the governments of the host countries are in a strong bargaining position and may get the lion's share of the income taxes arising from the international operations of the companies—as is the case in the international petroleum industry at the

present time. In other cases the situation may be reversed. In many important respects, however, the home government of an international firm has extensive power to reach across national frontiers and intervene in the operations of subsidiaries in other countries. The parent company is subject to the laws of its own country and to the pressures exerted on it by its own government with respect to the conduct of its international affairs, which includes by definition the conduct of its subsidiaries abroad. As we have seen, this may give rise to political as well as to economic actions which are reasonably interpreted as foreign interference in the affairs of host countries.

Even if left alone, however, the financial objectives of an international firm will cause it to discriminate among the countries in which it operates in order to minimize its tax payments. The organization of the firm may also give rise to other discriminatory policies that we have not discussed; for example, it is common practice for a firm to allocate particular export markets to subsidiaries in particular countries. This may well result in the restriction of the exports of these or other countries which might have developed an export trade in the commodities concerned in the absence of the division of markets imposed by the international firm and enforced by its control over subsidiaries in other countries.

Thus, in spite of the clear gains to be had from foreign investment —gains which are much greater than can be attributed to the increased supply of capital alone—few governments in receiving countries are willing to give international firms a free hand. Suspicion of foreign investment is especially prevalent in developing countries, partly because of the fact that such investment tends to come overwhelmingly from the economically more powerful countries of the world—at present particularly from the United States—and the vulnerability of the weaker countries to economic and political pressures from the stronger may be intensified, the larger the amount of investment that they accept. On the other hand, as pointed out above, any extension of commercial and financial relationships between the large and powerful and the weak and small countries may have a similar affect. The alternative to an extension of such relationships can only be an increasing economic fragmentation of a world made poorer all around, which is hardly an attractive prospect.

I have not tried to indicate solutions to the problems raised in this paper. Clearly the fundamental solution would be an arrangement under which the international firm would lose its nationality, so to speak, and thus be able to operate free from the control of a home government. This would require a procedure for the international incorporation of firms. International incorporation need not bring any important changes in a firm's ownership or financial position, but it might lead to a wider quotation of its shares on the stock markets of the world and the acceptance of payment in a larger number of currencies, and thus to a wider distribution of ownership throughout the world. If at the same time the firm became subject to a single international income tax, the proceeds of which could be distributed according to some formula among the countries in which it operated, its incentive to discriminate among countries for tax reasons would be removed. The difficulties would be many, but there is no reason to think that they would be insurmountable.

Developments such as these can be expected only in a far distant future, if at all. In the meantime the large international firm will remain the chief vehicle of direct foreign investment and lesser remedies may be attempted, including the adoption of 'international charters' of one kind or another designed to restrain the more undesirable types of governmental interference. So far, such attempts have been biassed in favour of the investing countries, for these countries have given insufficient attention to the defects of their own policies and to the less commendable behaviour of the private firms. At the same time, however, the firms themselves may offer solutions, for some of the more 'progressive' of the international firms are aware of the problems and make efforts of their own to adopt policies which reflect a more truly 'international' outlook in their operations.

One device that is widely discussed and has gained increasing popularity in many countries is that of the 'joint venture', which involves the establishment of a company owned in agreed proportions by foreign and local investors, who may be private individuals, the government or both. An already established foreign firm may be requested to accept local equity capital. Some international firms readily accede to such proposals; others resist them, or agree only for their less important subsidiaries. From the point of view of the foreign firms, the chief objections arise from a dislike of interference with their freedom of management, and particularly from a fear that political issues will be injected into their managerial deliberations from within. From the point of view of the receiving country, the chief advantages relate to the closer association expected between the firm and the local economy, leading to an enhanced contribution by the firm to the local economy, including a reduction in repatriated profits and of foreign control, since a firm with shared ownership of this kind may have a greater independence in making decisions than does a subsidiary wholly owned abroad.

Such arrangements are often useful means of reducing local antagonism toward foreign investment and may in consequence attract an increased volume of investment. But they may also result in a reduction of foreign investment by bringing about a substitution of domestic for foreign capital. When demands are made that an established firm admit local equity capital by selling shares to local investors, the effect is equivalent to a repatriation of part of its existing investment by the foreign firm unless, of course, the new local capital is used for expansion that would not otherwise have taken place. It is therefore commonly argued that for an underdeveloped country to divert its scarce capital to investment in enterprises that foreigners would be glad to undertake is bad economic policy. If a government demands a share in foreign enterprises, the great need for local investment in schools, roads, public utilities and other infrastructure, which can only be made by the government, is often put forward to dissuade it from using its resources to displace private capital; and even local private investors are urged to undertake activities that do not attract foreign investment, rather than to use their capital in an attempt to displace foreign investment.[13]

There is considerable force in this argument, especially as it applies to governments, and certainly the extent to which foreign investment is displaced by local capital that would have come forward for other uses must be taken into consideration. On the other hand, it is widely held that the supply of capital in developing countries is itself very much a function of existing opportunities for profitable investment, and that greater saving would take place, or that savings which otherwise would flow in non-productive directions (such as the purchase of gold ornaments) would become available for investment, if more secure productive openings for local investment were available. The shares of foreign firms might well provide very attractive investment opportunities for such savings, thus increasing the supply of local capital for investment. If this is in fact the case in some countries, then the insistence that foreign firms accept local equity capital—and especially private capital—may even increase the total supply of capital instead of merely forcing a substitution of foreign by local capital.

The 'joint venture' may be a useful vehicle for foreign investment and a helpful way of solving some of the problems we have been examining, but if it became the predominant form in which foreign investment took place, the role and structure of the large international firms would be drastically changed. No longer could they be looked on as reasonably cohesive administrative organizations operating in accordance with general policies laid down by their central management. Each of their foreign operations in which local capital played a significant role would have a much greater degree of independence both financially and administratively, and it might well be difficult to call the international group an *organization* in any meaningful sense.

The difficulties created by the operations of international firms for many of the countries in which they operate are already undermining the international acceptability of these firms. Solutions that compel an institutional partnership with the government or with private investors in host countries will undermine the autonomy and independence of the firms as international organizations. Increased government control might enhance the risk of conflicts between governments on economic matters. A strong case can be made for the decentralization of power that is inherent in the existence of large economic institutions relatively independent of governments. The question is clearly debatable, but in spite of all the defects of the present situation, the relative freedom of action of large international firms does have many economic and political advantages. It may be, however, that the only way of maintaining these advantages will lie in international arrangements which permit the firms to become truly independent of any one country, more internationally oriented in their policies, and more international in their personnel and capital structure, and thus more acceptable to countries fearing that foreign control, especially from the larger and more powerful countries, will undermine their independence.

## NOTES

1  It is misleading and in a sense even contradictory because it makes the government's *willingness* to accept the cost part of the definition of the extent to which the economy can be said to be controlled by outsiders. On the other hand, if an economy is so dependent on a few foreign firms that it would virtually collapse if the firms ceased operations, as for example might be the case with some of the oil-exporting countries, then in a very practical sense the government's choices are so severely limited that one could sensibly speak of foreign control. Even in such cases, however, the cost to the companies of ceasing operations may be so high that the government is nevertheless in a strong position.
2  It should be pointed out, however, that any increase in the trade or other commercial and financial transactions between countries will similarly augment the vulnerability of the weaker economies to pressures from the stronger.
3  Such instances are widely reported in the press but the type of pressure that is brought remains shrouded in diplomatic secrecy and is rarely precisely documented. Hence, it may be much exaggerated; nevertheless, the obvious inequality in bargaining power, and the fact that representations are in fact made in such circumstances, intensifies the general suspicion of foreign investment and leads to indiscriminate charges of 'neo-imperialism'.
4  E. T. Penrose, 'Profit Sharing between Producing Countries and Oil Companies in the Middle East', *Economic Journal*, June 1959, pp. 238–254. [See Essay VIII, below.]
5  Charles P. Kindleberger, *Economic Development* (New York: McGraw-Hill, 2nd ed., 1965), p. 334.
6  Kindleberger ignored the emphasis in my definition on the monopoly or

7 'It is possible, though unlikely, that one party gets all the gain, the other none. When this occurs it can be called exploitation', Kindleberger, p. 334. But why should zero gain be the criterion?

8 The scope for tax shifting of this kind is limited by a variety of considerations. It is often not practicable for an international firm to discriminate among countries on an individual basis, partly because most governments do watch for this kind of thing and the grosser and more obvious practices are more difficult to get away with. The international pricing policies of different firms differ substantially, but the discrimination that takes place is likely to be between groups of countries in the same categories rather than among countries selected individually. For a useful survey and analysis of some of the practices adopted see James Schulman, *Transfer Pricing in Multinational Business* (A thesis submitted to the Graduate School of Business Administration, Harvard University, 1966. Unpublished).

9 It is difficult to find any other interpretation of the high cost of materials and of imports in relation to the value of sales that a British Committee of Enquiry found was characteristic of the subsidiaries of Swiss pharmaceutical firms operating in the United Kingdom. See *Report of the Committee of Enquiry into the Relationship of the Pharmaceutical Industry with the National Health Service,* Appendix, I, esp. para. 23, p. 107. (Cmd 3410, 1967.)

10 See E. T. Penrose, *The Large International Firm in Developing Countries: The International Petroleum Industy* (Allen & Unwin. 1968), Chapter VIII. A more extensive discussion of many of the problems outlined can be found in this book, and I have taken most of the illustrations relating to the petroleum industry from it.

11 It should be pointed out, however, that the international companies were in a very difficult position, for they were caught between the governments of two groups of countries at this time: the countries of the Middle East from which oil was exported were demanding that f.o.b. export prices be raised, while the importing countries, like Ceylon, India, and Japan, were pressing for reductions in c.i.f. prices of imports. In the end the companies had to satisfy both groups by maintaining export prices for tax purposes in the exporting countries at higher levels than those they actually charged their importing subsidiaries or third parties. In effect this increased the total proportion of their consolidated profits that the companies paid the governments of the crude-oil exporting countries.

12 After the defeat of the government of Mrs. Bandaranaike, which had carried on the petroleum policies begun under the previous Bandaranaike government, the new government of Ceylon, faced with serious foreign exchange and economic difficulties, renewed negotiations with the companies. In June 1965 a compensation agreement was signed under which the companies received Rs. 55m., about half the amount originally suggested as the 'fair market value'. United States economic aid was subsequently reinstated.

13 For some time the policies of even the International Bank for Reconstruction and Development reflected the general belief, almost universally held in the United States, that public funds (including IBRD financing) should not be used to finance projects for which private foreign investment was available. This attitude of the Bank has now been modified.

political power of the company as the source of its bargaining power, but I do not think this destroys the validity of his criticism of my argument.

## VII

## THE STATE AND THE MULTINATIONAL ENTERPRISE IN LESS-DEVELOPED COUNTRIES*

*Rejects the view that the State is soon to lose its economic functions to the multinational firm.*

IN a recent series of lectures, Charles Kindleberger, an eminent American economist, predicted, rather hopefully, that the 'nation-state is just about through as an economic unit'.[1] This prediction comes just at the moment when large numbers of countries in the Afro-Asian world have become politically independent for the first time—members in their own right of the society of nations and under the impression that the era of the independent 'nation-state' has, for them, just begun. Professor Kindleberger's prediction implies that the peoples of these countries are under a grave illusion if they think that their newly-won independence will include any real influence over their economic affairs. Nevertheless, within Asia and Africa, which, after all, include China, reside the majority of the world's population, and it is of some interest to enquire whether this 'cosmopolitan'[2] point of view so vigorously put forward in different contexts by a number of North American economists, seems likely to prevail in the near future. True, Kindleberger does not specify the time horizon of his prediction, but one has the impression that he is not peering into far-distant historical vistas but into the foreseeable future. Already, he says, 'Tariff policy is virtually useless. . . . Monetary policy is in the process of being internationalized. The world is too small. It is too easy to get about. Two-hundred-thousand-ton tank and ore carriers and containerization. . . , airbuses, and the like will not permit sovereign independence of the nation-state in economic affairs.'[3]

True also, there is no hint that the 'nation-state' is 'through' as a political unit, which may be of some comfort for those countries under the illusion that their time has just begun. But historically political independence—like political dependence—has had very important implications for economic policy. No one, of course, assumes that any state can, by virtue of its political independence, be completely independent of the rest of the world in its economic affairs. Sharp

---

* This is a somewhat shortened version of a paper read to a conference at the University of Reading in May 1970, to be published in J. Dunning (ed.), *The Multinational Enterprise* (Allen & Unwin, 1971).

limits to the power of governments to shape the economic life of their countries as they would wish if no other country existed, are imposed by many types of circumstances, including the possibility of war, and are very different for countries of different sizes, with different endowments and in different locations. Such limits have always existed, but have not called into question the importance of the modern state as a significant economic unit.[4] Hence, the issue must turn on the extent to which contemporary developments are imposing such extensive new limits on the power of the state to adopt and implement independent economic policies that it will soon not be possible to consider the state as an important 'unit' from an economic point of view.

In this paper I shall deal with the question only so far as it relates to the countries of the 'third world'. First, I shall briefly recapitulate the contemporary developments that have led to the so-called 'nation-state controversy' and to conclusions which, from almost any standpoint outside North America, seem bordering on the absurd. Secondly, I shall examine the way in which the countries of the world are to become so closely 'integrated' into the international economy that they lose their economic identity; I shall then look at the relation of this process to conditions widely prevalent in the less developed countries. Finally, I shall suggest what seems to me a more likely course of events.

I

The new developments in the modern world that are to bring about the economic obsolescence of the state relate to technology and (or including) techniques of organization. These have reached the highest stages in their evolution so far in the United States and have enabled the large business corporations of that country not only to dominate, or if you like, 'integrate', the U.S. economy, but also to reach abroad on a hitherto unimagined scale and directly organize economic activity in other countries. It is not surprising, therefore, that American economists should take the lead in drawing attention to the prospect that these corporations (aided perhaps by some from other countries) will now proceed to accomplish in the international economy what they have accomplished at home. The emergence and spread of the large national corporation in the United States is held to provide a pattern, or prototype, not only for the evolution of the international corporation, but of the international economy as well.

The future economic organization of the world under the aegis of the large corporation is seen, therefore, as the culmination on a global scale of an evolutionary process that has been long in the making. Kuznets has looked at economic history in terms of 'economic epochs', the distinctive features of an epoch being determined by 'epochal innovations'.[5] He finds that the innovation distinguishing the modern economic epoch is 'the extended application of science to problems of economic production'.[6] It is within this framework that the modern corporation has developed. From small beginnings in the introduction of simple technology, the basic economic units in the organization of production—the factory and the firm—have grown steadily in size and in complexity.

This growth has been characterized by increasing division of labour, increasing use of capital, and continuous innovation in the techniques of both production and organization and in the nature of output. The potentialities of the corporate form of organization have been progressively developed as a means of raising capital, spreading risk, and expanding the scope of the enterprise through merger and acquisition. It became easily possible for one firm to integrate vertically entire industries within its scope and to spread horizontally over wide geographical areas. Because the potentialities for the growth of the firm tend to exceed after a point the growth of demand for a given product, the aggressive enterprise will forever be seeking new fields of activity, devising new products, and developing new markets.[7] This process has been especially prominent in the United States where 'product development and marketing replaced production as a dominant problem of business enterprise.'[8]

Such diversification brought with it an appropriate internal structure of organization, for it was found desirable to create separate central divisions to deal with the several activities of the firm, although the activities themselves were often carried on in separately incorporated subsidiaries. The 'multidivisional' form of organization is extremely flexible, for new activities can be taken on and old ones thrown off with relative ease, while the central activities of planning and directing overall corporate development and strategy can be concentrated in head offices with a very long view and a very wide horizon, which even curves round the earth. In these circumstances, and given the scope for profitable expansion abroad, the international spread of U.S. corporations, as well as of corporations in similar positions in other countries, is easily explicable. The chief older incentives to direct foreign investment—the development of new sources of supply of raw materials or control over old ones—were increasingly supplemented by the need to secure and control markets.[9] New methods of organization supported and induced by new technology in office machinery and data processing seem to have greatly extended the limits to size which had previously been imposed by the requirements of administrative coordination. Extensive decentralization of responsibility and authority are characteristic of the new forms of organization, but do not seem to have prevented the central planning of strategy and control.

Galbraith's vision of the emerging American society is relevant to the argument we shall be discussing, although it is possible that its adherents would vigorously reject any association of their views with those of Galbraith.[10] Nevertheless, it must be remembered that implicit in the discussion is the presumption that it is in the American image that the international corporation will re-shape the world, and Galbraith's picture is in broad outline plausible. It is of a society ruled—or rather managed—by the 'technostructure' of the large corporations. The members of this class—if it can be so called—are responsible for the profitable management of the vast resources of capital, skilled manpower, and technology at the disposal of the one or two hundred leading corporations which dominate the modern industrial economy. Successful management of resources on such a scale requires long-range planning. Successful planning requires, in turn, a high degree of control over the environment in order to reduce to a minimum unpredictable reactions to the corporations' activities.

This environmental control must be far-reaching. It requires extensive vertical and horizontal integration to establish and maintain control over sources of supply and of markets; effective use of advertising and other techniques of managing consumers' demand; the maintenance of close relations with government to achieve the necessary support for capital-intensive technological research and to ensure the required demand from the government as a consumer; the cultivation of close relations with academic institutions to ensure certain types of research and the required sources of technological recruitment; and the promotion of appropriate adaptations in the labour force and in the trade unions. The corporate 'technostructure' is thus the pervasive and effective power through the society.

As I understand the argument of those who predict that the economic future of the world lies in the hands of the large corporation, it is by an extension abroad of the same type of managerial control now exercised by the 'technostructure' in the United States, that the big corporations are supposed to integrate the world and reduce to insignificance the economic importance of 'nation-states'.[11] We now turn to a discussion of the way in which this state of affairs might be achieved.

II

I have seen very little concrete discussion of the process by which the world economy is to be integrated under the impact of the international corporations. The analysis seems to consist mostly of drawing analogies with the United States and sometimes with US/Canadian relations. In the United States the large corporations have not had to contend with independent states and national boundaries, for the local states have never had the type of power which could seriously interfere with interstate investment, trade, money flows, or migration. On the other hand, in spite of national boundaries, there has been considerable integration of the Canadian and U.S. economies through the agency of the international corporations. I shall return to this later.

It is true, as Kindleberger has emphasized, that the national corporation raised capital in those parts of the United States where funds could be obtained most cheaply and invested it where labour was relatively cheap, thus helping to equalize wages and cost of capital throughout the country,[12] and that through national advertising it created the same wants and brand attachments from coast to coast, thus creating a national market. In addition, its managerial and technical personnel (the 'technostructure') are highly mobile, further reducing regional attachments within the corporation. On all this type of thing is hung the generic label 'integration'—which, as the opposite of disintegration, is supposed to be a very good thing indeed. It is easy to see in broad terms that a similar process may take place internationally—in fact, that it already is taking place in some degree. But how is it to reduce the economic significance of the state?

A vivid picture of a possible structure for the emerging international economy matched against the structure of the international corporations that shape it has been painted by Stephen Hymer, so far as I know the only economist to go into much concrete detail.[13] Ingeniously combining location theory with Chandler and Redlich's 'three-level' scheme for the analysis of corporate structure, Hymer sketches a model of a world in which the international corporations dominate great industries on a global scale. Their head offices, on which the framework of control is centred and where global strategy is determined, are located in the great major cities of the world, to which the best men of all countries are attracted and which become the true centres of world power. Below these cities are arranged a hierarchy of lesser, though still large cities, where the activities requiring white-collar workers and technicians of all kinds are concentrated. These are concerned with communications, information, and all other functions involved in coordinating and supervising the managers of the production and distribution activities that take place at the next level below. The 'level III' activities are devoted to managing the day-to-day operations of the enterprise and are spread widely over the earth, their location being determined by 'the pull of manpower, markets, and raw materials'.

Thus the structure of the world economy itself reflects the corporate structure—cosmopolitan and truly international: a few great cities containing the richest and the most powerful who make the long-range plans and exercise over-all control, lesser cities, flourishing with lesser functions of coordination and supervision, and finally a hinterland of on-the-ground production and distribution personnel, and presumably even manual workers, in the smaller cities and towns below. All are 'integrated' within the hierarchical structure of the few hundred international corporations, and their development is shaped in accordance with corporate plans. Investment and production are spread widely over the world, but the centre of attention is on international *industries,* not on the development and growth of national economies.[14]

It is not very clear from Hymer's exposition how far he really believes in his own model, and certainly he is not convinced of its desirability. But if the type of development outlined is to 'integrate' the world economy so successfully that national frontiers become relatively insignificant, the international corporations will have to be able to carry on their activities—and especially those relating to their investment, trade, financial affairs, and labour and recruitment policies—with minimal interference from individual governments. It is envisaged, at least by Kindleberger, that the national interests of the countries involved would be protected by independent international bodies or through international agreements. The international arrangements would presumably be established partly as a result of free negotiations, such as those of the European Common Market, and partly also because the force of circumstances leaves little alternative to the governments concerned.[15] The truly international corporation, without national attachments, with no 'citizenship' and therefore indifferent to all national considerations, would attempt to equalize return on assets at the margin subject only to 'the discount for risk which applies realistically at home as well as abroad',[16] and would be free to do so within the internationally agreed constraints.

According to this 'cosmopolitan', or 'international' thesis, such corporations would do for the world economy what national corporations have allegedly already done for the United States economy: their capital would flow to raw materials and the cheaper sources of labour, with the result that the benefits of modern technology would reach the poorer and backward areas of the world, opportunities for all peoples, both at home and abroad, especially as internationally mobile employees of the large corporations, would be widened, and consumption standards would rise everywhere. In short, the classical advantages of the international division of labour and free international trade would bring about greater world output and greater efficiency in the allocation of world resources.

It is not my purpose here to discuss the merits of this argument.

Rather I now want to turn to an examination of the relationship that is envisaged between the international corporation and the state in the light of conditions and attitudes now widely prevalent in the less-developed countries.

## III

In discussing the problem from the point of view of the less-developed countries, I shall deal with three broad interrelated considerations: First, the emphasis that the new states place on 'nation-building', an integral part of which is the economic development of their own countries with a view to increasing the opportunities and standard of living of their own peoples in their own lands; second, the emphasis commonly placed on national economic 'planning' and the role of the so-called public sector—sometimes referred to loosely as 'socialism'; and finally, the ambivalent attitude toward foreign enterprise and the widely expressed fear of foreign 'exploitation'—again loosely referred to as 'imperialism'.

There is no doubt whatsoever that the less-developed countries in general take a 'narrowly-nationalistic' view of their situation. Economists from the industrialized country, the riches of which are spilling over the frontiers, may, with impeccable theoretical credentials and considerable practical force, point out to the poorer countries the folly of their ways, but one must deal with the world as it is. And the fact is that these countries are obsessed with their own economic problems and care not a bean for 'world allocative efficiency'. It follows, therefore, that their leaders must be convinced that the economic development of their own countries will be hastened if they give reasonably free rein to foreign corporations, or else they must be forced by a variety of economic or political pressures to 'liberalize' their policies if the multinational corporation is to prevail.

There are very great differences in the attitudes of the countries of the third world in this respect, but even in countries where foreign firms are reasonably welcome, restrictions are usually placed on their freedom of action, and these restrictions are often of a kind that force the foreign corporations to pay as much attention to national development needs as to international efficiency, if not more. There must be few underdeveloped countries in which foreign companies feel really secure and at ease, for suspicion of them is very widespread indeed.[17] Are such attitudes likely to change in the foreseeable future? I think not, for two reasons.

First, permissiveness in economic affairs generally is no longer as 'orthodox' as it once was, and economic planning, one of the most outstanding characteristics of the large corporation itself, is now widely accepted as an important function of the state in the interest of the development and stability of its national economy. The term 'planning' is only a generic one, covering many qualitatively different species of activity, but it carries with it the notion of establishing priorities among objectives, of setting targets and then of attempting systematically to achieve these targets.

Development planning is almost universal in the less-developed countries, aided and supported by international organizations and by developed countries of all political shades. In my view, development planning is not inconsistent with a large and growing private sector, nor with a high degree of autonomy for both private and public enterprises. But this is by the way, the fact is that many developing countries have conceived of planning as the antithesis of private-enterprise capitalism responding to the profit motive. Attributing much of their present lack of development to colonialist exploitation or capitalist imperialism, and with an eye on the alternative modes of economic organization seen in the communist world, many of them enthusiastically adopted—at least formally—a centralized type of planning, nationalizing their major industries and creating in the process a large 'public sector'. For the most part their reach exceeded their grasp, and their first enthusiasm has given way to much disillusion. A public commitment to 'socialism' usually remains, but in many countries reappraisal and retrenchment is the order of the day.

It would be a very grave mistake, however, to interpret retrenchment as a return to the older system, as a complete reversal of policy. More freedom and autonomy for enterprises there may be, and more sympathy and encouragement for the private sector. But governments will continue to be concerned with priorities, and will continue to intervene in attempts to ensure that the private sector, including foreign enterprises, will operate in ways consistent with overall government objectives. The priorities and objectives will differ for countries in different economic positions, but there is no reason to think that they will be consistent with the objectives of international corporations, were the latter free from state interference. They will continue to find that the state will remain a powerful and difficult economic unit. This is not to say that international corporations will become less important than they are now; on the contrary, their contribution should increase as countries develop, but this contribution will have to fit in with national priorities.

My second reason for expecting governments in the third world to continue to assert their economic independence *vis-à-vis* foreign corporations relates to the widespread dislike of foreign domination, ranging from obsessive fears of 'imperialist exploitation' to simple political pride in the ability to maintain a high degree of independence.

These attitudes will tend to persist as long as there are really glaring disparities in the economic position of the peoples of different countries.

As noted above, some countries trace many of their present problems to imperialist exploitation in the past, and their governments make great political play with anti-imperialist slogans in attempts to rally their people to support their policies. This is particularly common among governments which consider themselves 'revolutionary'. But even when governments are relatively conservative, there are always vocal groups which use such slogans in appealing to the people for support, and which governments can rarely afford to ignore. The roots of these attitudes, and the justifications for them, are far too complicated to explore here. They lie deep in grievances, real or imagined, inherited from the past, and in bitter feelings of inferiority directly traceable to the wide disparities in standards of living, technology and education that exist between the rich and the poor areas of the world. These feelings are exacerbated by sheer frustration and a sense of helplessness in the face of the overwhelmingly difficult problems inherent in attempts to force economic development and political progress at anywhere near the pace considered essential. They therefore become a potent political force in internal political struggles, and can be used to arouse class antagonism when conservative or 'reasonable' governments come too close to foreign business interests.

Because xenophobia is often used as a scape-goat for domestic troubles, there is a tendency among Western observers to put this entire complex of attitudes and behaviour into the box labelled 'irrationality' with no further attempt at analysis, or else to assume that the source of the trouble lies with 'communists' and the cold war. No one can deny that there is often much stupidity, illogicallity, and unthinking emotionalism to contend with, as well as genuine ideological controversy, but I think that there is a real problem underneath the froth worthy of serious attention—and this apart from the seriousness with which one must examine the ideological basis of conflicting views.

It is, of course, 'irrational' of economists to label other people's preferences 'irrational', provided that these preferences are consistent and that their implications are appreciated by those who express them. It has always been understood that 'welfare' encompasses much more than objectively measurable economic goods and services—workers (and businessmen) may prefer leisure to income after a point without being thought irrational, 'psychic utility' lies at the basis of the theory of consumers' choice. Even the notion of 'community preferences' is respectable in much welfare theory, in spite of the obvious difficulties, and surely it is not unreasonable to postulate that some peoples may prefer to run their own affairs as far as possible rather than have them run by foreigners. In other words, extensive and dominant foreign control of economic activities may in itself be a positive disutility to a community, and the community may be willing to incur a cost to avoid it.[18]

If 'disutility' of this kind exists, it may take the form of complaints that the country is being 'exploited' by foreign companies, who may then be accused of 'neo-imperialism'. It is very difficult to give a concrete meaning to such terms as 'exploitation',[19] but there can be no doubt that a feeling of being exploited is one of the compelling forces behind much of the hostility directed toward large foreign firms. Their size, international scope, political, managerial and technological expertise, and in general the apparently wider 'options' open to them in determining their policies, all give rise to a feeling in their less well-endowed host countries that bargaining power is grossly unequal, with the result that the foreigner obtains a disproportionate gain from his activities in the country. The notion of *disproportionality* rather than the notion that the country necessarily loses *absolutely* is, in my view, the important consideration on which to focus in explaining the prevalence of the belief that foreign investment tends to be exploitative.[20] The larger and more internationally powerful a foreign firm becomes, the greater may be its need to demonstrate its value to the host countries by contributing even more extensively to their development in terms of their own objectives.

Resentment, and difficulties, are sometimes intensified by attempts of dominant industrial countries to put pressure on less developed countries in order to force them to admit private investment more freely. There is a certain analogy here with what has been called the 'imperialism of free trade' in the 19th century. Britain was committed to free trade and she forced it on other areas under her political domination. It may be that she did so in the firm belief that, for all of the classical economic reasons, it was economically advantageous to the colonies as well as to the mother country. Nevertheless, one consequence was the destruction of some local industries with nothing to take their place. Today, infant industry (or, sometimes, infant country) arguments are almost universally accepted as grounds for making exceptions to free trade. But as Robinson and Gallagher have pointed out, the function of imperialism was to integrate new regions into an expanding economy, and it did so for a variety of agricultural and mining activities in an earlier period, but probably at an unnecessary cost to local development.

Today we have what might be called the 'imperialism of free investment', with precisely the same function, supported by arguments similar to those advanced for free trade, and enforced where possible by the dominant industrial power. From the point of view of the regions peripheral to the industrially advanced economies, however, 'free private investment' may have disadvantages similar to those attaching to free trade. Indeed, it is for this reason that even the most enthusiastic supporters of it admit the necessity for some sort of international controls.

It seems to me that these are powerful obstacles in the way of the new organization of the world under the aegis of the international corporation. Few will deny their importance, and the only question is whether or not the almost mystical forces of modern technology and organization are even more powerful. The vision of the American economists of an 'integrated' world may to some extent be shaped by their experience with Latin America and Canada, where the big American corporations are indeed making great strides in integrating many industries across national frontiers without serious opposition from governments. In Latin America, the organization of the bauxite-aluminium industry between the Caribbean, North America, and Latin America is a case in point:

'The resources of each region reach the other in a more finished form *via* processing plants in the United States. For example in 1962 Latin America imported 60,000 tons of aluminium, the bulk of its requirements, from the United States and Canada. But the United States and Canada in turn import nine-tenths of the raw materials needed to make aluminium (bauxite and alumina) from the Caribbean. That is to say, North American-based multinational corporations, four in number, mine and treat bauxite in the Caribbean and transfer the material to processing plants in the United States and Canada. Part of the output of aluminium and semi-finished aluminium products is then exported to Latin America. . . . What is . . . important to bear in mind is that the product and commodity flows are not so much between economies as between one plant and another within the multinational corporations and their own marketing agencies. Thus, intracorporate commodity transfers between plants located within different countries may satisfy the formal criteria of international trade but so far as resource allocation is concerned the flows are of an internal character, within frontiers which are institutionally, not politically defined.'[21]

It is possible, therefore, that the present position in Latin America gives much support to the American vision. I am not competent, however, to judge how stable this position is,[22] nor how much harmony there exists in the relations between governments and peoples in Latin America and the Caribbean on the one hand, and between both and their great northern imperial neighbor on the other. On this may depend whether or not experience with Latin America will contradict my thesis. The case of Latin America, however, must be distinguished from that of Canada. The essential difference lies in the fact that the peoples to the south are much less well-off by and large than are the Canadians. Here I have been concerned with the problem only with reference to the unindustrialized, undeveloped, poor countries of the world. The problems—and perhaps also the outcome—are very different when the relationships involved are less unequal and the sensitivities and needs less acute.

It is also possible, however, that my own views are excessively coloured by my own experience in the Middle East and certain parts of Asia, and with the international petroleum industry in all areas, including South America. In this industry, the necessity of international integrated planning has been put forward as persuasively as anywhere, yet events have been moving in a rather different direction since the war. The industry points to its own history in support of its arguments regarding the advantages of integration, for there is, and has been, a very high degree of international integration within the framework of a few large multinational enterprises. This integration includes a large part of the world, but excludes the Communist countries, as well as the United States, which has an independent and highly nationalistic policy of its own. Although from the point of view of overall economic efficiency the development of the industry has been seriously distorted in several important respects because of its oligopolistic and vertically integrated structure, very large supplies of oil have been brought to the consumers of the world at a really remarkable pace, and very large revenues have accrued to the less developed countries where crude oil is produced.

Before the war the international companies did not have much to contend with from the governments of the several countries in which they operated. It is probably fair to say that their greatest difficulties arose from the rivalries among their own home governments. Since the war there have been substantial changes, and, in addition to their increasingly effective demands for greater tax revenues, governments have interfered in a wide variety of matters, including transfer prices, product prices, production programmes and policies, exploration and exploitation arrangements, refinery locations, and a host of minor matters. Nevertheless, the unit of planning is still the international industry, and, strictly speaking, the crude-oil producing countries are still largely receivers of revenues, not exporters of oil.

Nevertheless, national enterprises in both exporting and consuming countries are becoming ever more important, and the international companies are finding their task of integrating and organizing the industry on a world scale (always excluding the US and the communist countries) increasingly complicated by government pressures. In my opinion, a process of international disintegration has begun in this industry, and among the major reasons for it are those I have discussed above. But I do not believe that this will be disastrous, or even a serious setback, to most of the international companies, or that it will be unfavourable to the development of the exporting countries, or disadvantageous to consuming countries. 'Adjustment problems', there will be, but, with the technical help of the oil companies, the exporting countries can find and produce oil. With cooperative arrangements between the companies and these countries, there is no reason why exports should not find their way in appropriate quantities, qualities and at appropriate times to refining centers, and, if international refining arrangements are required for particular markets, they can be worked out at least as efficiently as they are now. Arguments that integrated control by a few big western firms is essential for efficient operation of this industry are little more than assertions based on very flimsy assumptions. In a far less integrated and more open industry, the international companies could (and will) continue to grow, earn profits in a variety of ways, and in general adapt themselves to changing environments and diverse government policies.

\* \* \*

To conclude, I suggest that the state will remain an overwhelmingly important economic unit in the less-developed world, at least so long as that world is poor and technologically backward: (1) because the arguments designed to persuade it that its welfare would be increased by rapid and comprehensive 'integration' with the advanced countries via the international firm, are inconclusive and unpersuasive; (2) because their peoples seem to want to demonstrate their competence and independence and to establish their own economic priorities—and insist on doing so; and (3) because extreme inequality breeds fear of the more powerful. Where foreign domination has been extensive, as it has apparently been, and still is, in Latin America, it seems to raise political problems consistent with these considerations. It has even been suggested that some Latin American regimes are maintained in power only with the help of the great capital-exporting country whose firms have achieved dominant positions. In other instances, there are signs that national governments are attempting to assert their economic autonomy against international firms.

## IV

What then is the role of the multinational enterprise *vis-à-vis* the state? In spite of all the difficulties and attitudes outlined above, the less developed countries, by and large, are prepared to accept—even to welcome—foreign enterprise and foreign help on their own terms where they believe it is clearly advantageous. 'Socialistically' inclined governments tend to insist on partnership arrangements; countries more sympathetic to private enterprise may also do so in less restrictive ways. For their part, the great international enterprises are likely to invest wherever they consider it profitable to do so, without worrying overmuch about ideological considerations if only they feel reasonably secure—which means if expected profit will compensate for any additional risk. These enterprises have a great contribution to make to the less developed countries, but it is very different from what it is in the developed countries.

It has been persuasively argued that continual innovation in technology, and in methods of stimulating and sustaining consumers' demand with new products, new designs, new methods of marketing are necessary to maintain the momentum of the great industrial economies. Even if this is true, however, innovation and change of the kind that is characteristic of the United States economy, for example, is not the need, nor the popular demand, of the poor world. Many of their governments are aware of this, as are many international firms, some of which devote special efforts to research into the ways of meeting their real needs—agricultural development, the provision of water supplies, special health or nutritional problems, etc. Moreover, much of the nascent industry of many countries has been established with the help of the great international corporations, which have provided technical advice, granted licences and help under their own patents, given managerial assistance and even made training facilities of various kinds available for local technical people. If attention were focused on this kind of thing instead of on the more sweeping and spectacular generalizations about the 'integration of the world' through the multinational enterprise and the economic absorption of the 'nation-state', I wonder which function would really turn out to be generally the most important in the developing countries?

The multinational enterprise is a pool of managerial and technological expertise. It is also an efficient operating machine for both production and distribution. It can enter areas of inefficiency and set up and operate efficient units. It can also train others to do so, making much of its own expertise available in the process. Both of these are important activities, but I have not yet understood why the former will take precedence, or why world-wide planning on an industry level is the most efficient way to raise the standards of living of all peoples.

The notion that the multinational enterprises will integrate the industries, and thus the economies, of the world within their own administrative framework, and that this will maximize world welfare, rests basically on the notion that planning on a world-wide scale is not only the inevitable outcome of present trends and United States economic power, but is also the most effective way of bringing the peoples of the world closer to the United States levels of living. It seems to me equally likely that multinational enterprises will continue to function as important international organizations, complementing and aiding the economic efforts of independent countries to develop their economies, and that such enterprises will have an important independent role in all sorts of international economic relations, but will also have to conduct continuous negotiations with states whose economic sovereignty they must respect as they try to 'harmonize' their own interests with those of the countries in which they operate.

I have not tried to discuss 'welfare' considerations here or the extent to which integration by multinational enterprises would advance the development of any particular country, both economically and politically. My primary thesis has been that so long as the inequalities among the nations of the world are so great that a large proportion of the peoples of many are in real poverty, the governments of these peoples will be unwilling to give the rich and favoured foreign economic interests a dominating position in their economies. I also think it probable that, for as long as we can conveniently foresee, governments must insist on a high degree of sovereignty over their economic affairs in order to provide a national economic framework for the activities of their people on the one hand, and on the other to ensure that their economic needs are represented as identifiable claimants for international consideration. If the basic economic unit for planning is the independent country—or a group of countries where some of the national units are very small—it may be less likely that pockets of backwardness in the world will be forgotten and left to rot. It is argued that the existence of the economic sovereignty of the state leads to economic cleavages along national lines, but if national differences were eliminated through integration, would not cleavages of an even more intractable nature along class (or even colour) lines tend to be accentuated?

The concern of a government for the economic welfare of its country must inevitably encompass a variety of considerations which would not be important to a great international enterprise. Indeed, I suspect that multinational enterprises as we know them today would find their life intolerable if the 'state as an economic unit' really did disappear!

## NOTES

1 'The nation-state is just about through as an economic unit. General de Gaulle is unaware of it as yet, and so are the Congress of the United States and right-wing know-nothings in all countries.' Charles P. Kindleberger,

American Business Abroad: Six Lectures on Direct Investment. (New Haven and London, Yale University Press. 1969.) p. 207.
2 Cosmopolitan: 'Belonging to all parts of the world. Free from national limitations or attachments'. (OED). The selection of the term 'cosmopolitan' by those holding the views analysed in this paper to describe their position clearly has persuasive value and seems on the surface to be accurate. Whether it accurately describes the likely outcome of the events predicted is another question.
3 Kindleberger, *op. cit.*, pp. 207–8.
4 See, for example, the discussion by Simon Kuznets, 'The State as a Unit in the Study of Economic Growth', *Journal of Economic History*, Vol. 11 (Winter 1951).
5 Simon Kuznets, *Modern Economic Growth: Rate, Structure and Spread.* (New Haven and London. Yale University Press. 1966), pp. 2ff.
6 *Ibid*, p. 9.
7 See my *Theory of the Growth of the Firm*, esp. Chapters V & VII.
8 See Stephen Hymer, 'The Multinational Corporation and the Law of Uneven Development', in J. N. Bhagwati, ed., *Economics and World Order* (New York, World Law Fund. 1970).
9 Hymer, *op. cit.*
10 Kenneth Galbraith, *The New Industrial State*.
11 'The multinational corporation, if it succeeds, will reproduce on a world-level the centralization of control found in its internal administrative structure.' Stephen A. Hymer and Stephen A. Resnick, 'International Trade and Uneven Development', to appear in Kindleberger *Festschrift.* 1970.
12 C. P. Kindleberger, *op. cit.*, pp. 187–8.
13 Hymer, *op. cit.*
14 It should be noted that in spite of the terminology commonly adopted in these discussions, the 'world' referred to in fact excludes over a third of the population of the earth, for all of Eastern Europe and the largest country in Asia are presumably not expected to be amenable to integration by the international capitalist corporation. One might have thought that Kindleberger would have included 'left-wing know-nothings' among the ignorant. (See footnote 1, page 1.)
15 Both Kindleberger and Hymer suggest that the activities of international corporations have placed significant limitations on the freedom of action of even the United States government in the conduct of its foreign economic affairs.
16 Kindleberger, *op. cit.*, pp. 183.
17 The exceptions will be found primarily in countries which are straightforward political dependents of industrialized countries, such as, for example, Taiwan.
18 I have developed this argument in more detail in an essay, 'International Economic Relations and the Large International Firm', in *New Orientations: Essays in International Relations* (Peter Lyon and E. F. Penrose, (eds.). Cass 1970) pp. 114–22. [See Essay VI above.]
19 See my attempts to do this in 'Profit Sharing between Producing Countries and Oil Companies in the Middle East', *Economic Journal*, June 1959, pp. 251–2 [See Essay IX below]. See also Professor Kindleberger's criticism in *Economic Development* (New York: 2nd ed.) p. 334, and my further attempt in the article cited in the previous footnote.
20 'The notion is that the foreign company is obtaining a *disproportionate* gain—a gain greater than he ought in some sense to get when it is compared with the gain accruing to the receiving country. In other words, it is

a feeling of 'unfair' treatment, and this is what makes the notion so extremely difficult to analyse from an economic point of view, for it does not depend on any objective measurement of the absolute gain received by either side, but on their *relative* positions in the light of what seems fair to one side only'. 'International Economic Relations and the Large International Firm', *op. cit.*, p. 119. [See Essay VI above.]

21 Norman Girvan and Owen Jefferson, 'Institutional Arrangements and the Economic Integration of the Caribbean and Latin America', University of the West Indies (mimeo), pp. 13–15. The authors go on to say: 'The converse of the high degree of mobility within the company is a certain degree of rigidity so far as intra-regional product and capital flows between companies are concerned. For example, the [multinational corporation] with raw material facilities and processing capacity will not normally purchase raw materials from another producer. . . . This corporate integration can, and often does, result in regional fragmentation. This fragmentation does not only occur between the two regions, as already suggested. The experience of the Caribbean with its bauxite industry suggests that fragmentation can take place within a given region and even within one country. Thus that part of the bauxite output of Jamaica and Guyana which is produced by Reynolds Metals, the U.S. aluminium producer, reaches the fledgling aluminium smelting industry of Venezuela after being shipped to the U.S. and manufactured into alumina at Reynolds' plants and then re-exported to the South American mainland, in spite of the fact that Guyana borders on Venezuela and Jamaica is also nearer to that country than the United States. Conversely, Reynolds' bauxite output in Guyana and Jamaica has not so far been available to the existing processing capacity in both countries because this capacity is owned by a different company.' p. 15–16.

22 This paper was in press when Guyana announced partial nationalisation of its bauxite industry.

# SECTION II

# MULTINATIONAL FIRMS IN THE INTERNATIONAL PETROLEUM INDUSTRY

In the international petroleum industry very great changes took place during the decade over which the articles in this Section were written. The nature of these changes can be seen in the progression of the papers, the first of which was written in 1959 and the last in 1969. Essay VIII published in 1960 but written in 1959 on the basis of data available in 1958 was the first to bring out the significance of vertical integration in the industry. It was, of course, written before the creation of the Organization of Petroleum Exporting Countries, and for those who know the industry, the difference OPEC has made is evident by implication. One need only compare Essays VIII and IX with the last three in the Section to appreciate the extent of change in the relations between companies and governments in ten years.

# VIII

## MIDDLE EAST OIL: THE INTERNATIONAL DISTRIBUTION OF PROFITS AND INCOME TAXES*[1]

*Written at the beginning of the post-war turn in the fortunes of the international petroleum industry, this paper analyses the significance of the vertical integration of the industry. Future developments and problems are foreshadowed. Unless otherwise stated figures refer to 1958.*

THERE has been wide discussion of a number of problems associated with the distribution of income between countries chiefly engaged in raw-material production and countries chiefly engaged in manufacturing. In virtually all of these discussions the emphasis has been on the comparatively unfavourable position of the producers of raw materials, especially when the manufacturing firms, who are their customers and who also provide many of their industrial requirements, possess substantial monopoly power over the prices at which they sell their manufactured products. In one of the world's major industries, however, the situation usually presumed to exist is turned upside down. In the international petroleum industry it is now alleged that it is the raw-material producers who are getting disproportionate profits, while the refining and distribution activities in the industrial countries are carried on at a loss.[2] Nevertheless, the economists and political leaders in many of the countries producing crude oil believe that large profits are made in refining and distribution, profits which should properly be shared with the countries providing the basic raw material; and their demands for a full 'partnership' in the industry have not abated.

The demand for 'partnership' arises because the major international oil companies are vertically-integrated concerns, themselves using most of the crude oil they produce. Many people in the crude-oil producing countries believe that these countries should receive a share, not merely of crude-oil profits, but of the entire profits of the integrated companies, and some form of equity interest in the companies seems to them to be the best way of achieving this. But clearly the issues raised involve some interesting questions relating to the determination of 'profit' at different stages of production in the operations of vertically-integrated firms, questions which have con-

---

* Published in *Economica*, August, 1960.

siderable fiscal importance to a number of governments when the different stages of production are carried on in different countries. One may suspect that present methods of taxing profits in such companies lead to strange and anomalous results. The countries producing crude oil are in a unique position; but this fact alone points out, by contrast, the unfavourable position of other raw-material producing countries in which vertically-integrated firms are dominant.

I shall confine this discussion to the international petroleum firms operating in the Middle East, first briefly presenting the relevant facts about the industry, secondly examining the reasons why so much profit is attributed to crude oil production in the operations of the vertically-integrated oil companies, and finally analysing the relation between profits on crude oil and profits on refining and distribution with respect to the demand of the raw-material producing countries for a share in the latter.

## THE INTERNATIONAL PETROLEUM INDUSTRY

Seven international firms (the 'majors') dominate the international petroleum industry.[3] They produce over half the world's crude oil, and nearly 85% of that produced outside the United States and Soviet controlled areas; and they refine and sell the bulk of the petroleum products consumed outside these areas. They are responsible for almost 90% of the output in the Middle East and over 90% of the output of Venezuela.

In general, these firms (or 'groups') consist of a top holding company with numerous associated and subsidiary operating companies.[4] It is not necessary for our purpose to describe the organization of the major groups—suffice it to say that each is a recognizable entity, the boundaries of which, however, often overlap those of others; that the important activities of each are closely co-ordinated in one way or another; and that each is integrated vertically, controlling (or buying from one of the others on long-term arrangements) most of its own requirements of crude, owning or controlling through long-term charters most of its own transport facilities, owning its own refineries, and controlling its distribution outlets all over the world. The separate 'groups' are often closely associated through joint ownership of subsidiaries and through long-term supply or marketing agreements; but they are also in competition with each other, the competition largely taking the form of attempts to maintain or increase their shares in the markets for final products, or to obtain a position in new markets, through strategic acquisition of raw material supplies and refinery and distribution facilities, through improvements in products and service, and through advertising. Price competition is avoided as far as possible and market-sharing arrangements have not been unknown. Although these large, financially powerful, integrated firms produce the overwhelming proportion of the crude oil in the Middle East, the sale of crude oil is not, and has not been, their main marketing business, and obtaining revenue from crude oil sales is not their main concern. In spite of this we now find most of their profit attributed to the production of crude oil.[5]

PROFIT IN AN INTEGRATED FIRM

In an integrated firm, the total cost of any given amount of the finished product is the sum of the costs incurred at every stage of production, and profit for the firm as a whole arises when total sales receipts exceed this cost. If, at each stage of operation, the product is transferred to the next stage at its estimated cost at that point, no 'profit' appears until the final sale. Profit can only appear if the product is transferred at an accounting price greater than the cost of producing it; conversely, a loss appears if the transfer is recorded at a lower price. If the integrated business is, as a whole, a profitable use of funds, and if each stage of production is necessary to the whole, there are few circumstances in which a firm would be justified in claiming a profit or a loss on one of the stages in its vertically-integrated operations. A loss is incurred on a particular stage when the firm can buy its product at a market price lower than its own cost of production after making full allowance for any increased uncertainty of supply. On the other hand, when it can sell the product outside at a price higher than its own cost of production (or buy it only at this higher price) it can, for *accounting purposes*, impute a profit to the relevant stage of operations. Obviously, it makes a profit on any of the product actually sold in the market which it could not itself use in its integrated operations at a higher net return.

It is not, however, permissible from an economic point of view for a firm to attribute its profit to one particular stage of its integrated operations and then *for this reason* to consider all the subsequent stages 'unprofitable'. (Whether the subsequent stages are unprofitable in relation to an earlier stage depends on whether the total return on the firm's investment would be increased if more of its investment were transferred to the earlier stage.) It is possible that for some reason a firm may attribute so much profit to one stage of its operations that final sales cannot be made at a price which will cover costs including this profit. Clearly a loss must then be shown somewhere, for the combined profit of all parts of the firm must be shown as less than the profit imputed to the 'profitable' stage of production. But just where this loss is shown is a matter of choice for the firm and has no significance at all for the actual (pre-tax) profitability of operations in any particular stage of production.

Provided that a firm operates entirely within one country and under the jurisdiction of only one income tax authority, and provided that the tax laws do not themselves discriminate between different stages of the industry, it makes very little difference either to the firm's shareholders or to the government where it chooses to display the profits made on its total operations. But when the firm is an international one, it does make a difference not only to the firm but also to the countries in which it operates. The countries of the Middle East producing crude oil, the countries affording transit facilities for pipelines, the countries in which refining and distribution are carried on, and the country in which the parent corporation or its subsidiaries are chartered, are all interested in the profits of an international oil firm.

### PROFITS ATTRIBUTED TO THE PRODUCTION OF CRUDE OIL

Most of the oil in each of the Middle East countries is produced by a single national operating company or group of companies wholly or substantially owned by two or more of the seven major international firms. These subsidiaries—the national oil companies of the Middle East such as the Iraq Petroleum Company, the Kuwait Oil Company and The American Arabian Oil Company (Aramco)—sell the oil they produce, except for the 'royalty oil', to their shareholders, the majors.[6] The oil is valued at a 'posted price' which is changed very infrequently and then by fairly large amounts; this price is considerably above the cost of production although nearly all of the oil goes to refineries owned by the same groups which own the producing companies and set the price. Thus for each of the international companies a profit emerges on the crude oil produced on its behalf and then used in its subsequent operations.[7] This profit is divided equally between the companies and the producing country, royalty payments being included as part of the country's share. Thus, in the Middle East, whether the integrated companies use the oil themselves or sell it outside on long-term supply arrangements or 'spot' to independents, all transfers are presumed to have taken place at the relevant posted prices irrespective of whether they have or not, and profits are attributed to all crude oil produced.[8]

As the calculated profit on crude oil depends on the level of the posted price, and as the bulk of the oil is used by the integrated companies and is not sold to third parties, it is evident that the part of the 'profit' paid to the government of a producing country and relating to the oil used by the company itself, is really a charge on the *total* integrated operations of the company. This becomes especially obvious when we realize that 'profits' can be attributed to crude oil production while at the same time the subsequent operations of the company can be carried on at a loss. The question then arises why the oil companies choose to show so much of their profit at the level of crude production.

For a number of reasons relating to the structure and control of the industry, total profits before taxes attributed to the foreign operations of the integrated companies have been very high.[9] It is understandable that the countries producing crude oil would want a share of these profits and that they would not be satisfied with merely a share of the profit made on the relatively small amount of oil sold to third parties. Unless the major companies wanted *explicitly* to give the producing countries a share in the total profits on their integrated operations, some means would have to be devised of attributing a profit to crude production alone. The posted price of oil provided the means. Moreover it is easy to understand that the companies would attempt to sell crude oil to outsiders, including governments, at the highest price that would move the oil they wanted to put on the market. High prices for crude give them high profits on outside sales of crude, and also increase the costs of independent competitors in refining and distribution while having no effect on the costs to the integrated companies of using their own oil. Furthermore, the price of crude from the Middle East had to bear some sort of relation to the price of crude from the much higher-cost areas of the Western Hemisphere, and the interest of the majors in their Western Hemisphere production must have had some relevance to the price of Middle Eastern oil.[10]

These considerations are all part of the explanation of the level of posted prices. And once the principle of a posted price was accepted, difficulties would surely have arisen with the producing countries if crude used by the companies themselves had been valued at a lower price than that sold in the market. But there is another fact of decisive importance. The United States companies, which first adopted the device of posted prices and which first accepted the 50/50 profit-sharing arrangements, stood to gain under the United States tax laws by attributing large profits to crude oil production, in spite of the fact that they handed half of this profit over to the producing countries. Under United States law United States companies can deduct from their taxable income a depletion allowance equal to $27\frac{1}{2}\%$ of the gross income arising in crude oil production. The higher the posted price, the higher is the gross income attributed to crude and the higher the depletion allowance. Furthermore, under United States law the profits paid to the producing countries in the form of an income tax are considered a foreign tax and are treated as a credit against any

U.S. tax liability that remains. Thus the combination of '50/50', the depletion allowance, and a high posted price for crude oil virtually eliminates the tax liability of the American companies to the United States[11] and puts their shareholders (the majors) in a much better position with respect to taxes than they would have been had their profits been shown at some subsequent stage of production and hence been subject to the United States corporate income tax of 52%.

To summarise: the recorded profitability of crude oil production has been the result of the 'posted price' policy adopted by the international majors which has been advantageous both to the oil companies and to the countries producing crude oil. It provided a convenient way, without cost to the companies, of sharing the profits of the industry with the countries producing the raw material. A higher posted price merely meant that more was paid in taxes to the governments of the producing countries and less to the parent government;[12] and political relations with the producing countries were not strained by attempts to lower prices but instead were sweetened by the large profits paid over. It permitted higher profits to the majors on oil sold to outside users. It kept the raw material costs of competing refiners high; the lower the profitability of refining, the easier to prevent an influx of newcomers into the refining industry to compete with the majors. At the same time it reduced the impact of the increasing supplies of low-cost Middle East oil on the markets for the higher-cost production in the Western Hemisphere.

Thus the device of using the posted price to calculate profits on crude oil production is very largely an arbitrary, though convenient, bit of economic make-believe which has provided a means of solving some of the political and economic problems of the international oil companies; the profits so calculated have little economic meaning. This does not in itself condemn the system, but it has considerable relevance for our next problem: the desire of the producing countries to obtain a 'partner's share' in refining and distribution 'profits'.

### SHARE OF PRODUCING COUNTRIES IN REFINING PROFIT

If our analysis is correct, and apart from side effects on the competitive position of independent refiners, the level of posted prices has no effect on the profitability (before taxes) of the over-all *integrated* operations of the majors. For a given amount of profit to the firm as a whole, the higher the posted price, the lower the profits on refining and distribution. If no sensible economic meaning can be attached to profits attributed to the crude oil produced and used in an integrated firm, then any method of calculating such profits to be shared with the producing countries is merely a method of giving the countries a share in the *total* profits of the integrated enterprise. The oil companies insist that a 50/50 sharing of crude oil profits is 'fair' (presumably because equality implies 'fairness') in spite of the fact that the profits attributed to crude bear an arbitrary relation to total profits. A low posted price may mean that the producing countries receive only a small proportion of the total profits of the companies; a high posted price may mean that the profits received by the producing countries may equal or exceed one half of the total profit on all operations of the companies. In this case profits from all stages of operations have in fact been shared equally or to the advantage of these countries.

The latter is exactly what happened in 1958. In that year a number of factors, many of them long in the making, coincided to erode the price structure of the industry. In the first instance there was a slackening of demand for oil products largely related to the recession in the United States; inventories of finished products rose and prices weakened in world markets. At the same time it was difficult for the majors substantially to reduce their scale of operations.[13] Pressures became heavier in the individual countries producing crude oil for increases in their national outputs and increasingly delicate political conditions caused the companies to move warily, avoiding so far as possible actions that would reduce the flow of profits going to the governments of these countries, which were already granting new concessions favouring to some extent companies other than the majors. These circumstances, together with competition from Russian oil in European markets, and the imposition of import quotas by the United States, all worked to bring about a situation in which the supply of oil was increased in the face of weakening demand, and the majors in the Middle East, in order to sell their surplus oil, had to sell below the posted price. That is to say, increased supplies in relation to demand pressed down the market price for crude oil. Thus 'profits' attributed to crude oil production became even more fictitious; they were fictitious not only in respect of the oil used by the integrated companies in their own operations, but also as applied to outside sales. Since 'refining margins' became reduced by lower product prices (with crude oil inputs still being valued at or near the posted prices), 'profits' on crude became a greater percentage of total profits.[14]

Because of the widening gap between posted prices and market prices for crude oil, the majors finally reduced posted prices early in 1959. At the lower prices, the producing countries will receive not only lower profits but also a lower *proportion* of the total profits of the companies than they would have received if the higher prices had been maintained, since the reduction will affect little but the value put on crude oil for purposes of profit-sharing. Themselves pressed by competitive conditions beyond their control, the majors pointed to the exigencies of competition in order to persuade the governments that they too, must accept lower profits. It seems reasonable to agree that profits can hardly be increased or even maintained in a falling market; but it should be remembered that if the companies operating in the Middle East have oil which they must sell, they will if necessary sell it below posted prices, as they have actually been doing. In other words, while it is true that the effective price of crude oil may have to fall in response to supply pressures, price adjustments could conceivably be at the expense of the companies more than of the producing countries within the limits set by the bargaining strength of the respective parties.

This last is, of course, an important qualification; for the power of the governments of the producing countries to prevent reductions in posted prices is severely limited. They have no control over these prices except in so far as the companies think it politic to consult them, although they have always pressed for more control; and since the posted price is part of the technique of profit-sharing, it perhaps ought openly to become a subject for negotiation with the producing countries as part of the bargaining involved in the profit-sharing agreements as long as the present arrangements persist.[15] But each producing country is interested in expanding the sales of its own oil; hence if any individual government becomes particularly difficult, the companies have considerable scope for reducing output in that country while expanding it in the territory of a more compliant government; and the recent discoveries in North Africa have further strengthened the position of the companies. While the bargaining power of the companies is increased when it is evident to all that market conditions have deteriorated, it remains true that they cannot afford to ignore the political repercussions if the producing countries become very dissatisfied with oil arrangements.

While Venezuela has increased the rate of income tax to obtain a higher share of profits, and other governments have provided for a greater profit share in some of the new concession agreements,[16] a higher share of existing profits could also be obtained merely by maintaining the level of posted prices. This would have two major disadvantages from the point of view of the companies: it would worsen their financial position if the amounts payable to the producing governments exceeded the tax liability of the companies to their home governments or the tax relief obtained from them; and, perhaps above all, it would make more difficult the maintenance of an orderly structure of prices. If sales to outsiders in fact take place at posted prices (except for the more or less accepted discounts on the long-term supply arrangements), and in accordance with some recognized system, buyers and sellers alike know the general market position. But if each sale is subject to unknown discounts, the danger of more or less open price competition increases.[17]

One thing, however, is evident: if the producing countries could influence the manipulation of the useful, though artificial, device of the posted price as the basis for profit calculations, it would not, under present circumstances at least, be to their financial advantage to obtain the 'partnership' in the fully integrated operations of the companies that many of their spokesmen think they want; if they cannot obtain some control over the level of posted prices, the question is debatable. In 1958, however, they got considerably more than an equal share of profits, which to a large extent was merely at the expense of the foreign governments who granted tax relief to the international companies.[18]

NOTES

1  I want to thank all those who have read and criticized earlier drafts of this paper, and in particular Paul H. Frankel and his staff, who very kindly aided me in the collection of material.
2  For example, a recent study finds that in 1958, all profits in the international petroleum industry were made in the production of crude oil and that 'transportation, refining and marketing were carried on at a sharp loss'; furthermore, 'the oil companies would probably have had substantial losses' from these operations in 1956 and 1957 had it not been for the Suez crisis. William S. Evans, *Petroleum in the Eastern Hemisphere* (First National City Bank of New York), April, 1959, pp. 11–14.
3  They are, ranked by crude-oil output in 1958: Standard Oil of New Jersey, Royal Dutch/Shell, Gulf Oil, British Petroleum, Texas Company, Standard Oil of California, and Socony Mobil.
4  Royal Dutch/Shell is a special case. This group is not a legal entity but consists of two separate holding companies, one Dutch and one British, between which is a permanent contractual arrangement giving a 60% share of the group to Royal Dutch and 40% to Shell. The two organizations nevertheless form a co-ordinated firm in the sense used here.
5  I have been unable to secure statistics showing what proportion of the crude oil produced by the majors in the Middle East is sold to independents, but a brief look at the relation between the total crude production and the refinery input of each of the majors indicates the degree of integration. Standard Oil of New Jersey, Royal Dutch/Shell, Texas Company, and Socony Mobil are all net *purchasers* of crude—but they refine only slightly more than they produce. Royal Dutch/Shell has a long-term supply contract with Gulf Oil which nearly makes up the difference between its crude production and refinery runs, and in 1958, Socony Mobil took 67,600 barrels per day from British Petroleum under a long-term supply contract. Standard Oil of California produced about 9% more crude than its refinery input, and British Petroleum about 39% more. In addition to the contract with Socony, BP has recently made special supply arrangements with Sinclair (a United States independent) to supply its needs of Middle East oil for the United States market. Gulf Oil refined 55% of its crude output, an additional

(unreported) amount was processed by outsiders for Gulf's account, and over half of the rest was taken by Royal Dutch/Shell.

Of the majors, therefore, only Gulf and BP made substantial net outside sales of crude and both are endeavouring to extend their refining and marketing facilities. Because of the uncertainties surrounding oil discoveries, the output of petroleum available to a given company has not always been well adjusted to its refining and distribution capacity. Hence, in addition to the long-term supply contracts, there has been a considerable amount of re-shuffling between the companies of ownership both of crude-oil reserves and of refining and distribution capacity. Needless to say the effective price of oil under the supply contracts is not the posted price.

All the figures above are based on the 1958 Annual Reports of the companies. While it is true that figures of *net* sales of crude, based as they are on the world-wide integrated operations of each of the groups, give little information about their outside sales of Middle East crude oil, it is nevertheless obvious that sales of crude oil are only incidental to the main business of most of the companies.

6 The National Iranian Oil Company is an exception to the arrangement discussed here although effectively there is little difference in the outcome. The company is owned by the Iranian Government but operations are carried out on its behalf by a consortium in which the seven majors have 89% interest (British Petroleum 40%, Royal Dutch/Shell 14%, and the others 7% each). The members of the consortium receive oil in proportion to their share. The NIOC also enters into partnership with foreign companies in the exploitation of new concessions.

7 The method of showing this profit differs according to whether the national company was chartered in Britain or in America. The Iraq Petroleum Company, for example, is British and for tax reasons its accounting has been so arranged that the company itself makes no profit. Aramco, an American Company, has large profits attributed to it, also for tax reasons as we shall see. The fiction that Aramco operates as an 'independent' company making profits of its own is thoroughly irrelevant and should deceive no one. The price at which Aramco sells its oil is set by its board of directors, nearly all of whom represent the four major companies who own the company and who 'purchase' its oil at the price the board sets. The question still remains: why does it pay the majors, acting in their capacity as Aramco's board of directors, to set the price to themselves significantly higher than the cost of production?

8 The above exposition is a simplified, schematic representation of the principles involved in calculating profits. There are many more or less minor qualifications; '50/50' is a principle which may or may not work out in practice, depending on the particular calculations involved. In some of the concession agreements certain discounts from the posted prices have been allowed the companies in the calculation of profits. Furthermore, I have ignored the many arbitrary elements in calculating costs. And again Iran provides an exception with respect to the methods by which profits are calculated; but in effect a 50/50 profit-sharing arrangement exists.

9 For example, in 1955 stockholders' equity in Aramco was $491·5 millions and net earnings before Saudi Arabian income taxes were $465 millions, or 95% of net worth! In 1956, preliminary figures presented to a United States Senate committee showed stockholders' equity as $471·4 millions and net earnings before Saudi taxes as $480·3 millions. See *Emergency Oil Lift Program and Related Oil Problems*, Joint Hearings before Sub-committees of the Committee on the Judiciary and Committee on Interior and Insular

Affairs, U.S. Senate, 85th Cong., 1st Sess., pp. 1,237, 1,240 and 2,839. Nothing I say in this paper, however, should be construed as either approval or condemnation of the level of profit in the industry. Such an appraisal is not simple, and to make it is not one of the purposes of this discussion.

10 There is much controversy on how posted prices in the Middle East have in fact been determined and on why they have been as high as they have. The Federal Trade Commission of the United States has described complicated pricing formulæ of the basing-point variety which, together with the changing supply position of the majors in relation to demand, can explain for the most part the posted prices of Middle East oil. (See *The International Petroleum Cartel.* Staff Report to the Federal Trade Commission. Committee Print No. 6, 82nd Cong., 2nd Sess., pp. 352 ff.) These prices can also be explained in terms of a kind of non-price oligopolistic competition among the companies which together dominate oil production both in the Middle East and the Western Hemisphere. We have not space here to go into this complicated question; in fact, it makes little difference (except perhaps to the United States Courts) which explanation is accepted, for neither rests on the argument that the posted price of oil is the result of price competition among the majors, and in neither is it assumed that it would have been profitable for the majors to put all their efforts into crude production.

11 The effect of foreign tax payments on the domestic tax liability of an American company is well illustrated in the evidence presented by Aramco to the U.S. Senate Sub-committee and published in *Emergency Oil Lift Program and Related Oil Problems, op. cit.*, Part II, esp. tables pp. 1,442 and 1,471.

12 British Petroleum also now receives credit against its domestic tax liability for foreign income taxes paid.

13 'Each integrated concern must press forward with its programme of exploration and expansion of facilities of all kinds to ensure that it is in a position to secure a share of the future growth of trade. None can afford to hold back while its competitors capture outlets and forge ahead to its disadvantage.' *Petroleum Press Service*, October, 1958, p. 362.

14 It was estimated by the First National City Bank of New York (*op. cit.*, p. 12) that the seven majors paid to the producing governments on all accounts $260·1 millions more than the profits attributed in their accounts to their Eastern Hemisphere operations. (This figure includes relatively minor payments to Borneo, Indonesia and other governments in the Eastern Hemisphere.) Similarly, British Petroleum paid out in overseas income taxes more than half its net income in 1958. (*Annual Report*, p. 13.) This implies that losses were incurred on the rest of BP's activities; there is little published information on this, but in 1958, B.P. Benzin und Petroleum A.G., a German refining and marketing subsidiary, is reported to have made losses of DM 8·42 millions in spite of increased sales. (See A. M. Stahmer's *Erdöl-Informationsdienst*, September 1st, 1959.) Similarly the Deutsche Shell Aktiengesellschaft/Hamburg, a Royal Dutch/Shell subsidiary, showed a loss of DM 15·3 millions (see the company's *Annual Report*, 1958, p. 23); and Esso A.G., a Standard Oil of New Jersey subsidiary, reported a loss of DM 19 millions although turnover was up 10%. (*Petroleum Press Service*, July, 1959, p. 280). It seems that the refining industry of Germany, so far as it was controlled by integrated oil companies with a policy of selling oil to their own refineries at posted prices, yielded no corporate tax revenue to the German government.

15 See my 'Profit Sharing Between Producing Countries and Oil Companies in the Middle East', *Economic Journal*, June, 1959. [Essay IX below.]

16 It is true that agreements giving more than 50% of the profits to the producing government have provided that the government may bear some of the capital cost if oil is discovered; it is also true that oil has not yet been commercially produced under such agreements. Hence the nominal share of profits guaranteed to the government is sometimes misleading.

17 Cf., for example, the comment that the reduction of the posted price 'should also tend to eliminate, or at least reduce, the practice of discounting posted prices on marginal transactions, a practice which not only discredits posted prices as a commercial measure of value, but also leads to the payment of an undue tax . . .'. *Petroleum Press Service*, March, 1959, p. 87.

18 The hope of financial gain is not the only reason why partnership is often demanded by spokesmen of the producing countries. They are aware of the fact that their oil is valueless without outlets, and they feel at the mercy of the integrated companies so long as these companies have such a strong control over world refining and distribution. They therefore instinctively feel that if they had some control over these later stages, they would not be in such a vulnerable position. The fact is, however, that their profits have depended on the refining position of the majors and the subsequent limitation of price competition. If the present producing countries did have more 'say' in the policies of the companies, it is possible that they could prevent some of the expenditures now being made by the companies to develop alternative sources of oil, expenditures which may well be unnecessarily wasteful at the present time, but which may result in a reduced dependence of the world on Middle East oil to the disadvantage of the Middle Eastern countries. On the other hand, it is probable that each of the countries would press harder for increases in its own output and thus precipitate competition which would seriously reduce the revenues of all to the benefit of consumers. In addition, of course, the recent discoveries of oil in North Africa have already seriously weakened the bargaining position of the older oil-producing countries.

# IX

# PROFIT SHARING BETWEEN PRODUCING COUNTRIES AND OIL COMPANIES IN THE MIDDLE EAST[*][1]

*Analyses the nature of the bargaining between companies and governments. Written before the formation of the Organization of Petroleum Exporting Countries, the paper emphasizes the relatively weak position of the exporting countries at that time.*

IT is widely taken for granted that the oil-producing countries in the Middle East are being in some sense 'exploited' by the foreign oil companies; that their share in the revenues from the production and distribution of oil is 'unfairly' low; that their participation in the different operations of the companies is unduly restricted; and that foreigners are benefiting from 'Arab' oil at the expense of the rightful beneficiaries. The purpose of this paper is to explore some of the underlying considerations determining the sharing of oil revenues between oil companies and producing countries, and the related subject of nationalization, and to investigate what meaning can be given to the notion of economic 'exploitation' used in an invidious sense.

The vested interests in Middle East oil are so diverse and varied, the financial relationships between the various interests are so complicated, the formal terms and practical applications of the oil agreements are so difficult to comprehend, and the organization of the distribution of oil is so intricate, that it is no wonder that even experts seem at times confused, let alone the general public. I shall not be concerned here with any particular agreements or negotiations; I merely want to re-examine the fundamental principles involved in the matter and to point out a few of the significant economic relationships.

Some of the propositions advanced in the various discussions on the subject seem implicitly to assume that the sharing of oil revenues should accord with some principle of 'natural law'—this is often implied, for example, in the proposition that the 'Arabs' as such have a 'right' to all oil profits quite apart from legal arrangements. Here, I shall assume that all rights in the matter arise from man-made law, but that the governments of the countries under whose lands the oil lies have a legal right to deal with its exploitation as they see fit, with due respect for existing contractual arrangements. I shall also assume

[*] Published in *The Economic Journal*, Vol. LXIX, June, 1959.

that governments have the right to nationalize the property of foreigners as well as that of their own citizens upon payment of fair compensation, and that such action is justified when the economic interests of the country are thereby furthered.

At present the sharing of oil revenue is primarily a matter of bargaining between the different interests concerned, the bargaining being influenced by political as well as by economic considerations. Since I am not concerned with political controversies among the leading world powers over the 'control' of Middle East oil, I shall treat the oil companies as privately owned concerns interested primarily in making profits. The desire for an assured access to adequate supplies of oil for military and strategic as well as for economic purposes has made control of the sources of oil a major concern of many governments. Thus inter-governmental political manœuvring has been a dominant characteristic of the past controversies over oil concessions, and the present position of the oil companies in the Middle East is very largely the result of the past policies of and rivalries among certain Western powers. As a reaction against this, demands have been increasingly heard recently that the West should put its oil relationships with the Middle East on a 'business basis'; this presumably means that military and political considerations should no longer dominate the scene. In order to analyse the nature and probable consequences of oil relationships on a 'business basis', I shall therefore, neglect the military and political interests of the Western powers.

I shall deal first with the basic conditions affecting the bargaining process, secondly with the question of 'exploitation,' and thirdly, with some problems of international policy.

### I. THE PROFIT-SHARING BARGAIN

We start from the present position; we are not concerned with political and economic conditions relating to earlier times and earlier contractual arrangements, but only with the considerations affecting the granting of new concessions or the re-negotiation of present agreements. The governments of the producing countries have reserves of oil under their territorial jurisdiction. Some of the oilfields are actually worked by foreign companies under agreements which, on specified terms, enable them to explore for oil, produce and sell it. In other areas, not yet in production, exclusive concessions have been granted; in still others new concessions may be granted; and finally, some existing agreements contain provisions requiring the company, under specified conditions, to give up part of the area covered by the agreement and not yet in production. The governments of the producing countries may cancel existing concessions (we assume with fair compensation) through nationalization or otherwise and may re-negotiate existing agreements as well as grant new concessions. But whenever a foreign company is involved, the share of the revenue obtained by a government depends entirely on the agreement it can extract from the oil company. This is obvious in the case of new concessions; but even with respect to an established company the government is dependent on the willingness of the company to give up its profits, since the oil revenues arise in foreign currency outside the country, to which the government does not have access, and which it cannot even tax without the company's consent.

We have then, in essence, two opposing interests both benefiting from oil production, the benefit each obtains being largely determined by the terms of the bargain it can make with the other. One of the interests invests capital to start the industry and runs it, the other supplies the basic raw material, which is a 'wasting' asset—the extraction of the oil permanently reducing the resources of the country. Both are interested in promoting the most profitable long-run expansion of the industry, but each individual government is concerned with the profitable expansion of its own oil production, while the companies, with some choice of country in which to invest new capital, will be most interested in those countries from which the higher returns can be expected.[2]

Let us now examine the factors determining the outcome of the bargain that can be made. For the present I shall include royalty payments per ton of oil produced as well as all the various technicalities affecting the actual share of the revenues obtained by a producing country (as distinct from the nominal percentage share) in the category of 'profit-sharing,' although in fact there are many reasons why a country's actual share of the 'profits' may diverge from any announced percentage share.[3] I want first to consider the respective positions of an individual country and an oil company and the nature of the bargaining process between them when there is no competition in the industry and when the producing country cannot run it. It should be kept in mind that 'running' an industry includes selling the product.

Stripped of all complicating variations and special circumstances, the essence of the matter can be stated in its simplest terms as follows: The proportion of its profit that a company will be willing to give up depends on its estimate of the cost of meeting the government's final demands compared with the cost of resisting them, up to the point where the loss in either case makes the business unprofitable.[4] The demands of the government will depend *either* on the loss it believes it can inflict on the company by not giving or by cancelling the concession under negotiation, that is, on its estimate of the value of the concession to the company, *or* on the amount it thinks the company will be prepared to give up in order to avert political disturbances and to maintain political goodwill. The latter alternative applies in the case of negotiations with a company already established in a monopolistic position in the country and in circumstances where the government cannot run the industry.

In the absence of intergovernmental political manœuvring (which we have ruled out of our terms of reference), the problem with respect to new concessions is quite straightforward and need not detain us. The cost to the company of meeting the government's demands is the share of profit it must give up in one form or another; it need merely compare what is left to it under the terms proposed with the return that can be expected on alternative uses of its resources, making due allowance for differences in risk. It can always bargain for the best terms possible, but if the government sticks at too high a point, the company has no alternative but to refuse the concession. The cost of resistance is the loss incurred if the concession is not obtained. What the government demands depends on how badly it wants the oil production, how much it is aware of the value of its oil and how much it thinks it can persuade the company to give.[5] If the country cannot itself discover or produce the oil, it will gain by permitting the oil company to operate for nothing (if necessary it may even be worth the government's while to subsidize the company in some form), since exploration and production would alone bring considerable revenue and other advantages to the country in the form of wage payments and other outlays and of social and technological contributions of various kinds. A skilful government negotiator, however, may easily extract substantial royalties from the company if he realizes the value of the concession he has to give.

But with respect to the re-negotiations of existing agreements with companies already established in the country the matter is more complicated. Although the cost to the company of acquiescing in the government's demands is the additional profit lost in one form or another, the cost of resistance is not necessarily the loss of the concession and with this the future profits expected, for the company is in a strong position to precipitate an economic crisis which may drive the negotiating government out of office. The company need take no positive action; by simply refusing to budge it forces the government to take the risk of action which, once taken, cannot in the nature of the case, be easily retreated from. Furthermore, the government's action must be political—it may promote publicity which arouses the populace against the oil companies; it may condone or incite political harassment, such as acts of sabotage; it may and indeed is likely to threaten nationalization on terms unacceptable to the oil company. But where the economy of a country has become heavily dependent on the continuance of oil revenues, while at the same time foreign enterprise is required to run the industry, an existing oil company, if it has a virtual monopoly or acts in concert with other companies, is in an extremely strong position *vis-à-vis* any single government, for the latter courts political disaster if it gets into an all-out fight with the company. The progressive economic deterioration following the cessation of oil revenues is likely to cause greater and greater political difficulties for the government, its eventual repudiation and the substitution of a government pledged to re-establish the flow of oil revenues. The cost to the company of such a fight may be extremely heavy, but given the structure of oil interests in the Middle East, it is not likely to be disastrous for any of the big companies while it is likely to be disastrous for the government in power.[6] A company will naturally seek to avoid such an extreme show-down, but only if the expected cost of resistance exceeds the ultimate gain if the resistance is successful—and if the government's demands are very high in relation to existing royalties the company may stand to gain a great deal by resisting the government.

In deciding whether or not to refuse particular demands of the government, a company will not, in the first instance at least, be as much concerned with the probable cost of a show-down as with the question how much pressure it can bring on the government without forcing it to take serious action. This last question may be an extremely delicate one. Both the company and the government may be well aware of the government's fundamentally weak position if the battle lines are drawn, but the position of the parties is such that, when the company refuses to yield to its final demand, the only thing the government can do is to threaten nationalization. Once it makes such a threat publicly, it may easily become a prisoner of its own acts and be unable to back down, for in all the Middle Eastern countries popular feeling is easily aroused against the oil companies, and this puts great pressure on the moderate members of the government and strengthens the position of the more extreme members.[7] If the people can be appeased by the oil company's acceptance of the government's terms, the matter may end with a capitulation by the oil company, but there is likely to be a long-run enhancement of political antagonism on all sides. If the government is forced by popular pressure to go through with its threats of nationalization on unacceptable terms, the oil interests in the end may prevail through the downfall of the government in consequence of economic distress and in spite of the emotions of the masses.

The knowledge of most Middle Eastern governments that they really cannot afford a controversy that causes the cessation of oil revenues, together with the fact that it is they who must undertake positive action if the oil company cannot be persuaded by argument, puts them in a relatively weak position. A responsible government realistically attempting to increase its share of oil revenues must keep the company aware of the need to avert political action, but at the same time must not commit itself too deeply to an action it knows would be disastrous in the end. Paradoxically enough, this bargaining strength of the government may arise as much from its internal political weakness as from its political strength, as much from its inability to prevent disorder if the people become aroused as from its ability deliberately to interfere with the oil operations, as much from administrative incompetence as from Machiavellian policies.

On the other hand, the company's awareness of the danger of forcing the government into a position from which it could not easily retreat may induce it to accept even stiff demands in order to avert unpleasant and costly conflict. The dangerous aspect of the matter lies in the political repercussions over a wide area if the oil company miscalculates the pressures on the government and fails to give in before it is too late or if the government insists on more than the company is willing ultimately to give. So long as popular pressures are not great and the government's revenues are significant in absolute amounts negotiations may proceed reasonably peacefully, but clearly anything that increases popular agitation against the oil companies or against the terms of existing agreements, such as crises in other countries or even the mere knowledge that a government has obtained better terms in some other country, will reduce both the willingness and the ability of the government to compromise, and must therefore increase the company's estimate of the risk of loss if it makes no further concessions to government demands.[8]

In these circumstances there is every incentive for oil companies to adopt accounting and other techniques designed to permit a maximum effect on public relations with a minimum actual concession. For example, the prices at which the oil is sold as it leaves the control of the producing companies are posted by the companies. But for the most part the oil goes in pre-determined shares to the foreign oil companies who own the producing companies. These same companies, or their affiliates, also control the major distribution channels of the world, and it is extremely difficult to believe that they do not have a powerful influence on the prices with reference to which the profits to be shared are calculated and consequently on the revenue obtained by the producing countries. It has often been alleged that the prices are unfair and arbitrarily set to the advantage of the oil-distributing companies. It is not, therefore, so strange as it appears on the surface that one of the demands of the oil-producing countries has been that they should share in the profits made by the major oil companies

## PROFIT SHARING BETWEEN COUNTRIES AND COMPANIES 157

from the distribution of Middle East oil in world markets. Furthermore, the calculation of the 'costs' to be subtracted from the total revenue in determining profits is, as every accountant and economist knows, subject to any number of arbitrary judgments. Any announcement that a high percentage of profit is to go to the government has a popular appeal, but in fact the actual percentage granted may be much less significant than the method of calculating total profit.

This analysis of the essence of the factors underlying the bargaining process may seem extreme; and indeed for some countries, in particular the smaller autocratic sheikdoms, it would need considerable modification. Its chief pupose is to demonstrate that, in the absence of effective competition and when the producing countries are unable to run the industry (which includes selling the oil) there is no such thing as oil relationships on a 'business basis.' We are dealing essentially with a problem in bilateral monopoly slightly modified by marginal competition, and with a bargaining game in which there is a strong element of bluff; but one of the pawns in the game is the political stability of a sensitive area of the world. It seems highly probable that the Middle Eastern governments have obtained as much of the oil profits as they have with few overt crises precisely because the oil companies were willing to pay for political goodwill and because the profitability of oil in the Middle East has been so great that the companies could well afford to pay handsomely.

It remains to consider how the bargaining process would be affected by the introduction of competition, either competition among the oil companies for oil concessions or potential competition from the government made possible by growing ability to run the industry.

Again, the position with respect to new concessions need not detain us. If the government can itself exploit its own oil resources it naturally will not give a foreign company the right to do so unless it can obtain a share of the profit at least equal to the profit it could make by running the industry on its own account when all opportunity costs of the resources required are taken into consideration. This implies that the government is in a position to produce and sell oil as efficiently as the companies and only permits them to operate because it has equally or more profitable alternative uses for its own resources.[9]

If there were effective competition among the oil companies for a concession, this would plainly push the share of profits offered to the government up to the point where the company obtaining the concession retained only 'normal' profit which, for our purposes, may be defined as the amount of anticipated profit that would just induce the company to enter or continue to invest in the oil industry of the country. In these circumstances, the government should be able to extract nearly the full value of the concession from each company;

whether the introduction of effective competition would actually mean an increase in the government's revenue, however, is an open question which is discussed below. In the past, as noted above, competition for oil concessions was largely restrained by intergovernmental agreements, in which military and political considerations were dominant, as well as by the private agreements and financial inter-relationships among the oil companies themselves. Many if not most of these restraints still exist, and thus even for new concessions there is only marginal competition such as that recently offered by the Japanese and Italian oil companies, which are outside the combines at present controlling Middle East oil.

Even a small amount of competition for new concessions, however, is significant, for whenever any producing country obtains terms in an important agreement better than those prevailing elsewhere, an immediate pressure tends to be created for the revision of old agreements, because the amount the government can reasonably expect to get and the point at which it would be willing to compromise are both raised.[10] In addition, better terms obtained from a new competitor in the industry may encourage the rulers of the producing countries to attempt to withdraw concessions on some areas over which exclusive rights were granted under earlier agreements and which often include vast parts of the country not yet in production and relatively unexplored from a geological point of view when the agreements were made.

The companies are very much aware of these dangers to their positions and naturally seek to avert them—a difficult task if loose cartel arrangements had to be relied on; much easier when the various producing companies are jointly owned by a few dominant international concerns, which is largely the case in the Middle East.[11]

The government would also be in an extremely strong position with respect to established oil companies if it could run the industry without their help and obtain revenues not significantly below those previously obtained from the companies. In these circumstances, it would no longer risk political disaster if it nationalized the industry. Its threats of nationalization if the company did not comply with its demands would no longer be dangerous bluff but would have to be taken seriously, and the company's chief source of bargaining strength—the economic disaster for the country if the oil revenues cease—would disappear. If necessary to keep its concessions, the oil company would gain by continuing to give up part of its profits so long as the remaining return on its investment was at least equal to the return (appropriately adjusted for risk) it could expect if it invested compensation funds elsewhere,[12] and skilful government bargaining could conceivably force the company's share down to this level.

It would not be to the country's economic advantage, however, for the government to demand more than this, and thus force the company out, or to demand a share that would inhibit further expansion of the company's activities and create excessive emphasis on the short run, unless the government could operate the industry at a profit not only equal to but considerably greater than the revenue obtainable from the company over the life of the oilfields involved. A foreign company brings many indirect advantages to a developing country, including an increased rate of innovation, a freer flow of technical information leading to the more rapid application of advanced technology, and a greater supply of technical and managerial personnel. Most underdeveloped countries accepting foreign capital or foreign enterprise urge that the foreign companies should employ domestic personnel as far as possible, especially in technical and managerial positions. The primary purpose is to ensure that foreign enterprise provides opportunities for the nationals of the country to develop their own skills and executive potentials. But as such countries develop, many of the men and women so trained will be required in other fields, and it may well be desirable to use to the fullest possible extent opportunities for training made available by the foreign companies (which presupposes the presence of foreign personnel capable of providing the training) and also to withdraw trained and capable domestic personnel from the companies for other uses.[13] Nationalization may well mean that a larger proportion of domestic, managerial and technical skills will have to remain in the oil industry and thus will not be available elsewhere.[14] Paradoxically, many people in the underdeveloped countries complain of the shortage of capable technicians and capable management and at the same time urge the elimination of the foreign supplement to the available supply.[15]

In addition, the expenditures of the oil company in its ordinary operations in the country as well as for investment in further expansion and for taxes are an addition to the foreign exchange available to the country and these, too, would be lost on nationalization. Thus, even if we should make the assumption that a country could run its oil industry at a total profit equal to that made by the oil company before royalties and taxes, it still might not be advantageous to the government to buy out the company. The government, in calculating the advantages of nationalization, should take into consideration the alternative product that could have been produced by those of its resources which must be diverted into the oil industry (*e.g.*, the loss of productivity in other industries due to the reduced supply of managerial talent); the foreign exchange previously contributed by the oil industry in its operating expenditures, taxes and further investment expenditures; the amount of profit the company would have been willing to hand over to the country; and the compensation payments agreed on.

In fact, of course, even apart from problems of marketing the product, which we discuss in the next section, it is extremely unlikely that any of the producing countries will be able for some time to manage their oil industry as efficiently as the oil companies can, although some of them might well be able to produce oil with the help of individually hired foreign technicians. It would seem, therefore, that nationalization of the oil industry would rarely be profitable in any of these countries unless the alternative opportunities available to the oil companies or the rate at which they discounted the future made them hold out for such large shares of profit that the government would gain even after making allowance for less efficient operations and the considerations mentioned above.

Nevertheless, the ability to run the industry without risking economic disaster would put the government in a strong bargaining position *vis-à-vis* the oil companies, which would probably enable it in time to extract the lion's share of the profit from oil, and which would also help to reduce those political elements which now make it impossible to put East–West oil relationships on a 'business basis'. But inability to sell the oil is the enduring difficulty; this makes even the reasonable and reflective Arab nationalists feel at the mercy of the oil companies and they passionately believe that the companies are exploiting their country in an oppressive sense.

## II. THE QUESTION OF EXPLOITATION

The term 'exploitation' has no generally accepted meaning in economic analysis (except in the Marxian sense); it is full of ambiguities and is, like the word 'imperialism,' used more often as an undefined term of abuse than for serious analytical purposes. But the notion has a tremendously important influence on the actions of the people in the less-developed countries of the world, who often go so far as to treat all foreign investment as a form of exploitation.

Most of the serious charges that exploitation exists rest implicitly or explicitly on the assumption that foreign interests use their superior political or economic power to prevent the exploited country from making the most profitable use to itself of its own resources. This, broadly interpreted, provides a reasonable definition of exploitation for our purposes;[16] it does not imply that foreign investment is necessarily exploitative in itself, for if the profits of a foreign firm are high entirely because of its efficiency in production or in meeting demand, one cannot assume that the foreign firm is in any way preventing the domestic economy from using its resources more advantageously, even though the firm pays out a large proportion of its profit to foreign shareholders. On the other hand, even when the productivity of a foreign firm is higher than that which any domestic firm could hope to attain, a large part of its profit may nevertheless be traceable to factors other than its greater productivity, for example, to monopolistic restrictions enforced by the firm within the country, and it is then possible to inquire whether this leads to a less advantageous use of other resources and therefore whether exploitation exists. Thus, neither the actual amount of profit earned, nor the efficiency of the foreign firm, necessarily provide by themselves evidence of exploitation or the absence of it.

Apart from a number of popular fallacies which are not worth discussing, such as the notion that *any* oil profits going to foreigners imply exploitation, most of the substantial complaints that oil companies exploit the producing countries can, in the light of the preceding analysis, be reduced to the charge that the oil companies weaken the ability of the producing country to increase its share of the profits from oil by: (*a*) excluding competition for oil concessions, and (*b*) retarding the growth of the producing country's ability to run the industry, including its ability to market the product. In addition, there are charges of political interference and domination which are outside the scope of this paper.

(*a*) We have seen that, *other things being equal*, a producing country can be expected to obtain more advantageous terms in its oil agreements if there is substantial competition among the oil companies. Therefore, to the extent that competition is effectively restricted by the blanket nature of existing concession agreements (covering very large territories, if not the whole country), by understandings among the major oil companies with respect to the territories in which each will operate, or by the exercise of power to control distribution channels and to restrain the competitive entry of newcomers, the charge of exploitation *may* be justified. The past agreements among companies dividing the Middle East among them have often been pointed to by the Arab nationalists as humiliating examples of exploitation, and it can be safely said that competition, for all of these reasons, is still effectively restrained by the major international oil companies operating in the Middle East. But this statement does not yet indicate whether or not the restraint of competition actually prejudices the producing countries and therefore whether or not they are 'exploited'.

Effective competition presumably implies the existence of a considerable number of competing companies; no means has as yet been found, even by the United States anti-trust authorities, for making two, three or even four companies compete effectively if they prefer

not to do so. It can be cogently argued that because of the immense size and speculative nature of the original undertaking in the geographically and politically inhospitable countries of the Middle East, only big, financially responsible companies could enter the field with any assurance of success, and that this itself limited the possibilities of competition. On the same grounds, it can be reasonably held that restraint of competition was necessary to encourage such companies to undertake the risks involved. In addition, it is entirely possible that the producing countries might be worse off if oil concessions had been granted to many companies on a competitive basis, not only because the value of each concession would be substantially less and the royalties each company would have been willing to pay substantially smaller, but also because the utilization of their oilfields would have been more costly in the short run and wasteful in the long run. In other words, it can be plausibly argued that the same amount of investment would not have been forthcoming under a competitive regime in the industry, and that neither the efficient utilization of oilfields nor the organization of pipeline transportation and world-wide distribution would have been possible. Unlimited competition in oil production and distribution is held by many, if not most, oil experts to be not only wasteful but completely impracticable, for 'other things' do not remain equal when a competitive and monopolistic organization of an industry is being compared.

Thus the argument that the monopolistic position of the oil companies necessarily involves exploitation of the producing countries, in the sense that the countries would be better off if there were extensive competition, is by no means conclusive. Nevertheless, the fact remains that the absence of competition, for whatever reason, does give the oil companies a strong hand in bargaining and forces the producing countries to put the emphasis either on politics or on the necessity of developing their own ability to run the industry.

(b) It is charged that, by restricting the development of trained local personnel and by barring access to world oil markets, the oil companies interfere with the growth of the producing countries' ability to run the industry. In the first case, the complaint is that the companies fail to hire the nationals of the country for senior technical and managerial positions and to train local workmen for skilled jobs, thus effectively preventing the development of local skills and experience in the industry—that an important industry thereby remains in a very real sense an 'alien' in the local economy.

Plainly, even if there are very few nationals of the producing country in such positions, there is no warrant for assuming a deliberate attempt to harm local development, for if a given firm hired men only on the basis of their ability to do a job, the proportion of local employees in skilled and senior positions would tend to rise only slowly in underdeveloped countries. This is widely recognized, but critics insist that the foreign firms have not gone sufficiently out of their way, and at their own expense, to train local workers and executives. In fact, however, the modern oil company makes serious and costly efforts to develop local skills, partly because it needs such skills and partly for political reasons; there can be little question that this complaint is not well founded to-day, even if it were once valid, for there is strong evidence that the companies do go as far as is practicable in this respect.

The second aspect of the charge has more substance; regardless of the producing country's ability to produce the oil efficiently, the control of international distribution channels by the major oil companies can be used effectively to prevent the country from selling the oil.[17] The major companies not only can refuse to accept oil offered by a producing country but they may also be able to dissuade others from doing so, thus barring markets and forcing oil operations to shut down: the country is unable to run the industry.

Again, however, it is not clear that a concerted international attempt to break down the control of distribution channels by the major companies would, even if successful, benefit the Middle East countries; it is equally possible that any benefit would go primarily to the ultimate consumers of oil and oil products through a subsequent fall in prices. Perhaps consumers are also being 'exploited' by the oil companies, and it is possible that the disappearance of this 'exploitation' would bring in its wake a substantial fall in the oil companies' profits and a consequent fall in the revenues received by the producing countries whether or not they ran their own industries. Thus the elimination of this source of alleged exploitation might leave the producing countries with much less than they receive under conditions in which the oil companies can exploit both their suppliers and their customers.

\* \* \*

What, then, can we conclude with respect to the problem of exploitation? We have defined it as the use of superior economic or political power by foreign interests to prevent a country from making the most profitable use of its own resources. Clearly if the destruction of monopoly in the position of the oil companies would enable the producing countries to obtain greater revenue from their oil resources, we would be entitled to say that they are at present exploited by the monopolistic power of the companies. But if the facts should be otherwise, and the introduction of a substantial element of competition would raise current costs, or inhibit investment, or lead to wasteful utilization of the oilfields in the long run, or reduce prices and revenues to such an extent that total revenues to the producing countries became less than at present, we could not say that these countries were exploited *because of the monopolistic organization of the industry*, for the removal of the monopoly would not permit a more profitable use of its resources by the country.

Nevertheless, the superior economic power of the oil companies arising from their ability to inflict a disastrous economic loss on the producing country does give them a bargaining position which holds down the share of the profits the producing country can obtain in the oil agreement. Thus the producing countries can still complain of exploitation in the sense that their resources are being used to earn large profits which do not accrue to them. There is no reason from the producing country's point of view why the oil companies should earn greater profits than is necessary to induce them to continue investing in the industry and to compensate them for their initial risks. In other words, if excess profits are to be earned, a producing country can complain of exploitation to the extent to which it is unable to obtain them for itself because of the superior bargaining power of the oil company; if it could obtain them, the use of its resources would thereby become more profitable to the country. Whether this is reasonable, that is, whether, from some other point of view, the producing countries *should* properly have such profit is another problem entirely. Part of what appears as profit may really be economic rent attributable to the natural resources, and if the rest is basically due to monopolistic organization which is considered necessary for the efficient operation of the industry, a good case can be made for the claim of the producing countries. That excess profits, or at least large rents, do exist is plainly evident from the mere fact that the oil companies can afford to hand over large shares to the governments and still find it worth their while to expand their operations and continue their investments. Whether the oil companies are giving up all they could give up without making operations unprofitable cannot be decided here; it seems improbable.

### III. RECONCILING CONFLICTING INTERESTS

The countries consuming Middle East oil want an assured and adequate supply of oil at acceptable prices; the countries supplying it want to obtain the maximum possible revenue from its sale; the companies producing it want a profitable return on their investment. No one wants continuing political conflict, but each wants the maximum protection for its own interests. At present none of these aims can be achieved if the producing countries attempt to take over and operate their oil industries without the co-operation of the existing oil companies; perhaps as the countries develop they may be able in time to handle with reasonable efficiency the extensive scientific, technological and managerial problems of running this enormous industry, and to reach the point at which they themselves can supply their own oil for sale on acceptable terms to the companies controlling the world's distribution systems. That time has not yet come for most countries; perhaps it has nearly come for a few. But nationalization designed to eliminate foreign control would have consequences adverse to all interests at present; 'nationalization' as a formal act leaving the existing companies in effective control is, of course, always possible (and indeed even likely before very long in some countries) but would solve none of the difficulties.

There are three other possibilities: (1) matters could be left alone in the hope that moderation and good sense will prevail on all sides; (2) the Arabs could attempt to create some form of effective united front for consultation and perhaps action in dealing with oil companies; and (3) some form of international approach to the problem could be attempted.

(1) Although under present conditions there is danger of unexpected political explosion, actual conflict is not inevitable. It is true that the supply of oil is only uneasily assured as the contending groups attempt to maintain or improve their positions and that no amount of 'correct' behaviour on the part of the oil companies, of assurance that their intentions are good and their acts benevolent, nor demonstrations that the results of their operations are beneficial all around, will change the fundamental feelings of the producing countries that these powerful foreign interests are getting more than they should from the oil Allah has provided. On the other hand, although the oil companies must consider their 'duty to shareholders' and do not and cannot look upon their activities as simply a 'task of trusteeship'[18] for the producing countries, their tendency is to go very far to meet the demands and respect the feelings of the producing countries. They have placed nationals of the countries on their boards of directors, they have agreed to outside arbitration of disputes, and have virtually accepted the principle that existing contracts may be re-negotiated as circumstances change. It seems highly probable that increasingly greater shares of profit will be demanded by the governments. But if both sides have a genuine desire to avoid political conflict, it is possible that these demands will not be pressed too far and that the companies, for their part, will believe that political explosions are too dangerous to be risked, that it is preferable to continue to accept the government's demands so long as it is at all profitable to remain in the Middle East. Thus the various problems may be ironed out as they

arise by negotiation between companies and governments as they have been in the past, with grumbles on both sides but few serious disturbances.

(2) Advocates of the political 'unification' in some sense of the Arab world argue that concerted action by all oil-producing countries in the Middle East would remove their bargaining weakness and even give them the upper hand.[19] Apart from the very real difficulties in obtaining such unity between countries whose national interests with respect to oil production differ substantially in many respects, this argument has substantial validity in the short run; a single producing country is obviously more vulnerable than the group combined. But even for the group as a whole there is considerable risk in relying on the effectiveness of joint action, for though Middle East oil is low cost, fear on the part of the consuming countries of political instability or of extreme dependence on the whims of Middle Eastern governments would intensify the development, not only of new sources of oil, but also of substitutes for oil. This process has already gone a considerable way, and the producing countries would do well to remember that world supply conditions can change rapidly in times of crisis. Thus, even though concerted action would strengthen the bargaining position of the producing countries, it would not resolve the political tensions nor ensure that either their own economic interests or those of others would be advanced.

(3) In the minds of the people of the Middle East, the Western governments cannot escape responsibility for the position or actions of the oil companies whose relations with the producing countries must inevitably become in some degree entangled with their politics. Many of the difficulties arise because the oil companies are themselves international organizations, their activities transcending national boundaries, their interests and powers of control involving many peoples and governments, yet subject to the effective scrutiny, let alone supervision, of none. It is unrealistic to think that oil relationships with the Middle East can be put on a 'business basis' and the question may be fairly asked whether the Western governments should stand aside and leave their oil relationships with the producing countries solely to such business and commercial interests. Perhaps the more hopeful long-run solution to this type of problem lies in the creation of some form of consultative international body representing all groups, including producing countries, oil companies, the consuming public, the interested as well as 'neutral' governments (including perhaps some of the non-producing countries in the Middle East), and well staffed with economic and technical experts empowered to investigate all aspects of the operations of the companies—their costs, their financial interrelationships, the nature and effects of their

agreements and operating policies, the way in which their prices are fixed at all levels of operation. Such a body could have many advantages: it could provide a consultative group where disagreements could be discussed and also an independent source of information; it could reduce the feeling of dependence in the producing countries on the goodwill of the oil companies; and it could certainly help to make information more readily available; for at present the serious student of the problem finds important facts missing.

### PROFIT SHARING BETWEEN PRODUCING COUNTRIES AND OIL COMPANIES IN THE MIDDLE EAST: A REPLY*

#### By H. W. Page

As one who has been intimately associated with most of the major oil negotiations in the Middle East during the past decade, I was very interested in Dr. Penrose's article, 'Profit Sharing Between Producing Countries and Oil Companies in the Middle East' in the ECONOMIC JOURNAL of June 1959. It is apparent that Dr. Penrose has given a great deal of thought to these very complicated problems, and she is to be congratulated on her lucid presentation. However, for the record, I think I should point out a few things that are factually incorrect and several inferences with which I do not agree.

Despite its qualified language, the thesis of the article is unmistakable: There is lack of competition in the bargaining processes by which concessions are granted and agreed upon, and the decisive weight is on the side of the companies in their negotiations by reasons of their monopolistic power.

It is clear that the term 'monopoly' as used in this article has several meanings. It is, of course, correctly used when applied to an oil concession. Any oil concession is a monopoly, *i.e.*, the company involved is granted exclusive rights to explore for, develop and carry away oil from a specified area for a specified time. Obviously, nobody would risk capital in exploring for oil unless they had some form of guarantee—call it a monopoly if you wish.

I find it extraordinary, however, that Dr. Penrose assumes that there is not competition for concessions. A high degree of competition has been going on in both Venezuela and the Middle East, not to mention other areas. As a matter of fact, the competition became so keen that the terms offered for recent concessions has caused speculation as to the business acumen of some bidders. There are indications that several of the successful bidders are already regretting their success.

The implication in Dr. Penrose's article also is that there was no

* Published in *The Economic Journal* Vol. LXX, September, 1960.

competition in the past. It is perhaps forgotten—although well known in oil circles—that the Saudi Arabian concession could have been in other hands if the others had thought it worth a down payment of £30,000. Dr. Penrose states: 'There is little doubt that some of the early agreements were made by rulers who knew little of what they were signing away.' Conversely, the companies knew little of what they were receiving. I am confident that a study of early agreements would show that they were very close to the maximum terms which anyone would pay at the time. The rulers and governments demanded and received gold payments—most of them 4s. gold per ton. This royalty, of course, was in addition to other payments, such as bonuses, dead rent, etc.

It should be recalled that during the 1930s crudes similar in quality to the Middle East crudes being produced were selling in Texas for less than $1.00 per barrel at the well and even lower than $0.50 per barrel in 1931 and 1933. The cost of delivering Texas oil from the well to Northern European markets was a little less than the tanker cost from the Persian Gulf. Residual fuel oil was available in seemingly unlimited quantities at the large Caribbean refineries at $0.60–0.75 per barrel f.o.b. Although there were no posted prices in the Middle East pre-war, it is not difficult to estimate the maximum value of the crude for delivery into Europe in competition with Texas and Caribbean crude and products. It can easily be seen that 4s. gold (about 13 c. per barrel until February 1934, when gold was revalued at $35 per ounce, and thence about 23 c. per barrel) plus other payments to the governments came to an average of nearly 50% of the profit, and at times was even more.

After the War, with inflation and the rapid increase of crude prices, higher United States and United Kingdom corporate income taxes and frozen gold prices, the oil companies operating in the Middle East recognized that the fixed royalty payment no longer gave the equitable division originally intended, and that additional royalty payments were neither economically practical nor a permanent method of maintaining equity between the parties. The 'rapid spread' of 50/50 was the result of this recognition of the need to restore the equity which had been frustrated by drastically changed conditions, beyond the control of either party and unforeseen at the time the agreements were negotiated.

There is no relationship between this adjustment of old agreements and the signing of new concession agreements. New concession agreements should bring the Governments whatever they are able to negotiate, whether it be better terms for the Government or less favorable. If some of the early agreements, made before the potential of an area was known, appear more favorable to the oil companies than new ones negotiated after the risk-takers have demonstrated the area was rich in oil, is this not the proper reward for risk-taking?

Even in Iraq, where Dr. Penrose states incorrectly that five American companies now participate, it should be realized that three of these companies pulled out more than thirty years ago because they thought the terms too stiff.

There is the implication in Dr. Penrose's remarks that there is no potential competition from governments—that they are restricted in marketing. In fact, practically all producing countries can receive royalty oil which they are free to sell anywhere, at any price they are willing to accept. They find, however, just as the oil companies do, that a lot of capital investment is required in transportation, refining and marketing facilities before crude can be disposed of.

Dr. Penrose writes that 75% was obtained by the 'Emir of Qatar' from the Italians and Japanese. The correct facts, which were widely reported in newspapers and oil journals, are that the Italian agreement was made with Iran, and the Japanese agreement with Saudi Arabia and Kuwait for the offshore area beyond territorial waters of the Kuwait/Saudi Arabia Neutral Zone.

The Italian agreement, although incorrectly reported as a 75/25 deal, actually involved a 50/50 agreement with the National Iranian Oil Company, which included liability of half of the development costs. This group in turn has a 50/50 concession with the Iranian Government. In the case of the Japanese company, the agreement with Saudi Arabia amounts to 56% profit payment and 57% to Kuwait.

As for the question of sharing profits made outside the country, it is important to note that those who suggest doing so are unwilling to share in the capital investment required to generate these profits, if any. The financial statements of the oil companies indicate that the total profits in 1958 in the Eastern Hemisphere on all operations, including producing, were appreciably less than the payments made by them to the governments of producing countries. (See recent report of First National City Bank of New York by William S. Evans, *Petroleum in the Eastern Hemisphere*.)

Dr. Penrose writes that it is extremely difficult to believe that the oil companies do not have a powerful influence on prices and, consequently, on the revenue obtained by producing countries. I think it can be shown that world competition sets the price. If the price is unfair—and the implication here is that it is too low—then two things happen. First, the price is available to all and if too low, competitive refineries will buy and take away product markets by price competition while still making a satisfactory profit. Secondly, if the price is too low the Government could take its royalty oil in kind and sell it for a higher price. In actual fact, most of the complaints that the oil companies receive is that posted prices have been too high.

Dr. Penrose refers to the 'transportation monopoly of the major companies.' If the major companies have a transportation monopoly they are certainly doing a poor job with it. At present there is a 20% surplus of tankers, and owners are ready to charter at below out-of pocket costs to save tie-up charges.

In Dr. Penrose's mind there is clearly much confusion between nationalization and the cancellation of concessions. I doubt if it is generally realized that oil concessions in the Middle East are already nationalized. That is, all the facilities are either the existing property of the Government, or a government entity, as in Iran, or become the sole property of the Government under terms of the contract after termination. Rights are then granted for a period of years in return for the performance of certain obligations. If a country were to seize the properties and prevent a company from exercising its rights under such a contract it is incorrect to dub this act as nationalization. It is nothing more or less than the abrogation of a solemn contract which is part of the law of the land.

One of Dr. Penrose's important assumptions is that governments have the right to nationalize the property of foreigners as well as that of their own citizens on payment of fair compensation and that such action is justified when the economic interests of the country are thereby furthered. The important point here is that although the countries may have the power to abrogate a contract they do not have the right unless they pay prompt, adequate compensation, plus liquidated damages for loss of profit during the remaining life of the contract.

In so far as principle, however, is concerned, it is important to keep the record straight. If there is one thing that has been laboriously achieved in Anglo-American history over past centuries it is that the King or the Government is under the law and not outside it. What protection is there for either private citizens or private corporations if we allow in principle that governments are free to break, unilaterally, solemn contracts ratified by the highest authority in the land? It is true that in some countries there are overriding government rights, such as to achieve national security, but they are exercised only under careful restrictions, including fair compensation not determined unilaterally but, if necessary, by the courts. We must never forget that such safeguards are not often available in international affairs. The lack of them, however, does not warrant acceptance of the assumption that a government is superior to its contracts. Of what use then would be a contract? Would it permit a company, too, to unilaterally abrogate its contract if its economic interests were thereby furthered? No, a government is, equally with the company or individual, bound to the provisions of a contract freely entered into, and if revision is necessary, then it is a matter for negotiation and not unilateral action.

*Standard Oil Company*                           H. W. PAGE
*(New Jersey).*

---

PROFIT-SHARING IN MIDDLE EAST OIL—REJOINDER

ALTHOUGH my article on the sharing of profits from Middle East oil production[20] was painstakingly designed to analyse as objectively and impartially as possible the fundamental considerations governing the bargaining positions of oil companies and governments in the Middle East, it has (perhaps for this very reason) come under attack from both sides. My friends among the Middle Eastern economists have privately objected to my contention that the oil-producing countries would not gain by 'nationalizing'[21] the companies, and also to my conclusion that these countries are not 'exploited' by virtue of the monopolistic position of the companies. On the other hand, the distinguished director of the Standard Oil Company of New Jersey, Mr. Page, objects to the notion that the oil companies have monopolistic power with respect to concession agreements, marketing, price or transport, while Professor Salera insists that in any case it is the governments that often enjoy 'the superior economic power.'[22] In addition, Mr. Page raises the difficult and as yet unresolved question under what conditions a sovereign state can legally abrogate its contracts with foreign nationals.

With this last problem international lawyers have been grappling for some time and I have no special competence in the matter.[23] Mr. Page and I evidently agree that adequate compensation should be paid, but it seems to me that there is ground for argument over the question whether compensation should include 'liquidated damages for loss of profit during the remaining life of the contract,' when much of the profit can be attributed to the monopolistic market position of the private producers. Clearly we cannot undertake a full discussion of this problem here; there is no way in economics of determining what is a 'proper reward for risk-taking' or an 'equitable' division of profits, and therefore economic analysis provides no criteria for deciding how much 'damage' due to loss of profit should be included in an 'equitable' compensation. I was concerned neither with the legal issue nor with the problem of determining appropriate compensation but only with the circumstances in which it might be to a country's advantage to abrogate its contracts on payment of compensation.[24]

When we come to the question of the amount of competition in the international oil industry we have to analyse a very mixed bag. Mr. Page is certainly correct in stating that there has been intense competition for the new concessions recently granted in the off-shore areas and in Iran; there was nothing to the contrary in my paper, and I certainly would not quarrel with the implication that the governments have done extremely well as a result.[25] As to the original concessions, I suppose that the political manœuvring and behind-the--scenes bargaining could be classed as a kind of competition, although surely Mr. Page has not forgotten the governmental policies and private restrictions affecting such competition. Whether or not the terms of most of the original concessions were the 'maximum terms which anyone would pay at the time' is very difficult to judge, and there is little point in disputing the matter; the important issue with which I was primarily dealing related to the problem of renegotiating existing agreements, and here Mr. Page does not appeal to competition but to the recognition by the oil companies of 'the need to *restore* the equity which had been frustrated by drastically changed conditions' (italics mine).

I always find myself slightly bewildered by the tendency of the spokesmen for business firms to emphasize at one moment competition as the most significant force determining a firm's behaviour and at the next moment to emphasize equity, the public interest and fair play. Equity becomes an important factor only when competitive forces are largely inoperative; rarely can the same phenomena be explained on *both* grounds.[26] My discussion of the bargaining problem was based on the assumption that both the oil companies and the governments were concerned to do as well as they could for themselves, and I still suspect that this assumption comes fairly near to the truth. The companies insist that a 50/50 split of the profits is fair; this may be so, although I have not seen any attempt to show *why* it is fair. Presumably to everybody except economists anything split equally is *ipso facto* split fairly!

Professor Salera, on the other hand, holds that the 50% profits tax imposed on Aramco by the Saudi Arabian government was a 'unilateral decision' by the Government which made Aramco very unhappy. I have yet to see any evidence that it was unilateral, and Mr. Page implies that, equity having been served, Aramco would not have been unhappy. As a matter of fact, to give 50% of its profits to Saudi Arabia in the form of an income tax should not have distressed Aramco, since it cost the company practically nothing—it merely meant that the company, which is incorporated in the United States, paid taxes to Saudi Arabia instead of to the United States Government, whose corporate income tax is now 52% in any case.[27]

I would agree with Professor Salera that political considerations complicate the oligopoly bargaining problem, but it certainly is not necessary to postulate anything so pseudo-scientific as 'the dynamics of the international oil industry' (p. 313) to explain the likelihood that the governments will be able to obtain progressively larger shares of the profit. The simple facts are that the governments are no longer easily dominated, they are becoming increasingly aware of their strategic political position (although North African oil will reduce their economic bargaining power), and their peoples, imbued with highly charged feelings of national consciousness, are sometimes fiercely (and often irrationally) determined to reduce Western influence in their economies to a minimum. Ricardian rent analysis, to which Professor Salera appeals, is a strange and lonely traveller in such company, and it has little relevance to the problem of how profits (or rents) are to be divided between companies and governments.

So far as the price structure of the international oil industry is concerned, it must be remembered that the seven international major companies control around 90% of the crude oil produced in the Middle East and in Venezuela, and refine and sell the bulk of the petroleum products sold outside the United States and Soviet controlled areas. Crude-oil production in the United States, the third major source of supply, is regulated by government agencies. Estimates of the control of the commercial tanker fleet by the integrated companies through ownership or long-term charter range from 72% to as much as 90%.[28] It is true that since 1958 price competition in crude oil outside the United States has for a number of reasons become a serious problem for the industry, and that the established pattern has been breaking up. But the argument that 'world competition sets the price' of crude oil has, I think, been effectively exploded, even if we were to use a very loose notion of 'workable competition.' There is not space to discuss the matter here, and I can only refer the interested reader to the important literature on the subject.[29] I do not intend to imply, however, that price competition in the economists' sense *should* be the primary determinant of crude oil prices; this is an entirely different and rather difficult question.

But surely it is disingenuous of Mr. Page to say, on the one hand, that the governments of the oil-producing countries are not restricted in marketing their royalty oil, since they 'are free to sell anywhere, at any price they are willing to accept,' and that they would do so if oil prices were too low, yet on the other hand to indicate that they *cannot* do so because they do not have the 'transportation, refining and marketing facilities' necessary before 'crude can be disposed of.' Is this not an admission of my contention that the integrated companies have extensive control over these later stages of the industry and are unwilling to provide markets for oil available from other sources?

Indeed, the most significant fact about the international petroleum industry is the vertical integration of the major companies. For the most part, the majors produce their crude oil for their own use; selling crude is not their primary business, and relatively little is sold 'spot' to independents on the open market. Nevertheless, as Mr. Page points out, their financial statements 'indicate that the total profits in 1958 in the Eastern Hemisphere on all operations . . . were appreciably less than the payments made by them to the governments of the producing countries.' In other words, in 1958 *all* of the profit from the *integrated* operations of the companies was put at the level of crude-oil production. As any economist would immediately suspect, there are tax advantages in such an arrangement, at least for the American companies. The device by which this is achieved is the posted price, which is the basis for calculating payments to be made to the governments of the producing countries on all oil produced, but which has no necessary relation to the price at which the majors actually sell the crude oil they do not use in their own subsidiaries and certainly no relation to the value of the oil used in their own operations. Whether the posted price is 'too high' or 'too low' depends on the point of view; there are no economic criteria. It certainly has been higher than genuine price competition would have produced—but then it is not necessarily the actual 'market price'; it is naturally too low from the point of view of governments desiring a greater share in the oil companies' profits.

NOTES

1  I would like to thank for their criticism and suggestions all those who read earlier drafts of this paper, especially those from the University of Baghdad, the Iraqi Government service, and the Iraq Petroleum Company, Dr. Yusif Sayigh of the American University of Beirut, Professor Fritz Machlup of the Johns Hopkins University and my husband, Professor E. F. Penrose.
2  The large oil companies control oil properties in numerous countries throughout the world and their production policies in the different countries are subject to many influences, including tariff arrangements, the domestic policies of national governments, questions of the political security of their investments and reliability of their supplies of oil, and private agreements aimed at 'regulating' supplies of oil. Most of the companies control both low- and high-cost sources of oil but do not necessarily put the greatest emphasis on expanding low-cost oil.
3  Many of the demands made by a government often affect the costs of the company rather than the profit-sharing arrangements. But since increased costs reduce the profits available to both parties, we can, for our purposes, treat such demands as part of the profit bargain. They are usually quantitatively unimportant in any case.

4   It is possible, indeed even likely, that an oil company would concede some of the demands of a government simply because it considered the demands to be fair, and not at all because it believed that to do so was necessary to safeguard its position. But it would be extremely difficult to separate concessions made solely in a 'spirit of fairness' from those made because the company was convinced that a 'spirit of fairness' would in the long run assist in safeguarding its position.

5   I must emphasize that we are only concerned with present-day conditions. There is little doubt that some of the early agreements were made by rulers who knew little of what they were signing away and who were attracted as much by the immediate cash for themselves as by any calculation of the future returns to the country.

6   This point is well illustrated by the conflict between the Iranian Government and the Anglo-Iranian Oil Company in the years 1948–53. To be sure, both the British and American Governments intervened in the dispute, but without much avail, and the dispute ended with the fall of the Iranian Government and the practical victory of Western interests. The company suffered heavily; but during the years 1952 and 1953 when it had no Iranian oil its profits were greater than ever, derived of course from other sources than Iran. See the *Annual Reports and Accounts* of the company and also the discussion by Benjamin Shwadran, *The Middle East, Oil and the Great Powers* (London: Atlantic Press, 1956), Chapter VI. Stephen H. Longrigg's account in *Oil in the Middle East* (London: Oxford University Press, 1954), Chapter X, is more sympathetic to the company than that of Shwadran. The latter implies that nationalization as such was the primary issue; Longrigg places the emphasis on the terms of the nationalization. Both accounts should be read.

7   The ordinary people of the Middle East, knowing nothing of the oil industry, are easily convinced that their government could run it as well as the foreign company. The relative success with which the Egyptians have been able to operate the nationalized Suez Canal has enormously strengthened the belief in the more developed countries such as Iraq that similar success would attend the nationalization of oil.

8   The Iraqi agreement, for example, made express provision for re-negotiation if better terms were obtained in neighbouring countries.

9   As we shall see below, this last consideration means that it will not necessarily be advantageous to a developing country to exclude foreign companies merely because it can produce as efficiently as they can.

10  The rapid spread of the so-called 50–50 arrangements after the Aramco agreement with Saudi Arabia in 1950 is an example of this; and almost immediately after the announcement that 75% had been obtained by the Emir of Qatar from the Italians and Japanese, increased pressure was evident in Iraq for a revision of that country's agreement with the Iraq Petroleum Company, which is jointly owned by British Petroleum, Royal Dutch-Shell, Compagnie Française des Pétroles and a combination representing five American companies. [See correction below.]

11  This is not intended to pass judgment on whether or not competition would be desirable from other points of view. See II below.

12  I do not want to discuss the difficult problem of how 'fair' compensation can be determined. Clearly the amount of compensation the company expects to get would be an important consideration in its calculations of the profitability of remaining in the country. For our purposes it is not necessary to examine this question; we merely assume that some agreed formula can be found.

13 The Iraq Petroleum Company, for example, has an excellent training programme which not only provides skilled people for the company but for outside firms as well. In particular, those who are unable to reach the high standards necessary to complete the programme successfully often take positions in other firms where the training they have had gives them substantial advantages over others and benefits their employers as well.

14 It need not do so if the nationalization were carried through reasonably amicably and if the new owners were able to retain or obtain the foreign personnel they needed. But it is almost certain that the ease with which qualified foreign personnel could be obtained would be substantially reduced.

15 On the other hand, there are sometimes acute problems arising in an underdeveloped country because a growing group of middle class 'educated' young people have difficulty finding suitable employment, and the government tends to look to the oil companies to help solve this unemployment problem. To some extent the problem arises because of an unwillingness of the young people to seek employment in agriculture and other occupations of low prestige value and because of the failure of the government to encourage small-scale private industrial enterprise.

16 It should be noted that to accept this definition of exploitation is to look at the matter from the point of view of the producing country and not from the point of view of the economist analysing the optimum utilization of the world's resources. [See Essay VI, above for further discussion of this problem.]

17 As noted earlier, it is also alleged that the control of distribution permits the oil companies to set arbitrarily low prices at which Middle East oil is sold as it leaves the producing areas and enters the world distribution system. This, however, is essentially a question of how the profit to be shared should be calculated and becomes part of the profit-sharing agreement. Whether or not the producing countries can alter such conditions in their favour depends on their bargaining power at the conference table.

18 Stephen H. Longrigg, op. cit., p. 48. Mr. Longrigg, in his able and sympathetic exposition of the history of the oil companies in the Middle East, swings uneasily between these two points of view.

19 At present the Arab League is promoting conferences and discussions among the Arab countries with a view to obtaining a 'unification' of their oil policies. One of the most discussed projects involves the construction of an Arab fleet of tankers which is presumably partly intended to weaken the transportation monopoly of the major companies.

20 'Profit Sharing Between Producing Countries and Oil Companies in the Middle East,' ECONOMIC JOURNAL, Vol. LXIX, June 1959, pp. 238–54.

21 Mr. Page is, of course, quite correct in pointing out that strictly speaking the issue is not one of 'nationalization' but of 'abrogation of contract,' since the companies operate only under concession contracts. The former term is the more commonly used, perhaps because it has less invidious connotations nowadays, but Mr. Page is within his rights to insist on the latter.

22 Virgil Salera, 'A Note on Profit Sharing in the Middle East Oil,' ECONOMIC JOURNAL, Vol. LXIX, December 1959, pp. 812–13.

23 For a convenient survey of the issues see George Lenczowski, *Oil and State in the Middle East* (New York: Cornell University Press, 1960), pp. 94 ff.

24 As I have pointed out in an earlier paper, it would rarely pay a country to nationalize a foreign firm if it had to pay compensation including the full present value of expected monopoly profits. See my 'Foreign Investment

and the Growth of the Firm,' ECONOMIC JOURNAL, Vol. LXVI, June 1956, p. 234. [Essay IV above.]

25 I am indeed grateful to Mr. Page for giving me the opportunity to acknowledge in print three unforgivable errors with respect to the new Italian and Japanese concessions, and to the composition of the American group in the IPC. All of the errors, which he corrects, occurred in the same footnote of my original article (p. 245), and I have been busily correcting them by hand in offprints. My excuse would only be understood by people doing research and writing in out-of-the-way parts of the world, and I can only offer apologies for my failure to check more carefully. Fortunately, the footnote errors have no bearing whatsoever on any part of my argument.

26 Some of the difficulty arises, of course, because business-men and economists use the term 'competition' differently. A delightful example recently came to my attention: the Chairman of the Board of the Standard Oil Company of Indiana, after pointing out that in the first six months of 1960 demand could be expected to rise while production controls would prevent 'senseless overproduction of products' in the United States, went on to say, 'In this economic climate prices should firm and strengthen, but the oil industry is so intensely competitive that improvement in the ratio of demand to supply does not automatically cause strengthening of price'! First National Bank of Chicago, *The Outlook for Business: the First Six Months of 1960*, December 8, 1959.

27 In the *Hearings* cited by Professor Salera, Mr. Davies, Chairman of the Board of Aramco, attributed the 50% profits tax of Saudi Arabia to the *benevolence* of the Saudi Arabian Government, stating, '. . . the Saudi Arabian Government was insisting on an income tax which would result in no additional burden on the company.' *Emergency Oil Lift Program and Related Oil Problems, Joint Hearings Before Subcommittees of the Committee on the Judiciary and Committee on Interior and Insular Affairs* (U.S. Senate, 85th Cong., 1st Sess., 1957) Part 2, p. 1430.

Mr. Davies also explicitly agreed that there were no income taxes paid to the United States on the operations of Aramco in Saudi Arabia because of the tax credit against foreign income taxes which is granted by the U.S. Treasury. See *ibid.*, p. 1421; also pp. 1442 and 1471.

28 See Melvin G. de Chazeau and Alfred E. Kahn, *Integration and Competition in the Petroleum Industry* (New Haven: Yale University Press, 1959. Petroleum Monograph Series, Vol. 3), p. 343. Incidentally, as Mr. Page suggests, there is a severe surplus of tankers at the present time; furthermore, if one can accept Professor Nielsen's analysis of the past record in forecasting the need for oil tankers, one can also agree that the oil companies as well as the independent tanker companies have often done a 'poor job with it.' There is nothing in economic analysis suggesting that errors of business firms provide proof of the existence of competitive conditions! See Robert S. Nielsen, *Oil Tanker Economics* (Bremen: Weltschiffahrts Archiv, 1959).

29 I can only mention a few of the significant discussions:
*The Price of Oil in Western Europe*, Economic Commission for Europe, March 1955.
*The International Petroleum Cartel*, Staff Report to the Federal Trade Commission, Committee Print No. 6 (82nd Cong. 2d Sess. 1952), pp. 352 ff.
Michael Laudrain, *Le Prix du Pétrole Brut* (Paris: Genin, 1958). Contains extensive bibliography.
M. G. de Chazeau and A. Kahn, *op. cit.*, pp. 211–17. One of the more recent, though brief, discussions.

# X

## MONOPOLY AND COMPETITION IN THE INTERNATIONAL PETROLEUM INDUSTRY*

*Briefly summarizes the monopolistic and competitive aspects of the industry in historical perspective.*

OF all the world's major industries, the international petroleum industry seems to get itself into more political hot water than any other. Although it provides a rich source of revenue to the governments of exporting and importing countries alike through oil revenues or excise taxes, the public in most countries look upon it with distrust, and not a few governments have acted harshly against it at one time or another. It has had to live under almost continuous scrutiny nearly everywhere, and is constantly subject to accusations of 'monopolistic exploitation' of consumers, or crude-oil producers, or both at the same time.

In recent years both economic and political difficulties have been accumulating for the industry which are bringing about very significant changes in the pattern of competition and the structure of control. The immediate cause of the difficulties is reasonably straightforward—the amount of crude oil that is readily available and could profitably be put on the market exceeds demand at ruling prices. The excess supply has both prompted and made possible increasing price-competition in world oil markets with different effects on different companies and different oil-exporting countries. Oil revenues are of overwhelming economic importance to the exporting countries, which naturally resent recent cuts in the posted prices of crude oil on which the revenues of most of them depend. They are demanding a larger share in oil profits at a time when these profits are under pressure. On the other hand, the cost of oil is a very significant item in the balance of payments of the importing countries, and some of them, especially the less developed, are increasingly concerned about it. Hence the international oil companies are caught between governments with fundamentally conflicting economic interests.

Politically the international companies are sitting targets. They are very large international organizations with considerable political and economic power. They have contractual control of most of the crude-

---

* Published in *The Year Book of World Affairs, 1964*. London Institute of World Affairs (London: Stevens. 1964).

oil production in the Middle East and Venezuela, and together they produce nearly 85% of the world's supplies outside the United States and the Soviet Union, and own around 65% of the refining capacity and all of the major pipelines. By virtue of their wealth and position they are natural objects of suspicion and obvious political scapegoats when difficulties arise. Of the seven largest, five are American, one is British, and one is Anglo-Dutch. Naturally, all three countries gain from the operations of the companies, and therefore it is not surprising that in many eyes the oil companies are presumed to serve the political as well as the economic interests of their governments.

Outside of the United States and the Soviet Union the oil that became vital to the economies of the great industrial countries has so far been found in largest quantities in under-developed and sometimes politically unstable countries, and its exploitation has required the managerial control, capital and technology of the advanced countries. Hence, as the oil-producing countries pursued their uneven course up the slippery path of political and economic development in a stormy environment of growing nationalism and suspicion of 'capitalist imperialism,' some political tension would have been likely regardless of the particular form of organization and control prevailing in the industry. Similarly, some economic difficulties would have been inevitable in view of the rapidly changing conditions of both supply and demand, and of the problems created not only for the older producing centres by new discoveries of large and cheap reserves, but also for the producers of the chief competitive fuel—coal. Nevertheless, the structure of organization and control that in fact developed in the international industry, and the practices associated with it, have themselves created many of the special difficulties of the present which are, in turn, likely to force substantial changes on the industry before very long.

### 'MONOPOLY' AND 'COMPETITION'

There has been much unproductive argument over the question whether the oil industry is 'monopolistic' or 'competitive.' In the end such arguments usually turn on definitions and it makes precious little difference (except perhaps from the point of view of public relations) which label we attach. Whether the structure and operations of the industry are considered desirable or undesirable in some way depends on whether we think that better overall results would have been obtained under different conditions. But one of the disadvantages of analysing the industry using the concepts of monopoly and competition is that each of them has strongly controversial associations which evoke judgments on the part of the reader and prejudice his appraisal of the results. For this reason I want to avoid these terms as far as possible in spite of the title of this paper, falling back on them only when there is no suitable alternative.

No one denies that the companies do work together in many respects; there are all sorts of enterprises carried on jointly by two or more companies in marketing, refining, crude-oil production, and exploration, in the success of which the participants naturally have a common interest. Moreover, all of the companies also have a common interest in maintaining the profitability of the industry as a whole at as high a level as possible consistent with the full exploitation of the opportunities available to each. Each company may endeavour to maintain or improve its own position *vis-à-vis* the others, but each accepts the restraint that arises from the recognition that some actions, perhaps profitable in the short run, would in the long run be defeated by the retaliation of others. The particular restraint most often exercised is in respect of prices, although at the same time non-price competition can be alarmingly vigorous. It is this latter kind of competition, and the sense of rivalry behind it, together with occasional price-competition and the numerous and real limitations on their freedom of action, that makes the companies feel that the charge of monopoly is absurd and that they are essentially competitive; it is the restraint each exercises in the common interest of all, together with the high degree of positive collaboration in many markets and many producing areas, that makes large numbers of outsiders feel that the industry is essentially monopolistic.

But firms are in business to make money, and their actions are in general taken with a view to enhancing their long-run profitability. Wherever possibilities for making profits exist, it may be assumed that some firms will take advantage of them, restrained only by the general ethics of the society in which they operate, and by the law. It is often argued that firms should take care to operate in the 'public interest,' but this seems to me to be a dangerous notion. It is not the business of private firms to set themselves up as judges of 'the' public interest. In particular, this requirement would put an international firm in an impossible position, since it is more than likely that the interests of the different publics to be considered would be incompatible.

The business of the private firm is to seek out legitimate opportunities for profit; it is the business of governments to define legitimacy and to attempt, however imperfectly, first to determine and then to protect the interests of their respective publics. One of the real political advantages of the economist's notion of 'competition' —or 'impersonal market forces'—as an adjudicator between conflicting interests is that no government or private group is required to shoulder the responsibility of such an adjudication, which in the nature of the case is likely to be a thankless job.

Moreover, there are severe limits on the extent to which private companies can act in the public interest if influential groups among the shareholders or the company executives feel that the private interest of the firm will suffer. There is wide scope for disagreement over whether or not any given action is in the long-run interest of 'the' public, and a notorious tendency for any group to identify, quite sincerely, its own welfare with the public interest. The difficulty becomes especially acute when the immediate outcome of any action proposed in the 'public interest' will almost certainly affect profits adversely in the short-run if at the same time there is considerable scope for disagreement over the long-run consequences.

## SOME HISTORICAL CONSIDERATIONS

It will not be possible in this paper to go into the historical development of the industry, and show how the present structure evolved as a result of a complex interaction of all sorts of forces—rivalry among producers, technical peculiarities of the industry, strategic and political interests of governments, the vagaries of tax laws, United States attitudes towards monopoly, economic and political conditions in the areas in which oil was found, the effect of the Great Depression on the United States industry, the phenomenal increase in demand, and many others. One of the dominating features of the industry up to the Second World War was the intense rivalry and numerous mutual accommodations of the chief contestants, but it would take many pages to present even the bare outlines of the story, which began with the activities of John D. Rockefeller in the 1870s and the rapid rise to dominance of the Standard Oil Trust in the United States. As the United States quickly became the world's largest oil producer, so did Standard Oil become the world's largest company, with a leading position abroad as well as at home, although it was faced with sharp competition in Europe and the Far East from groups drawing their supplies primarily from Russia.

But oil was also soon discovered in other parts of the world and a small Dutch company was chartered in 1890 to exploit Indonesian oil—the Royal Dutch—which, under able and aggressive leadership, soon began to hold its own with Standard Oil, and eventually joined forces with Shell Transport and Trading, a British company with extensive European and Far Eastern markets. Then came the British discoveries in Iran in 1908 and the emergence of the Anglo-Persian Oil Company (now British Petroleum) with the financial help, first of the Burmah Oil Company, and then of the British Government. In the jockeying for position in the Middle East after the First World War the American companies pushed their way into Iraq, and obtained a secure position in the Persian Gulf where vast reserves were discovered. As a result, all the reserves in the Middle East ended up in the hands of eight companies.[1] Oil was also discovered in Venezuela just before the First World War, and by 1932 Standard Oil of New Jersey, Shell and Gulf had obtained control of almost all South American production.

Were the story of the rise of the major companies to be told in detail, it would show clearly that superior efficiency in production and distribution, in invention and technological advance, could not account for the dominant position they achieved. Their record in finding, producing, and distributing oil and its products is indeed impressive, but efficiency in this respect would not have been enough to ensure their dominance. Hence the story of the rise of the great companies deals as much with financial power, commercial and political negotiations and intrigue, with cartel agreements, marketing alliances, price maintenance arrangements, price wars and armistices, mergers and combination, actions to avoid taxes, and the national and international political interests of governments, as it does with the economics of production and distribution. This statement does not necessarily imply any condemnation of the companies. Several were managed by brilliant and ambitious men in an environment in which commercial greatness could be achieved in no other way. Moreover, and more important for our purposes, it has been cogently argued that quite apart from the specific methods used to achieve the end, economic conditions and political circumstances, together with the technological relationships characterizing the production and distribution of petroleum and its products, made the emergence of a few vertically integrated organizations large enough to exercise extensive control over the industry not only an historical necessity but an economic desideratum. And there is considerable evidence that the men who brought it about believed this.

### ARE DOMINANT FIRMS NECESSARY?

The argument was perhaps most persuasively presented in 1946 by P. H. Frankel in his *Essentials of Petroleum*[2] and can be summarized as follows: Because of the uncertain results of exploration, the high overhead costs at all stages of the industry, and a high inelasticity of demand in the short run, the industry is not 'self-adjusting' in the sense that a fall in prices chokes off supply significantly or stimulates demand. Therefore the industry is subject to continuous crises in the absence of reasonably strong control over supply. The uncertain results of exploration create the likelihood that either too little or too much oil will be discovered in relation to the amount that will be bought at profitable prices; the fact that crude oil production, transportation, refining, storage and distribution all require large amounts of capital leads to heavy fixed charges and forces competitive producers to produce as much as possible as quickly as possible in an attempt to recover their sunk costs and to cover at least part of their overheads;[3] the fact that demand is not readily expanded when prices fall means that prices are easily pushed below costs of production by relatively small surpluses. Thus 'hectic prosperity is followed all too swiftly by complete collapse, and redress can be hoped for only from the efforts of "eveners," adjusters and organizers.'[4] And '. . . for technical reasons alone the formation of paramount oil concerns was inevitable; their role could not be taken over by a welter of smallish firms.'[5]

This essential role of the large firms could only be played if an important bottleneck of the industry could be brought under their control. For Standard Oil in the United States in the last decades of the nineteenth century the relevant 'bottleneck' was transportation, but for the international industry the key proved to be crude oil reserves. Vertical integration is held to be necessary because efficient operation requires a continuous and secure flow of supplies, and big firms with large and far-flung markets and heavy investments cannot afford to be entirely dependent on others for their supplies unless these are securely tied up under long-term contracts. Hence, '. . . whatever we may do, the fundamental factors come to the surface; the oil industry to exist at all, calls for concerted effort and, however often a co-operative structure may have been disturbed or broken up, it will soon begin to form again.'[6]

This chain of reasoning is supported by an appeal to history which relies largely on the experience of the United States, where very special legal conditions prevailed. In the international industry both the legal and political environment were different, and the emergence of oligopoly[7] occurred in a very different way. In the United States the 'law of capture'[8] almost forced instability on the domestic industry, and outside the United States the interplay of international politics, and the terms on which concessions could be obtained in the Middle East gave the already well-established companies powerful advantages.

But apart from historical consideration, the essential theoretical foundation of the argument has been powerfully challenged, notably by Professor Adelman. He makes a strong case for the proposition that the industry is at all stages subject to increasing costs except where inappropriate policies have led to the development of large excess capacity. Where Dr. Frankel puts the emphasis on the uncertainties of *exploration* as a cause of alternate excess and deficient supplies, Professor Adelman puts the emphasis on the costs of *developing* discovered reserves. He denies that an exceptionally high ratio of fixed to variable costs is characteristic of the industry; rather he finds an 'extraordinary variability.' He shows that costs of development may be expected to rise as output rises after perhaps a short initial period of flush production, and since it will only pay to produce up to the point where this rising marginal cost equals price the industry is, in fact, 'self-adjusting': 'The crude oil industry, contrary to common belief, is inherently self-adjusting; more precisely, it has a strong adjustment mechanism for determining the level of output and its division among various sources of supply, by the price acting upon the cost which must be incurred to bring up more output. This incremental or marginal cost, for every individual unit and *a fortiori* for the system as a whole, increases rather than decreases with greater output.'[9]

It cannot be denied that the mechanism of adjustment may at times be severely strained, and that in the process of adjustment to changed conditions some producers may be hurt severely. It is true that at times the conditions of supply (or demand) may change abruptly, and that the adjustments required of the industry may not be acceptable to the public or private groups who have the power to prevent them, at least for the time being. It is also true that it is possible to devise inappropriate legal arrangements, such as the 'law of capture' in the United States, which virtually create instability. But none of these is sufficient to resolve the question at issue, which is whether the oil industry in private hands and subject only to the technical regulations required to prevent uneconomic waste (compulsory exploitation of each reservoir as a unit, proper well spacing, etc.) would necessarily suffer worse 'crises' than other industries, or whether the 'self-adjusting' mechanism could in fact take care of most of them without creating self-destructive instability.

We cannot decide this question here, nor do we need to, for we are not attempting to determine the optimum type of organization and control for this industry but rather to analyse some of the economic consequences of the particular organization and control which have in fact prevailed. Even if it could be conclusively shown that stability is desirable, and that some sort of regulation is necessary for stability, or at least for short-period stability, it would not follow that the specific pattern of the past was either necessary or sufficient for long-term stability. It is undoubtedly true that the postwar impact on world oil markets of the development of unprecedentedly large low-cost discoveries in the Middle East was relatively painless precisely because they were in 'strong hands,' and were marketed with due regard for the interests of powerful political groups in all parts of the world. But it is also true that the policies pursued have aggravated, rather than corrected, the underlying disequilibrium of supply and demand with consequences that may yet prove explosive. To understand the nature of the problem we must examine briefly the circumstances surrounding the evolution of oil policy and prices.

## OIL IN 'STRONG HANDS'

Between the First and Second World Wars there were numerous attempts to establish cartels both nationally and internationally on the part of the dominant producers. The best known is the 'As Is' or 'Achnacarry' agreement of 1928 which expressed the desire of the companies to limit competition and to agree on markets and prices, and which laid down the broad strategy to be followed. The general principle of 'as is' was more or less the basis of a number of later agreements—few of which were entirely successful. By 1928 large oil discoveries had already been made in Venezuela, and in the 1930s exports from Iraq began, and oil was discovered in Saudi Arabia, Bahrain and Kuwait. The problem facing Jersey Standard, Shell and British Petroleum, the leading international companies, was how to ensure that the newly discovered reserves would be in 'strong hands,' i.e., in the hands of companies who had market outlets, and would not have to go after other companies' markets and cut prices to dispose of their oil.

There was nothing sinister in this—it was sound business practice—and since some companies needed more crude for their markets and others had more than they could use, the obvious solution was to join in a somewhat polygamous wedlock—or at least in respectable long-term liaisons—those in surplus and those in deficit either through the joint ownership of producing companies or through long-term supply contracts. (There was also, incidentally, the question of dowries, some companies having capital that others lacked.) Joint producing companies and long-term contracts were both effective, and the oil in Venezuela and the Middle East did not get into the hands of groups who found it necessary to undermine the existing price structure in order to sell it—at least until the late 1950s.

And so it came about that a pattern of ownership was evolved which made necessary the close co-operation of the major marketing groups in plans to develop and produce oil from the great new discoveries outside of the United States. At the very least, therefore, each group had considerable information about the plans of others. An essential condition for ensuring 'orderly' price movements—control over the rate of supply—seemed, for the moment at least, to have been established. If supply can be reasonably adjusted to any assumed level of demand there is no need at all to worry about direct control of prices. Oil company spokesmen have often indignantly denied that they control prices, insisting that prices are determined by 'supply and demand,' and there is no reason to quarrel with them. The point is that in the interest of 'orderly' markets and of stability in the industry the companies have until recently been able to do a fairly good job of controlling total supply.[10]

On the other hand, supplies can never be perfectly adjusted to demand at ruling prices at all times, and whenever there are a number of producers in an industry, some at least of whom would find it profitable to sell more, the situation will inevitably 'deteriorate' if producers act independently. In general, procedures for dealing with the problem fall into three broad categories: (a) the establishment of machinery for continuing and close consultation on prices; (b) the acceptance of recognized rules for calculating prices to be followed by all; and (c) the publication of prices which all are expected to observe. The first was probably used in the oil industry to some extent before the last war,[11] and the second certainly was. When exports began from the Middle East there were no published prices at which oil could be bought by outsiders, since almost all oil was produced by the companies for use in their own refineries or to supply their long-term special contracts. The sales that were made, however, were at prices related to the price of oil exported from the United States Gulf. The delivered prices of oil and oil products everywhere were equated to the f.o.b. price United States Gulf plus transportation costs from the Gulf to the export destinations. In other words, oil was sold to independent buyers everywhere, including those in the Persian Gulf itself, at the price it would have fetched had it come from the United States Gulf. This basing point system ('Gulf plus') provided the rules for the calculation of prices up to the war.

During and just after the war substantial changes took place, largely under pressure, apparently, from the British and United States Governments and from the Economic Co-operation Administration, which financed oil for European reconstruction. To meet the sharp increase in demand the companies rapidly developed the low-cost Middle East reserves, each striving for its share of the growing market, but avoiding price competition in the process. Nevertheless, as exports increased, the price of Middle East oil was reduced relatively to that of the United States, until, around 1946–47, it was no longer directly linked to the United States Gulf price, and 'basing-point' rules finally gave way to posted prices. The public posting of prices was apparently started by Socony, which has a 10% interest in Aramco in Saudi Arabia, in November 1950,[12] one month before the agreement to share profits with Saudi Arabia in the form of an income tax was made. The profit-sharing arrangements made with the governments of the crude-oil producing countries required that the companies establish a common standard for the valuation of crude oil, but the posting of prices was also to be a method of ensuring that all the oil companies followed a common price policy in their own operations.[13] For a while the posted prices were apparently observed by the companies in their 'arm's length' transactions, but there is little evidence that attempts to form cartels similar to the pre-war ones have been made since the war or that the companies have acted under any explicit agreement, or have even consulted, on price policies. In fact, most of the evidence is to the contrary, for it points to a developing price competition, and it was not long before first indirect and then direct discounts were offered by sellers including the major companies themselves.

### THE SIGNIFICANCE OF THE PRICE OF CRUDE OIL

Crude-oil production in the Middle East seemed so very profitable that the necessity of increasing the rate of payment to the governments of the producing countries became increasingly pressing.[14] Some writers choose to treat the high profitability attributed to Middle East oil as the rent which low-cost producers always earn, and some treat it as a monopoly profit deliberately created by the majors who wanted to maintain high crude prices in order to protect their Western Hemisphere investments from a too rapid expansion of Middle East oil. The relative decline in the prices of Middle East oil as exports expanded has been used as evidence that there was 'real competition' between the high- and low-cost sources of supply, and elaborate theories have been constructed showing a competitive price moving Westward as the 'watershed' of competition pushed the boundary of the area in which Middle East crude could undersell oil from the United States nearer and nearer to the East Coast of the United States.[15]

Doubtless there is some truth in all of these ways of looking at the movements in the prices of crude oil, but I think they all attach too much importance to the price of crude as itself a source of profit to the companies, and as a determinant of the competitive position of high- and low-cost crude. Apart from the incidence of taxation, which is discussed below, the market price of crude oil was of relatively little importance for the profitability of most of the major companies simply because they sold very little oil in a free market. Most oil produced was used by the companies themselves, the 'sales' involved being transfers from one affiliate to another, or was disposed of under long-term contracts at special prices. The cost of production and transportation would be an important consideration in a company's decision whether to lift from Venezuela, the Middle East, or the United States; it would make no sense for a company to make such a decision with reference to the relative market prices of crude when it was neither obtaining it at these prices nor wanting to sell it. Many economic and political factors would inevitably influence the choice of source of supply, but the market prices of oil could not have been one of them. For the same reason, the profit directly made on crude sales must have been relatively minor. It is true that most companies showed very high profits on crude operations in the Middle East, but this was because they chose to value the oil they used as if they had in fact sold it.[16] There was no particular reason why they should not have done so since it accorded with the book-keeping arrangements between their affiliates. But profits are the difference between total costs and total sales receipts, and when the accounts of a group of affiliates are consolidated to show the total profit of the group, transfers between the affiliates cancel out. Thus, if an extraordinary profit was shown by the companies on crude production alone this was largely the result of arrangements deliberately made by the international integrated companies under which a high price was put on crude oil sold in the market, and an equally high value on that used in their own operations. There were good reasons for this: under United States tax-laws the American companies could claim $27\frac{1}{2}\%$ of the gross income from crude oil operations as a depletion allowance to be deducted from taxable income; higher profits could be made on outside sales; and finally the higher the price at which oil is available to independent refiners the higher their costs, and the less dangerous they are as competitors in product markets.

Moreover, a high common price for crude oil would put a floor to the extent to which affiliates of the groups themselves would be tempted to cut prices in product markets. That this consideration did not entirely escape the notice of oil policy-makers in the early days at least, has been shown in the records of the negotiations in the 1930s over the price policy to be pursued with respect to the oil produced by the Iraq Petroleum Company. The French wanted IPC oil to be made available to the companies at a low price. This was resisted by the other groups for a number of reasons, one of which was that the French might use a low price 'as a basis for influencing product prices in France.'[17]

In the short run, therefore, a higher rather than a lower price for crude oil was in the interests of the integrated companies because it meant higher profits on any amount that was sold at or near that price, and it raised the costs of competing refiners and marketers, thus restraining to some extent competition in product markets. It did not make a very important contribution to total consolidated profits except through the tax advantages thus obtained, nor could it have determined the sources from which they would lift crude oil. As time went on, however, other powerful forces tending to support crude oil prices became increasingly important—the interests of governments.

### THE INTERESTS OF GOVERNMENTS

Whatever the price policy of the majors, and the means of implementing it, had been before and after the war, perhaps the most significant post-war change was the extent to which the companies increasingly lost their freedom to manoeuvre. Restrictions on this freedom came from the governments of both exporting and importing countries.

An exporting country would, in general, be expected to have a direct interest in the prices of its major exports. But when these exports are in the hands of foreign companies which themselves receive the export proceeds and remit only part of them in royalties, taxes or other direct payments to the government of the exporting country, the latter will concern itself with export prices only to the extent that they influence the foreign-exchange revenues received. The maximum price a company will pay for the privilege of producing and exporting a commodity is set by the expected value of that commodity to the company, and when the commodity is a raw material used in the manufacture of finished products by the company producing it, the maximum is determined by the profit the company expects from its full range of operations, including final sales of the products. Clearly the higher these profits, the greater are the maximum revenues a government could expect to obtain from the companies, and an arrangement between companies and governments that gave the latter a stake in total profits would provide a strong incentive for them to co-operate in measures designed to maintain prices of final products as distinct from 'prices' of the raw material itself.[18]

But the arrangements actually made between the companies and the Middle Eastern governments in the early 1950s provided for an equal division of the profits attributed to crude oil, which were determined with respect to the price at which crude oil was offered for sale—the so-called 'posted price'—and bore no necessary relation to the total profits of the companies. This was a way of giving the producing countries a share in the total profits, since integrated companies make their profits from using rather than selling crude oil, but at the same time it gave the governments a vested interest in the way in which the crude oil itself was valued. It must have been assumed when these arrangements were made that movements in the posted prices for crude oil would be related to movements in product prices, at least in a downward direction, and therefore that any decline in the profitability of the integrated companies would be shared by a similar decline in the profits to be shared with the producing countries. In principle, the companies have the right to alter posted prices in accordance with changing market conditions; in practice they have lost the power to do so without risking damaging retaliation from the governments concerned.

It may be that the companies had little choice in view of the profitability of their oil operations in the Middle East. The governments of the producing countries under any circumstances would have demanded an increasing share of these profits. The alternatives were some form of royalty payment per barrel, some form of profit-sharing, or some combination of the two. But the governments would have compared their revenues with the total profits of the companies regardless of the form in which they received the revenues. Profit-sharing which took the form of an income tax imposed by the governments of the producing countries had substantial tax advantages for the United States companies, since the taxes paid could be offset against their tax liability to the United States treasury. For some companies at least this meant that they paid no taxes at all to the United States Government and that, on balance, their tax payments to the producing countries involved them in little or no additional cost.[19]

From the governments' standpoint, too, the arrangement was a good one so long as prices were not falling, for it assured them a minimum income per barrel probably greater than they would have received if the payment had taken the form of a royalty which would not have had the same tax advantages for the companies. But this method of paying the governments of the producing countries at once introduced powerful political support for keeping up the general level of crude oil prices. Between 1951 and 1957 there were four changes in the posted prices for crude oil from Ras Tanura, all of them upwards, but a reduction in 1959 evoked immediate adverse political consequences for the companies.

From the consuming countries also, there were dangers to the position of the international oil companies from lower crude oil prices. In the United States, domestic producers, especially the smaller high-cost producers, were a vocal and determined political force, and the international companies dared not push too far their imports of foreign oil for their own uses for fear of unpleasant political consequences. Furthermore, sales to independent refiners and marketers even at posted prices could also have hastened the imposition of import quotas which, in the event, proved politically inevitable in any case. In Europe the coal producers, also a powerful political force, were becoming more and more concerned over the impact of increasing imports of crude oil, and the companies were anxious to avoid provoking legislation adversely affecting the terms on which oil could be imported and sold. For other countries, notably the United Kingdom, the price of crude oil did not press on the balance of payments so long as the oil was paid for in sterling to British companies. In other words, no consuming country of consequence was willing to raise any objections to the ruling prices for crude oil, although some, notably Italy and the Scandinavian countries in Europe, and some of the underdeveloped countries in Asia and South America, were becoming more and more restive.

#### THE TEN 'GOLDEN YEARS'—AND AFTER

So, for some ten years after the end of the war little occurred to disturb the structure of oil prices or fundamentally to alter the pattern of control of the industry. The competing oil companies had established a *modus vivendi* permitting 'peaceful co-existence,' and relations with the governments of the exporting countries were reasonably amicable except for the conflict with Iran, which may to some extent have been due to the difficulties facing Anglo-Iranian in adjusting to the changed conditions created by the profit-sharing arrangements inaugurated by the United States companies (including the inflexibility of the United Kingdom tax authorities, who at that time did not permit the United Kingdom companies to set foreign income taxes off against their domestic tax liability—that came later). The developing pressure of increasing capacity was masked by the consequences, first, of the Iranian conflict, which kept large amounts of oil off the market for three to four years, then of the Korean war, and finally of the Suez crisis, which quickly followed and further disrupted oil supplies for a short period. Hence, from the end of the war until around 1956 supplies of crude oil were for one reason or another remarkably well adjusted to demand at ruling prices in spite of the fact that already developed excess productive capacity existed and that prices were well above the cost of finding and developing additional supplies of oil. These have been called the 'ten golden years,'[20] but the glitter was only on the surface. For by 1957 the pressures of supply that had been steadily building up began seriously to be felt, and the oil companies found themselves entering rough water, both politically and economically, largely because prices had been too high too long.

There has been much argument over this question whether or not prices were 'too high.' Spokesmen for oil interests point out that oil has been substantially cheaper than other fuels, that prices have not risen as much as general price indexes and therefore that the 'real' price of oil has declined, that profits of the integrated companies are moderate, and that even if they were not, large amounts of capital are needed for investment, etc. All this may be true, but what it amounts to is an attempt to *justify* price levels, and whenever one finds attempts to do this it is safe to conclude that prices are *not* being set by keen competition. One of the chief results of competition is to squeeze profit margins, and when this happens producing interests do not spend their time insisting that prices are not too high, rather they complain bitterly that prices are too low.

In any event, we are not here concerned with the justification for any given level of prices but with an analysis of economic consequences, and these are not affected by arguments. It is clear that the price of crude oil to those who did not own it, or were unable to obtain it on near to 'cost' terms, was high enough to provide a strong incentive for them to search for it and, having found it, to develop it. Some of the major companies were short of cheap (*i.e.*, 'cost') crude. A number of minor companies were attracted into crude oil production abroad and began to explore, many because they realized that future supplies, particularly in the United States, would be had only at rising cost and they did not want to depend on cheap supplies being available in the market. Independent refinery and consumer interests, sometimes backed by governments, saw that the costs of exploration and production, if successful, might be handsomely repaid through a saving in the cost of crude. Many were successful—notably the Japanese-owned Arabian Oil Company and some of the lesser companies in North Africa and Venezuela. Moreover, as the foreign exchange bill for oil of the underdeveloped countries rose with increasing imports, the desirability of intensifying the search for domestic supplies became stronger; to protect their markets the majors joined the search, and again were often successful. Finally, the Russians, who had also been developing increased productive capacity, decided to export crude oil to world markets and, like good capitalists, cut prices just enough to secure orders. As this new capacity was being developed for crude oil, natural gas supplies were found, and gas became an increasingly important competitor of fuel oil. All this began to affect the very large firms, whose position as suppliers of products or producers of crude oil is not and never has been impregnable, either individually or as a group.

First, in order to dispose of their own surplus crude as well as to protect their markets from the incursions of others, they began discounting the posted prices, in the beginning by offering lower prices, special freight arrangements, or other terms to large buyers, and later by straight discounts to almost all comers, including their own affiliates.

Secondly, as the prices at which oil moved diverged more and more from posted prices, the companies decided to cut the posted price in 1959 and again in 1960, since they had to pay tax on the basis of posted prices regardless of the price at which the oil was actually sold. The cuts were moderate but provoked immediate adverse reactions from the producing countries, who proceeded to establish in 1960 an organization of their own to strengthen their collective hand *vis-à-vis* the companies—the Organisation of Petroleum Exporting Countries. OPEC at first tried to make the companies restore the previous level of prices, but now seems largely to have abandoned the attempt, and is endeavouring instead to increase the oil revenues of its member countries by other means. It is not yet clear exactly what role this organization will play in the international industry but it has undoubtedly been an important force in preventing further price cuts so far. Posted prices may remain for a long time to come completely separated from market prices and, if so, they will represent nothing more than a way of calculating a payment per barrel of oil to be made to the producing countries.[21]

Thirdly, falling prices were not entirely offset by increased sales, and profits were threatened. The companies not only promptly reduced their investment expenditures but also their operating costs, and in one annual report after another it was made clear that 'economy and efficiency' had had an important role in holding the level of total profits. Evidently, a good deal of fat had accumulated in their corporate bodies which competition was effectively removing.

Fourthly, in spite of the efforts of the major companies, independents (including State-owned enterprises) gained ground. The share of the eight majors in the world's crude production outside of North America and the Soviet bloc fell from 91% in 1957 to 84% in 1961. According to estimates of Jersey Standard the majors accounted for only 35% of the refining capacity under construction or planned in the area through 1965, and at the end of 1965 they are expected to have 57% of the total capacity as compared to 65% in 1961.[22]

And finally, changes have been taking place in the attitudes of the underdeveloped importing countries of Asia and Africa, partly because of their policies with respect to economic development, partly because of the increasing competition in world oil markets, and partly because Russian competition has brought home to some of them the extent to which the vertical integration of the major companies had narrowed their choice of supplier to the cost of their balance of payments. These changes may have a greater effect on the structure of the industry in the long run than any we have so far mentioned.

### UNDERDEVELOPED IMPORTING COUNTRIES

As noted above, the only time when the governments of the industrialized importing countries of the West exerted effective pressure on the oil companies to lower the prices of crude oil was during and immediately after the war, when oil imports to Europe were financed under the Marshall Aid plan and pressure from the Economic Co-operation Administration forced a reduction in prices of Middle East oil. For the most part the largest Western consuming countries acquiesced in the price policies of the oil companies, each for reasons of its own, with the result that the profitability of the industry before taxes not only permitted but made necessary a very large transfer of income from the consuming to the crude oil producing countries in the form of oil revenues.

The share of the total profits of the industry that the governments of the crude oil producing countries can obtain depends on their bargaining skill in negotiations with the oil companies up to the point at which their demands exceed the maximum the oil companies believe they could give and still profitably continue oil operations in the countries. It is probable that the oil companies are not entirely agreed on what this maximum is, and certainly the governments cannot know it in advance; hence dangerous mistakes can easily occur.[23] At present there is still considerable bargaining leeway, since profits are being maintained through increased sales and the improved efficiency that the recent increase in competition has stimulated. But if market prices continue to fall under the impact of excess supplies and adversely affect profits, while at the same time the governments of the crude producing countries continue to press for larger revenues, a point must be reached at which the governments can get no more unless prices can be raised and profits increased.

But in such circumstances prices cannot be raised unless something is done to reduce the pressures of supply which push them down, and this is precisely why pro-rationing—the control of supply through the imposition of export quotas on crude producing countries—figured prominently in the early proposals for possible action when OPEC was originally formed. This has not been followed up for a variety of reasons, but it is now argued by some that the revenues demanded by producing countries will in themselves so raise costs for the companies, and particularly for those who are less well established in the international business than are the majors, that the rate of supply will be reduced because of rising costs. This could come about if

(a) the 'uncontrolled' supply has been coming from producers who would be forced to cut back by higher costs; *and* (b) if the lower cost producers with large developed reserves were willing and able to restrain the rate of increase of their own supplies; *and* (c) if the individual governments concerned would accept a slower rate of exploitation of their own reserves rather than reduce their demands on the companies. But regardless of whether the excess supply is eliminated by deliberate control, or by rising costs to the companies as a result of increased demands for revenues, the result would be the maintenance of higher prices for crude oil in order to provide greater revenues to the producing countries. Among those who would pay these prices are the underdeveloped importing countries of Asia and Africa whose *per capita* incomes are by and large below those of the oil-producing countries, who have launched extensive development programmes and who are in consequence finding their balance of payments under continual strain.

There is no doubt that the cost of imports of oil and oil products is a cause of concern to a number of these countries as their fuel requirements increase, and they are casting around for ways of reducing the price they pay. They know that they benefit from lower prices, and their officials are fully aware of the nature of the conflict of interest between the importing and exporting countries with respect to the price of crude oil. But their immediate problem is how to maximize their bargaining position *vis-à-vis* their present suppliers— the international oil companies—in order to take the greatest possible advantage of the existing supply position, for they are not at all sure that the prices they in fact obtain are as good as could be had. From this point of view, the extent to which their domestic refining and distribution facilities are owned by the international companies is an important consideration.

Up to fairly recently nearly all of the petroleum requirements of most of the countries of Asia and Africa were met by the import of products from the large export refineries in the Persian Gulf at f.o.b. prices roughly equal to those prevailing in the Gulf of Mexico. Refining margins were therefore substantial,[24] and provided a prima facie incentive for the importing countries to seek savings in foreign exchange by establishing domestic refineries to process imported crude oil. The international oil companies were not very enthusiastic about this development, since the larger export refineries were in many cases not only more economic, being able to take advantage of certain substantial economies of scale, but were more suited to the nature of their integrated oil operations.[25] But when it became clear that the various governments were not only determined to have refineries, but that they would also be able to obtain financial and technical assistance for the purpose, notably from the Russians and certain independent oil companies,[26] the international majors found it necessary themselves to construct or to finance the construction of local refineries in order to protect both their product markets and their outlets for crude oil.[27] Until recently most of their refineries were thus wholly-owned affiliates of one or more of the international companies.

As more and more crude began to move at prices openly discounted from the posted prices, the importing countries began to look harder and harder at the prices being paid by the domestic refining affiliates of the international companies supplying their crude oil. The refining companies had little choice in the price they paid, since the crude was supplied to them by trading affiliates of the parent companies at what the companies chose to call the 'world competitive price'—in other words, the posted price. 'It was hoped that the general acceptance of a procedure along these lines would be an assurance to the governments of consuming countries of the acceptability by others of the price basis of their oil imports,' states Shell International.[28] Clearly this hope could not be realized once the consuming countries began to suspect that they could get better prices, and that the transfer prices of the companies were putting an unnecessary burden on their balance of payments.[29] Offers by the Russians to supply crude at lower prices brought the matter to a head during 1960, particularly in India where it seemed for a while as if the Indian Government might attempt to require the international majors to take Russian crude in their own refineries. The majors then discounted the posted price to their own affiliates, and India received cheaper crude.[30]

Such experiences also demonstrated to the underdeveloped importing countries that an important element in bargaining may be the existence of some refining capacity that is not tied to a particular source of crude oil. With 'free' capacity available a country can shop around in order to get the best terms available for crude. In some countries the companies have attempted to meet this problem by agreeing to supply their affiliated refineries on terms at least equal to those offered by any other supplier (excluding the Russians, whose terms the oil companies have insisted are 'political' rather than 'commercial'). But even this does not necessarily provide adequate protection for the importing country, because independent suppliers are in some cases loath to offer better terms when they know that they would have little chance of obtaining a contract since the existing companies need only match the terms to retain their position. In other words, independent suppliers may be understandably unwilling to have their offers used merely as bargaining counters to force better terms from existing suppliers.

These considerations reinforce the prevailing suspicion of extensive foreign ownership of domestic industry, and there is an increasing unwillingness to permit the international oil companies to build wholly-owned refineries and an increasing tendency to insist on some participation by the local government or by local interests or even to prefer loans (on favourable terms) to finance locally-owned refineries. Such developments, together with the growing significance of companies other than the 'seven sisters'—as Signor Mattei of ENI (the Italian State company) labelled the seven largest companies—are rapidly causing far-reaching changes in the position of the international majors in the underdeveloped countries of Asia and Africa, and are reducing the ability of the companies to establish by their own actions a uniform 'acceptable price basis' for oil imports. Among other things, this brings nearer the day when the governments of the crude oil producing countries will have to recognize that they are attempting to improve their own terms of trade and foreign exchange receipts at the expense, not only of the rich industrialized West, but also of the poorer developing countries of the world who are importers of crude oil. The dilemma is a real and potentially serious one.

Nevertheless, the fact remains that some 85% of the world's crude oil production (outside the United States and the Soviet bloc) is under the control of eight major international oil companies and, barring unforeseeable upheavals, both producing and consuming countries must therefore negotiate primarily with them for a long time to come. So long as the profits of the companies are large enough to leave scope for concessions without either an increase in the price of oil or a reduction in the per barrel revenues of the exporting countries, the problem can be solved by bargaining skill and good judgment with respect to the real limits beyond which demands must not go, and resistance must not be offered. But no one can negotiate with market forces, and these may in the end have such a decisive effect that entirely new approaches must be attempted. Shell International has argued in the pamphlet already cited that 'The pricing mechanism has efficiently co-ordinated economic forces and reconciled divergent interests in the past, and should be allowed to continue to perform these important functions in the future.'[31] If the 'reconciliation' in the future is to be like that of the past, however, considerable skill may have to be exercised by someone in order to persuade the underdeveloped importing countries that the arrangements made by the oil companies within which the 'price mechanism' is allowed to operate take adequate account of their economic interests.

I would like to thank Professor M. A. Adelman, Dr. P. H. Frankel, Professor Fritz Machlup, Mr. John Trimmer, Mr. Bickham Sweet-Escott, Mr. Brandon Grove,

Mr. S. Cody and Mrs. Rose Greaves, for criticism and suggestions. In addition, the manuscript was read and commented on by other officials in the oil companies, to whom I am grateful. I am also grateful to the London School of Economics for financial support of my research. Needless to say, however, I am alone responsible for the final product, for I was not always able to accept the advice given to me.

## NOTES

1 Standard Oil of New Jersey, Royal Dutch/Shell, Gulf Oil, Socony Mobil, Texaco, Standard Oil of California, British Petroleum and Compagnie Française des Pétroles.
2 P. H. Frankel, *Essentials of Petroleum: A Key to Oil Economics*, 1946.
3 It should be noted that with respect to crude oil, this argument is valid only if producers are extremely stupid, or forced to produce, or if the development of the reserves discovered *and* the subsequent production are subject to constant or decreasing marginal cost. It is the latter that Frankel implies. With respect to the other phases of the industry, the argument assumes that the minimum economic unit is very large in relation to the size of the market.
4 *Op. cit.* in note 2 above, at p. 67.
5 *Ibid.*, p. 85.
6 *Ibid.*, p. 97.
7 The term 'oligopoly' is used to describe a situation in which producers are so few that each is aware of how the others would be likely to respond to his own competitive actions. Oligopoly leads to stability of price so long as all producers avoid price cuts because they fear retaliation from others, and so long as newcomers do not upset the balance.
8 Under the United States 'law of capture' oil belongs to him who brings it to the surface. Since oil migrates under ground to areas where pressure is reduced, the drilling of a hole may attract oil from an entire reservoir the extent of which bears no relation to the property lines drawn on the surface of the earth. Thus there may be many properties lying over a reservoir, and many producers, each of whom has an incentive to produce as fast as possible to get a maximum share of the oil before it is 'captured' by others.
9 M. A. Adelman in M. Clawson (ed.), *Natural Resources and International Development*, 1964, p. 32.
10 It has sometimes been argued that supplies in the Middle East were developed as fast as technical conditions permitted, and were not determined by the marketing policies of the companies. This argument hardly merits discussion, but a revealing example of how efficiently supplies can be regulated is provided by the change in oil production from Kuwait and Iraq as Iranian oil first disappeared and then was brought back into the market from 1951 to 1956.
11 United States Federal Trade Commission, *The International Petroleum Cartel*, 1952, Chap. VIII.
12 *Ibid.*, p. 369, note 59.
13 That the publishing of prices was intended to accomplish both of these objectives has been explicitly stated by Shell, according to whom the 'main reasons for the establishment of posted prices outside the United States' are the following:

'The offer to sell to buyers generally was intended to provide an assurance to the governments of producing countries that oil being exported at publicly posted prices to affiliated companies was not being under-valued.

'It was intended that posted prices plus appropriate long-term freights would provide the basis on which oil would be imported into consuming areas.' See Shell International Petroleum Co., Ltd., *Current International Oil Pricing Problems*, August 1963, p. 6.

14 See C. Issawi and M. Yeganeh, *The Economics of Middle East Oil*, 1962, Chap. 5, for a discussion of the profits shown on crude production in the Middle East.

15 See, for example, Wayne Leeman, *The Price of Middle East Oil*, 1962, Chap. 4, for a discussion on this type of theory.

16 Shell International in the pamphlet mentioned above, and the Organisation of Petroleum Exporting Countries (OPEC), in a reply to it *'Pricing Problems: Further Consideration,'* have recently exchanged views on the question of whether or not profits attributed to Middle East crude-oil production can be called 'high,' and each produced figures to support its point of view. I think one has to accept the judgment of the *Economist* (September 21, 1963, p. 1042): 'Few people who have ever looked at Middle East oil objectively are in doubt that the return on investment there is exceptionally high . . . ; the figures are also emotive, but attempts to explain them away generally turn out to be emotive too.'

17 Letter from H. G. Seidel of Jersey Standard quoted in United States Federal Trade Commission, *op. cit.*, in note 11 above, at p. 76.

18 I am not suggesting that this would have been a practicable arrangement.

19 See E. T. Penrose, 'Middle East Oil: The International Distribution of Profits and Income Taxes,' 27 *Economica*, August 1960, pp. 203–213. [Essay VIII above.]

20 P. H. Frankel, *Oil; The Facts of Life*, 1962, p. 12.

21 Not all of the producing countries receive revenues calculated with respect to posted prices. In both Libya and Venezuela there seems to be considerable doubt about the prices that the governments will ultimately accept for tax purposes.

22 *Petroleum Press Service*, June 1963, p. 208.

23 For a discussion of some of the relevant aspects of the bargaining problem, see E. T. Penrose, 'Profit Sharing Between Producing Countries and Oil Companies in the Middle East,' *Economic Journal*, June 1959, pp. 238–254. [Essay IX above.]

24 See the tables and discussion in C. Issawi and M. Yeganeh, *op. cit.*, in note 14 above, at p. 69 *et seq.*, and Wayne Leeman, *op. cit.*, in note 15 above, at p. 103 *et seq.*

25 These points are well discussed in the papers presented by P. H. Frankel and W. L. Newton to the Fifth and Sixth World Petroleum Congresses.

26 Not, however, from the United States or the International Bank, who took the view that aid should not be given where private enterprise would do the job.

27 See P. R. Odell, *An Economic Geography of Oil*, 1963, p. 128 *et seq.*

28 Shell International Petroleum Co., Ltd., *op. cit.*, in note 13 above, at p. 6.

29 Anyone reading the Damle report on oil prices in India cannot help being struck by the suspicion and frustration of the government investigators when they tried to discover how the prices at which crude oil was being imported were set.

30 See J. E. Hartshorn, *Oil Companies and Governments*, 1962, p. 218.

31 Shell International Petroleum Co., Ltd., *op. cit.*, in note 13 above, at p. 16.

# XI

## VERTICAL INTEGRATION WITH JOINT CONTROL OF RAW-MATERIAL PRODUCTION: CRUDE OIL IN THE MIDDLE EAST[*][1]

*An economic problem not entirely peculiar to the oil industry is discussed.*

WHEN vertically integrated firms which are competitors in markets for end products jointly control the production of their raw material, the rate of expansion of the output of the raw material will be influenced by the competitive strategies of the competing firms, and may well be reduced below the rate that would be attained if the requirements of every one of the firms sharing the ownership of production were met. In the normal course of events shareholders of a company expect its management to make as much profit as practicable and expect to share in profits in proportion to their shareholding. But the ordinary notion of profit as the difference between costs of production and sales receipts is not of much relevance for a company that is little more than a legal entity in a vertically integrated chain of operations and whose shareholders are primarily interested in obtaining the raw material for use in their own manufacturing operations rather than for sale to outsiders. This is particularly true when the entire industry is dominated by vertically integrated concerns and the 'independent' market for the raw material is very thin. In such a case the shareholders are not chiefly interested in maximizing profit from the sale of the raw material in the open market and in obtaining a proportionate share of profits; their main objective is to secure supplies of the raw material at cost. Of course, shareholders will gain from any outside sales of the raw material that can be made at profitable prices, provided that such sales do not increase competition with the sellers' products in product markets.[2]

In an international industry vertical integration may extend across national frontiers and the different stages of production may be concentrated in different political areas, with the result that the pattern of organization and the policies adopted to facilitate the operation of the integrated firms may have side effects that raise difficulties for one or more groups of countries. The problems created

[*] Published in *The Journal of Development Studies*, Vol. 1, No. 3, April, 1965.

are clearly evident in the international petroleum industry, but may well exist in less striking forms also in other industries.

The purpose of this paper is to analyse certain aspects of the organization of production and the structure of the market in the international petroleum industry, describe the way in which the problems arising have been dealt with by the major firms in the industry, and discuss some of the wider effects of the policies pursued by them. I shall deal only with the major crude-oil producing centers of the Middle East from which about 56% of Europe's oil imports come.

## The Crude-Oil Producing Affiliates

The great international oil companies are all vertically integrated in varying degrees, and although they compete with each other in the markets for refined products, they cooperate in the production of crude oil through joint ownership of crude-oil producing affiliates. Almost all of the crude oil from the Middle East is produced by companies owned or jointly controlled by two or more of the international major companies.[3] The cost of production of crude oil, and thus the cost of oil to the integrated companies that own it, has been very much below the price at which oil can be bought in the market, and companies with access to crude at or near 'cost' are in a more favourable position than those that have to buy their crude in the market. Indeed, most of the major companies which have searched for crude oil abroad have done so in order to ensure supplies for their own refining and marketing networks. One of the exceptions to this was Anglo-Iranian (the predecessor of British Petroleum) which proceeded to integrate forward when it acquired supplies of crude oil in Iran before the First World War.

The arrangements for a shared ownership of the crude-oil producing affiliates in the Middle East were made for different reasons in the different countries. In Iraq, a combination of the historical position of the British and British/Dutch companies, the political settlement with the French after the first world war, and pressure from the State Department for an 'open door' for United States companies resulted in the ownership of the Iraq Petroleum Company being effectively shared on equal terms between Anglo-Iranian (now British Petroleum), Royal Dutch/Shell, Compagnie Française des Pétroles, and the Near East Development Corporation, which is in turn owned by two United States companies—Standard Oil of New Jersey and Socony Mobil.[4] In Kuwait, British Petroleum and Gulf Oil hold 50% each of the Kuwait Oil Company as a result of complicated political and commercial negotiations in the interwar period between American and British interests in which both the Foreign Office and State

Department were involved, as well as the British-protected Shaikh of Kuwait.[5] In Saudi Arabia, Standard Oil of California, the company which originally obtained the concession and discovered the oil in the 1930's, had neither the marketing organization nor the capital adequately to develop the concession. This was partly remedied by an agreement before the second world war whereby the Texas Company with marketing outlets East of Suez obtained a half-interest in the Arabian company, and was further remedied after the war when Standard Oil of New Jersey and Socony Mobil, with marketing outlets West of Suez as well as additional capital, were brought in.[6] In Iran, the Anglo-Persian Oil Company, which had been sole owner of the great oil concession in Southwest Iran, was nationalized in 1951 and replaced by a consortium of international and U.S. companies in 1954[7]

Thus it came about that for a complicated set of commercial and political reasons the great oil-producing companies of the Middle East were set up as the producing affiliates of the integrated majors, owned jointly by them in unequal shares. But because the crude oil produced by these affiliates was almost entirely destined for the refineries of the integrated companies,[8] the expansion of output of the affiliates was influenced by the competitive relationships among their co-owners and *a fortiori* by the conflicts of interest among them.

*The Conflicting Interest of Joint Owners*

The competitive position, and in particular the profitability, of different sellers of refined products is affected by the terms on which crude oil is available to them. If crude-oil producers were independent of refiners and were competing to maximize returns from crude-oil production, then similar crudes from the same sources would tend to be equally available at the same f.o.b. cost (subject perhaps to quantity discounts) to all companies engaged in refining. In fact, when the major companies began in 1950 to publish common prices for which crude oil f.o.b. the Persian Gulf was not only to be sold to independent buyers but also transferred to their own refining affiliates, one of the effects was to provide an illusion of this equality of treatment. But since the relevant competitive unit is the integrated group, not the refiner or marketer, we need to look behind the illusion. Vertical integration partially destroys the independent market nexus between crude-oil producers on the one hand and refiners on the other, for the cost to an integrated company of its own crude oil is not a uniform market price but the cost of producing the crude. Since there are wide variations in the costs of production of crude oil from different sources, it follows that the raw-material component in the cost of oil products will be different for different integrated companies depending on the distribution of their ownership rights in crude oil.[9]

Joint ownership of a crude-oil producing company goes only a little way toward establishing equal access to crude among co-owners because in the absence of special agreements each owner is entitled to crude oil 'at cost' only in proportion to its share in the ownership of the crude-producing affiliate. This follows from the perfectly acceptable principle that investors are entitled to share in the benefits of a joint venture in proportion to their investment.[10] The pattern of ownership of the crude-oil resources of the Middle East that emerged in the interwar period left the international companies in different positions with respect to the amount of crude oil they owned in relation to the requirements of their integrated operations, some having abundant supplies of owned or 'cost' crude and others owning less than they used. British Petroleum and Gulf, for example, have access to very large supplies of owned crude in relation to their requirements, while Shell and Socony Mobil are short of crude in spite of the fact that both are part-owners of some of the largest oil fields in the world.

The major companies are in fact competitors in product markets,[11] and those in a position to do so stand to gain if they can establish arrangements which will put any of their 'sister' competitors at a disadvantage in obtaining supplies of crude oil. Except in Kuwait, the arrangements made both with respect to investment in new capacity and to the terms on which current output can be 'lifted' by the different companies, inevitably create precisely this disadvantage for those companies whose requirements are disproportionately large in relation to their ownership shareholdings in the affiliates. This situation arises, not only because of the restriction that is imposed by the ownership shares on the amount of oil that can be taken 'at cost' from a given affiliate, but also because the arrangements that have been adopted in some of the producing affiliates impose a limit to the amount of oil an individual company can effectively request the company to produce when investment plans for the expansion of new capacity are being developed. This limitation, together with the extra cost of acquiring oil in addition to the ownership share, not only has the effect of discriminating among the owners of the affiliates with respect to the terms on which oil is available but may also retard the development of productive capacity and strengthen existing inducements for at least some of the owners to search for new supplies of oil elsewhere.

*The Arrangements Governing the Expansion of Output*

The arrangements that have been adopted for planning the amount of investment to be put up by the parent companies for the expansion of the crude-oil output of their affiliates in the Middle East differ among the affiliates, but all of them are solutions to a simple problem: if a parent company is entitled to a share in output only in proportion to its share in the ownership of the affiliate, what happens if the company wants, and is willing to put up the funds for, an amount of crude oil, that, given the demands of other owners, exceeds the amount it is entitled to according to its ownership share? To illustrate the problem with an imaginary example, suppose we have a producing affiliate owned by parent companies A, B, and C in the proportions 20:35:45, the output of which is being planned for a future period. Of the new increment in production, Company A, with little crude oil of its own elsewhere wants 90 'units' of crude oil annually, B wants 40 units, and C, having plentiful alternative supplies available, wants only 20 units. To satisfy the demands of all parents, the additional output would have to equal 150 units, but B and C would have to put up funds for the production of more oil than they themselves wanted. A problem clearly arises for which there are a number of solutions.

The output of oil would be largest if ownership shares were ignored and A allowed to put up 60% of the investment funds necessary to produce 150 units, taking 90 units, or 60% of the output; but this would either create problems with respect to the ownership of the new fixed assets or would involve a *de facto* lending and borrowing among the companies. The most rigid arrangement would be to require that funds for additional investment be accepted in the same proportion as the original shares, observe C's unwillingness to invest for more than its 20 units and limit total output to 44 units, 45% of which would give C its 20 units of crude oil, with A obtaining only some 9 units instead of 90. Within these two extremes are numerous opportunities for arrangements requiring owners to compromise on investment plans and permitting those who want 'too much' to pay more in order to induce their co-owners to make the requisite investments. The particular arrangements that have in fact been chosen in the Middle East reflect a number of factors, important among which have been the political need to expand crude-oil output of the affiliate concerned at some minimum rate, the actual or potential competitive relations among the various parent companies, and perhaps the amount of oil the affiliate was expected to be able to deliver during the life of the concession.[12]

*Kuwait*

The most liberal arrangement has been adopted by the owners of the Kuwait Oil Company, which has the richest oil fields in the Middle

East. Here the fact of joint ownership seems in no way to restrict the amount of oil that can be taken by the parent companies, British Petroleum and Gulf Oil. Funds for new investment are apparently put up by the owners in proportion to ownership shares (which are 50% each), but each of the parents is allowed to ask for and to lift as much as it wants; if one of them lifts more than its 50% share it pays the other the capital cost of the investment required for its production.[13] Apparently neither company exercises a veto over total investment, and thus over output, in order to restrict the amount taken at cost by the other.[14] In making this liberal arrangement, however, the companies originally agreed not to use the oil so freely available to each to injure each other's position in product markets. This part of the agreement has been abandoned in view of, among other things, the attitude of United States governmental authorities to such agreements. The fact is that at the time of the original agreement between Gulf and BP setting up the arrangements for the exploitation of Kuwait Oil, the two companies did not look on each other as potential competitors. If they begin to do so now, with competition in product markets growing, the arrangements could come under strain.

*Iraq*

In Iraq the arrangements are more complicated and the companies with large reserves available elsewhere in relation to their requirements are in a position to set a limit on the amount that can be taken by others; moreover, oil lifted in excess of a company's ownership share is made more costly to the company lifting it. The investment programme for the Iraq Petroleum Comany is set for five-year periods five years in advance, and this long planning period in itself reduces the flexibility of the company and has probably resulted in a rather slow response to the enormous post-war increases in demand.[15] Investment in the company is made by the parents in proportion to ownership, and the regulations governing the post-war programming of output were laid down in a Heads of Agreement document of 1948.[16]

For the period up to 1952, when presumably post-war readjustments were taking place and output was primarily determined by the speed with which physical capacity could become available, there were no stated restrictions; for the period 1952–56 definite quantities were set forth in a schedule to the Agreement.[17] There is no way of knowing how these quantities were determined, but they show Compagnie Française des Pétroles to be a heavy 'overlifter', i.e., obtaining much more oil than her ownership share entitled her to; the excess was bought from the other shareholders at a price halfway between cost and the sales price, chiefly, apparently, from the Anglo-Iranian Oil Company (BP), since AIOC was listed as requiring very much less than she was entitled to. For the period 1957–61 (which had to be planned in 1952) there were two options for planning output: each of the groups owning the company was limited *either* to a stated increase (126,000 tons for each 1% of ownership) *or* to a quantity calculated in accordance with what is known as the 'five-sevenths rule', whichever was the higher. According to the 5/7ths rule the total amount to be programmed is determined by adding up the amounts wanted by the various companies and then reducing this amount if the highest demand exceeds five-sevenths of the sum of the two lowest demands. The revised total was then allocated to the companies according to their ownership shares, and arrangements were established so that companies wanting more than they were entitled to could in effect buy from those wanting less. Thus a limit was set to the total amount that could be obtained by any company, but no company, by setting its demands very low, could arbitrarily thus keep down the amount available to the others below the fixed quantities stated in the first option.[18] In subsequent periods, the first option was replaced by the '125% rule', which in effect enabled the companies which had been allocated the largest quantities in the previous five year period to obtain at least a 25% increase, again preventing one or two companies from unduly restricting total output by putting their demands very low.[19]

The arrangements with respect to lifting from current output, as distinct from planning the investment to provide for future output, permitted each company to contract to buy oil in excess of its ownership share from others willing to sell, at a price half-way between cost and the 'posted price',[20] thus permitting the selling company to make a profit on sales to its partners. We do not know how the system has actually worked and the companies no longer publish the amounts of oil they lift from the IPC, but it seems almost certain that had Kuwait's type of agreement existed in Iraq, Iraq's output would have been developed very much faster, and perhaps the pressure to explore elsewhere felt by CFP, Shell, and Mobil, in particular, would have been reduced.[21]

*Saudi Arabia*

The way in which the investment programme of Aramco is determined has never been published, although two writers have described the terms on which companies are allowed to lift current output.[22] Aramco, unlike the IPC, was set up as a profit-making company, and its profits are distributed to the parent companies on a complicated system which in effect makes oil that is lifted by a parent in excess of its ownership more expensive than the owned oil—how much more depends on the prices realized by Aramco from sales of all oil. Thus, although the parents of Aramco, who control its Board of Directors, have apparently been willing to expand capacity rapidly and each can increase its offtake progressively from year to year, they do not make the same profits on 'overlifted oil' and, as in the IPC, it is therefore more expensive. So it can come about that Socony Mobil, one of its owners, has less owned oil in relation to its requirements than any of the other majors. Although it is not clear how expansion of capacity is determined and whether the system under which profits are distributed provides sufficient inducement for each parent to invest its proportionate share of the total capital required to meet all demands, regardless of the amount of oil it is itself prepared to lift, it is difficult to escape the conclusion that there must be some restriction on the amount of oil an owner can get at cost, since at least one of the owners is acutely 'short' of owned crude oil.

*Iran*

Finally, another variation on the theme is to be found in the operation of the Iranian Consortium. Again the details have not been published, but according to one informed observer the total programme is determined by the lowest 'nomination' of the shareholders holding some given percentage of the shares in the consortium. Each shareholder proposes a total programme for the output of the consortium for the following year; these are ranked in order, and the programme chosen is the 'lowest total nominated that will cover the estimates put in by shareholders representing a given percentage majority of the shares in the consortium. This total is then redistributed into entitlements proportionate to the holdings of each shareholder in the consortium' and those who exceed their entitlement ('overlifters') have to pay the full posted price.[23] In other words, the proposals ('nominations') by each shareholder for the consortium's *total* output are ranked in order of size and that output is chosen which is the *lowest* of the nominations of companies representing some unknown (to outsiders) percentage of the shareholders.[24]

Whether or not a very few companies have the power to determine the total output of the consortium thus depends on what this critical percentage is, for a low nomination by a few companies can be defeated only if those that want more own a percentage of shares in the consortium that equals or exceeds this critical cut-off point, which is probably between 60% and 79% of total shareholding.[25] The arrangements made for determining output of the Iranian Consortium must undoubtedly have been the result of a compromise between the desires of the Iranian government (which wanted United States companies to come in), British Petroleum, and the United States companies themselves, many of which were not at all eager to take on the obligations associated with joining the Consortium. There are no provisions for 'overlifting' except at the 'posted price', however, and consequently there presumably is none, at least under present circumstances. This high cost of overlifting would in itself reduce the amount taken below what it otherwise would have been.

*Effect on the Rate of Development of Production*

Whether or not the arrangements I have described have in fact retarded the development of any of the producing affiliates cannot be determined unless one knows how they have been operated, but clearly it is possible that they may have done so, and it may be useful to consider the conditions under which conflicting interests of parent companies would restrict the amount of investment made in crude-oil producing affiliates and thus retard the expansion of capacity. As we have noted, the problem arises because of the vertical integration of the parent companies, for if the crude-oil producers were selling their crude to outsiders, rather than transferring it to owners, all the owners would have an equal interest in expanding output to a point which maximized the total revenue to the producers. Profits would be distributed according to ownership shares, and presumably any owner having better use for its investible funds could dispose of part of its share. But when part ownership in an affiliate is obtained, not for the profit directly made on sales of that affiliate's output, but to provide raw material for the owners' further activities, then some sort of inducement will be required to persuade an owner to invest in expanding capacity of the affiliate to an extent greater than necessary to meet its own requirements. Such an inducement may be provided if sales of the affiliate's output can be made between the parents at prices which provide a return on the capital invested by the seller that is better than any alternative return and high enough to offset any competitive disadvantages to the sellers of providing the oil to the buyers. Provided that the necessity of paying such prices does not reduce the amount demanded of the affiliate by any owner, then the arrangement will not result in a total output lower than would have been produced had each company been entitled to obtain its requirements at cost.[26]

Thus, there are two conditions which must be satisfied before we can safely conclude that joint ownership of a raw-material producing affiliate by vertically integrated parents will not retard the rate of expansion of the affiliate: (1) The parents as a group must be willing to invest in expanding capacity to the full extent necessary to supply all of the demands on the affiliate by all owners. (2) The price that the parents which want extra supplies must pay must not reduce the demands of these parents on the affiliate.

If there had been no question about the first condition being satisfied there would have been no need for rules regulating the amount of oil any company could lift at some price, and Kuwait-type agreements would have been made. Hence some owners must have wanted restrictions (a) because they feared that they would be required to invest more than they wanted to unless restrictive regulations were adopted, or (b) in order to put some co-owners at a competitive disadvantage, or (c) as a means of regulating total output to prevent surplus supplies, or (d) to protect reserves.

For vertically integrated companies operating in markets where demand was expanding very rapidly, large investments in refining and marketing were required which made heavy demands on capital funds; moreover, most of the companies most of the time have strongly preferred self-financing. Given the very large difference between costs of production and posted prices, the 'half-way price' charged the 'overlifters' undoubtedly provided a very nice return on investment for the 'underlifters'—the sellers among the parent companies—but there may well have been some reluctance to invest in order to provide oil to co-owners, who were also competitors in product markets. It is clear from the record, for example, that some of the companies at least, wanted to restrict the amount of cheap oil the French could get from the IPC for fear that the French would use it to cut prices in product markets.[27] It seems unlikely that fear of early depletion of reserves was a major consideration in the regulation of output in view of the very large reserves available, but there can be little doubt that the very size of these reserves made it imperative from the companies' point of view to do the best they could to ensure that the rate of exploitation was kept more or less in line with expected rates of increase in demand at ruling prices. The arrangements discussed here are, in effect, the mechanism through which the companies attempted to adjust increasing supplies to increasing demand.

As to the second condition—the willingness of owners to buy extra oil—assuming that crude oil was not available elsewhere at lower prices, and that the expansion of an individual company's sales of products was not determined by incremental costs but by other considerations, then the higher price charged to 'overlifters' for crude would not significantly affect the amount taken. If this were the case, however, it would also follow that the competitive position in product markets of rival parent companies would not be altered by differences in the costs of crude oil, except in so far as high costs reduced the rate of investment of some companies by reducing profits and retained earnings. There is no way of knowing how significant considerations of these kinds have been. It is perhaps likely that the companies with higher average crude-oil costs hesitated to expand in some markets which were only marginally profitable in any event. But there can be little doubt that a company having to pay higher prices than its competitors for its crude oil would feel at a long-run competitive disadvantage and would have every incentive to search for a cheaper source of oil. Thus the offtake arrangements in the Middle East provided for some companies a very strong incentive to find new oil of their own.[28] The incentive was the stronger the greater the gap between costs and posted prices, since lower posted prices meant lower 'half-way prices'; in recent years, however, the situation has changed considerably; market prices of crude oil have continued to fall while posted prices have remained up; oil has become available at prices even less than a half-way price calculated on the basis of present posted prices. Unless the purchase arrangements are modified, they could, in present circumstances, significantly reduce 'overlifting' in some Middle East affiliates.

The only way to avoid the difficulties discussed here, given the vertical integration of the parent companies, and to ensure that the development of existing crude-oil producing affiliates is not retarded by the desire of some owners to profit at the expense of co-owners by raising the cost or restricting the availability of oil to them, is to allow all owners who want to take oil at 'cost' to do so, providing either that the overlifters pay the interest and depreciation costs on the capital investment to the partners putting it up (as is evidently done in Kuwait), or put up the capital themselves. The former arrangement preserves the effect of the existing ownership shares on the distribution of capital investment, but removes it on the distribution of oil to the parent companies; the latter ignores ownership shares in both the distribution of capital investment and in the availability of oil, which becomes determined for each company by the proportion of actual investment it contracts for.

That procedures of this kind are feasible has been demonstrated by the working of the Gulf-BP agreement in Kuwait. Nor would they reduce the chief alleged advantages to the major companies of vertical integration—an assured supply of crude oil from numerous sources. One of the consequences would probably be a more flexible pattern of investment and oil supplies among the various companies,[29] but at the same time competitive pressures in product markets might well increase, bringing effects on prices thoroughly unwelcome to the oil companies and producing countries alike. Moreover, to insist on arrangements that would permit companies to gain from joint ventures benefits which were not proportionate to their shareholdings would be inconsistent with the accepted notion that investors 'ought' to benefit in accordance with their investment; and to promote a change in effective ownership shares through disproportionate investment in new capacity could be looked at as depriving some of the original investors of ownership rights.[30]

But in a system of private enterprise extremely profitable ownership rights are, among other things, a challenge to the ingenuity of those whose interests are adversely affected to find a way around them. Among those adversely affected were countries which are predominantly importers of crude oil, such as Japan, Italy, India, and Pakistan, who owned no Middle Eastern oil, as well as the companies 'short' of cost crude such as CFP, Shell, Socony Mobil, and numerous 'independents'. Moreover, oil exploration is a competitive process, and if some groups have a strong incentive to search for more oil, all others must do so in self-defence, not to mention the fact that if some of the dominant producers would like to restrict the availability of low-cost crude to others, it behoves them, if possible, to find it first.

Thus both the oil companies and the older producing countries of the Middle East are on the horns of a dilemma. In the short run it pays them to restrict output and maintain prices; in the longer run this policy is likely to increase supplies elsewhere not only bringing down prices but leading to a permanent loss of market shares. There can be little doubt that the control exercised over the output of crude oil from the Middle East, imperfect though it was, made it possible for the companies to pay the very large revenues they did to the governments of the oil-producing countries. But the high prices thereby maintained, as well as the high degree of concentration in the hands of the majors, were undoubtedly important factors in speeding up the development of oil supplies elsewhere.[31] From some of the effects of this the oil companies are partly protected to the extent that they have obtained a share in the new sources of supply; the producing countries are less well protected, although they too may mitigate the effects on themselves by granting concessions and encouraging exploration in their own territories, especially by those companies in greatest need of cheap crude. The newer producing countries benefit.

Nevertheless, the policies that were pursued in the past, and that brought such enormous short-run benefits to a few countries and companies, may yet yield bitter fruit, especially for the companies with the largest stake in the older producing centres. For example, British Petroleum has a weight in the determination of the distribution of output as between Iran, Iraq, and Kuwait, greater than her percentage shareholding in the producing affiliates would imply, and for political as well as commercial reasons may well wish to maintain a high rate of increase in the output of these countries. At the same time some of the companies, for whom supplies of oil have been made expensive in the Middle East, may be reducing their demands in favour of other sources and may well want substantially to cut their 'overlifting' as soon as present obligations expire. Hence, as alternative supplies become increasingly available, a few of the larger companies may be left with the political burden of attempting to maintain the relative position of the older centres of production which their own policies have helped to undermine.

NOTES

1. I would like to thank for reading and criticizing this paper Professor M. A. Adelman, Mr. J. E. Hartshorn, Dr. Paul Frankel, and Dr. P. R. Odell. I should also like to thank the numerous officials from the oil companies who were good enough to discuss with me, or to comment in writing on, some of the issues raised. Since many of the arrangements discussed in this article are made by agreement among several companies, no one of the companies is, quite understandably, in a position either to comment on the accuracy of my description or to give me additional information. Nevertheless, the willingness of their officials to discuss the problems with me helped me to appreciate the very real dilemmas and difficulties with which they are confronted in matters like those I am analysing and to understand their point of view. Needless to say, the companies do not necessarily, or indeed presumably, agree either with my argument or my point of view.

2. The major groups owning the crude-oil affiliates have, of course, always to balance the gains to be had from selling crude at good prices against the (possibly long run) competitive consequences in product markets. In fact, considerable sales outside the group of integrated majors have been made, especially by companies with very much larger supplies of crude oil available than they could use in their own operations. These sales may have been increasing in recent years and may have helped to contribute to the continuing pressure on crude-oil prices.

3. The major international companies are Standard Oil of New Jersey, Royal Dutch/Shell, British Petroleum, Gulf Oil, Texaco, Standard Oil of California, Socony Mobil, and Compagnie Française des Pétroles.

4. Each of the four groups owns 23·75% of the IPC, 5% being owned by the Gulbenkian interests. There were originally a greater number of companies in the American group, but now only the two listed above remain. The negotiations to set up the IPC took place in the 1920's, its name being changed from the Turkish Petroleum Company in 1928. It is a British Company.

5. The Kuwait Oil Company was established as the outcome of an agreement in 1933. It is also a British company.

6. The Arabian American Oil Company (Aramco) is now owned 30% each by Jersey, Texaco and Standard of California (Socal), with Socony Mobil holding 10%. It is a United States Company.

7. The oil of Iran is owned by the Government through the National Iranian Oil Company (NIOC) and is produced on behalf of the NIOC by Iranian

Oil Operating Companies which is owned by British Petroleum (40%), Royal Dutch/Shell (14%), Compagnie Française des Pétroles (6%), Jersey Standard, Socony Mobil, Socal, Texaco, and Gulf (7% each), and Iricon (a consortium of 9 United States Companies) (5%).

8   The two owners of the renowned Kuwait oil fields (Gulf and BP) have disposed of a significant part of current production on long-term contracts to other major integrated companies, for the amount of oil discovered was vastly in excess of what either company could use. At the time when the agreement were made restraints were placed on the extent to which either company could use oil sold under the agreements to compete with the other. The contract between Gulf and Royal Dutch/Shell gives Shell virtually the equivalent of *de facto* ownership rights in a portion of Kuwait oil.

9   Although the present discussion is concerned with the cost of crude oil, it should be remembered that oil is not a homogeneous commodity and similar considerations apply with respect to the quality and location of crude oil.

10  There is, therefore, no presumption in the analysis that follows that different arrangements 'ought' necessarily to have been made; nevertheless one may still find that the type of arrangement analysed has consequences which would support a case for modifying it. Clearly, however, this will depend on the point of view.

11  There are exceptions to this statement in some markets. For example, Shell and British Petroleum market jointly through Shell-Mex and BP in the United Kingdom, British Petroleum sells through Burmah Shell in India, and Shell sells for BP in some 15 countries in the Far East which together take almost as much oil as the U.K. market. Socal and Texaco are associated in production, refining and marketing in many areas through Caltex and until recently Stanvac operated for Jersey Standard and Socony Mobil in the Far East. These joint operations have been under considerable strain in some areas in recent years as competition throughout the industry has increased.

12  In view of the very large reserves of oil in the Middle East, oil in the ground had virtually no future value, particularly since supplies were expected to far outlast the period of the concessions. This consideration was therefore probably of very little significance for the companies.

13  Since the capital cost includes depreciation and amortization as well as interest, the 'lending' company recovers its capital in time. The Kuwait agreement is described in U.S. Federal Trade Commission, *The International Petroleum Cartel*, (Washington, D.C., 1952) pp. 133–34.

14  But BP is permitted under the agreement to supply Gulf's requirements from Iran or Iraq if it so wishes. *Ibid.*

15  For this reason steps are being taken to shorten it, for no other company in the Middle East has such an inflexible programming policy. [The arrangements in the Iraq Petroleum Company have been radically changed since this paper was written.]

16  This, and other documents, are published in *Current Antitrust Problems*, Hearings Before Antitrust Sub-committee (No. 5) of the Committee on the Judiciary, House of Representatives, 84th Cong., 1st Sess. Part II. Serial No. 3.

17  Schedule B in *ibid.*, p. 962.

18  The text of the '5/7ths' rule is given in footnote (19) below. To illustrate the nature of its restrictive effect if one or two companies tried to use it to keep down total output, consider the following example. (I ignore for

simplicity the 5% interest of Gulbenkian and assume equal shares for the four major groups.) Suppose that the two low companies asked for 150 and 200 'units' of oil; the highest amount any company could get would then be restricted to 5/7 x 350=250. Imagine, then, that the following set of demands existed: company A, 150, B, 200, C, 220 and D, 350, a total of 920. D's demand would be reduced to 250 and total output to 820. Company A, by further reducing its demand to 80 units could restrict all others to 200 units and total output to 680 units. Company A could conceivably consider this a useful way of restricting the competitive power of the others, while at the same time selling its own excess at the 'half-way price' to others since, out of a total output of 680, A would be entitled to 1/4 or 170 units. If in fact A takes only 80, having plenty of oil elsewhere, it can dispose of 90 to the others, but even if D takes all the excess oil it will still be unable to satisfy its original requirements.

19  The following is the text of the two alternatives—the '5/7ths rule' and the '125% rule'. The phrase 'basic proportion' means the ownership share; the phrase 'adjusted requirements' refers to the requirements of each company as worked out by the Managing Director of the IPC to conform with the upper limits permitted each company, sound oilfield practice, and reallocation of surpluses of some to others according to basic proportions. The Managing Director of the IPC is directed to make all the calculations.

After the period 1957–61 each of the Groups owning the IPC was limited to the 'higher of the two following quantities:
the quantity resulting from the following computation:
   (1) divide the last notified requirements of each Group for the year in question by that Group's basic proportion:
   (2) take 5/7ths of the sum of the lowest two figures obtained under (1);
   (3) multiply the resulting figure by the Group's basic proportion'.
or
'the quantity resulting from the following computations:
   (a) determine the annual average of the adjusted requirements of each Group in the immediately succeeding five-year period (e.g. in 1957 in revising requirements for 1962–66, the adjusted requirements for 1957–61 would be used);
   (b) divide the average for each Group as determined in (a) above by that Group's basic proportion as of the date on which the adjusted requirements used as a basis were fixed (e.g. in 1957, in revising requirements for 1962–66, and using adjusted requirements for 1957–61, the average would be divided by basic proportions as of 1952);
   (c) multiply the largest figure so obtained in (b) above by 1·25;
   (d) multiply the resulting figure in (c) above by the basic proportion of each respective Group at the date of such revision (e.g. in 1957, in revising requirements for 1962–66 basic proportions as of 1957 would be used.)'

The first of these (the 5/7ths rule) is found in Clause 12 of the Heads of Agreement of 1948, *Current Antitrust Problems, op. cit.*, p. 919; the second the 125% rule) in Clause 13, which came into effect after 1957–61, *ibid.*, p. 922.

20  The 'posted price' is the published f.o.b. price at which oil is offered by the companies. It is also the price used to calculate the 'profits' attributed to crude-oil production for income tax purposes. In recent years the companies have had to sell below this price under the pressure of competition, but for political reasons they have been unable to lower 'posted prices' which still determine the 'profits' subject to tax. According to recent press

reports the 'half-way' price is now calculated with reference to average prices actually received from crude-oil sales instead of to posted prices.
21 Some of the parent companies of the IPC, and particularly Anglo-Iranian, apparently were very much concerned to arrange for production to develop slowly, (see Federal Trade Commission, *op. cit.*, p. 105) while others very much wanted more oil. CFP, for example, was largely responsible for pushing up the rate of development of Iraq oil, but nevertheless has had to buy large amounts at the 'half-way' price. See M. Laudrain, *Le prix du pétrole brut* (Paris, 1958), p. 268.
22 J. E. Hartshorn, *Oil Companies and Governments* (London, 1962), pp. 164–5; Wayne Leeman, *The Price of Middle East Oil* (Cornell Univ. Press, 1962), pp. 21–23.
23 J. Hartshorn, *op. cit.*, p. 163.
24 To illustrate, suppose the following are the proposals by the shareholders for the total output of the consortium in the following year, and suppose the percentage of shareholding of which the lowest proposal will be accepted is 65%. Then the output planned for would be 60.

| Shareholder | % Shareholding | | Proposal for total output (ranked by size) |
|---|---|---|---|
| Socony Mobil | 7 | | 85 |
| Shell | 14 | 66% | 80 |
| BP | 40 | | 75 |
| Iricon | 5 | | 60 |
| Esso (Jersey Standard) | 7 | | 50 |
| Texaco | 7 | | 45 |
| CFP | 6 | | 40 |
| Socal | 7 | | 35 |
| Gulf | 7 | | 30 |

It should be noted that this type of arrangement *need* not result in an output lower than could have been obtained if each company were able to nominate and get just the amount it wanted. It might even have been a way of trying to ensure that total output would be maintained at an acceptable level.
25 Since British Petroleum was the sole owner of the old Anglo-Iranian Company before the nationalization, yet for political reasons could not be given a majority interest in the consortium, it is probable that the critical percentage of total shares which determines the output of the consortium is such as to ensure considerable weight to BP. BP owns 40% of the consortium, and if the critical percentage is less than 60%, then all others together could possibly determine the output of the consortium without BP, although BP would be entitled to 40% of the output; if this percentage is more than 60% BP alone could determine the maximum output. When we consider the circumstances surrounding the negotiations between the companies to set up the consortium, it seems improbable that the percentage would be less than 60%, even though a greater percentage would give BP very much more weight in the decisions determining total output than would appear from its 40% shareholding. However, in spite of the fact that BP has abundant oil available elsewhere, it would be surprising if its nominations for the consortium's output were unduly low in view of its political interest in ensuring a level of output satisfactory to the Iranian government.

Shell and Mobil are short of 'cost oil', while Gulf and Socal have very large supplies in the Eastern Hemisphere; hence it is probable that Gulf,

Socal, and perhaps Texaco would be the 'low bidders' for Iranian oil, but low nominations from those companies would be defeated if the critical percentage is less than 79%, since together they own shares equal to only 21% of the total. Therefore, since BP would not have wanted Iran's output to be held down by these companies, the critical percentage is probably less than 79%, and more than 60% and the total output of the Iranian consortium is probably determined by the bids of either Jersey Standard, Iricon, Texaco, or CFP, although CFP now probably requires less Iranian oil than earlier in view of the supplies now available in Algeria.

26 It should be noted that if we pare away the accounting smokescreen, the Aramco-type arrangement where a profit is attributed to the affiliate has no different effect from the IPC-type for the problem discussed here except that the Aramco arrangements take account of sales at discounted prices. See Leeman, *op. cit.*, pp. 21–22.

27 Federal Trade Commission, *op. cit.*, p. 76.

28 In a brief presented in a Civil Action brought against the company by the United States Government in 1953, Socony-Vacuum (now Socony-Mobil) made the following statement:
'Socony began to embark on a systematic foreign exploratory effort primarily for two reasons: *First*, because substantially all foreign production was in the hands of large integrated companies which were competitors of Socony and, therefore, could not be depended upon for supplies. *Second*, because a long period of time would elapse between the inception of such an exploratory effort and the development of commercial production. . . .'
*Current Antitrust Problems*, *op. cit.*, p. 879.

29 Quite apart from the context of this discussion there is, I think, much scope for making more use of the flexibility that could be provided by arrangements which permit a change in ownership shares by altering the proportions in which new investment is made. This could be a particularly useful device for developing countries wanting to attract private foreign investment for a period, but wanting over a longer period to reduce the extent of foreign ownership in their industrial economy.

30 But strictly legal considerations should not be allowed too much weight when other considerations point to the desirability of change. See for example, the well-argued criticism by F. R. Parra of the 'legalistic' attitude of the companies in *Exporting Countries and International Oil* (Organization of Petroleum Exporting Countries, Geneva, 1964, EC/64/I). On the other hand, the oil companies have more than once showed willingness to re-negotiate agreements in the interest of the producing countries.

31 There are, of course, forces inducing intensive exploration in spite of the present 'surplus' of oil other than the ones discussed here: the desire for 'security' through diversification of sources of supply, the hope of finding very low-cost oil (or gas) very close to the great consuming areas, the desire of a government to discover oil within its own boundaries, etc., are all important. But all of these motives are powerfully reinforced by prices significantly higher than the expected costs of finding and developing new supplies.

## XII

## GOVERNMENT PARTNERSHIP IN THE MAJOR CONCESSIONS OF THE MIDDLE EAST: THE NATURE OF THE PROBLEM*

*Explains why the internationally integrated companies may prefer to withdraw from their major concessions in the Middle East rather than concede 'partnership', or equity participation, to the governments of the exporting countries if their managerial prerogatives are threatened.*

THE question of government equity participation in the major crude oil producing companies in the Middle East is clearly moving to the center of the stage in international oil affairs. The statement by the Oil Minister of Saudi Arabia in a seminar on the international petroleum industry at the American University of Beirut in June, to the effect that the Saudi Arabian Government intended to obtain an equity interest in the companies exploiting its oil reserves has been given considerable publicity everywhere. It seems to have caused something of a flutter in the dove-cotes of the oil industry in the Middle East—and Aramco can, with much justification, claim a history of 'dovishness' in oil matters. To be sure, the Minister's claim was more evolutionary in tone than revolutionary, for he stressed a willingness to begin modestly—perhaps with only 10% of the equity —although the eventual desideratum would be at least a half share for the government. Even making full allowance for Saudi bargaining techniques, there was no hint in the Minister's talk of confiscatory threats, but rather the assertion of a faith that reasonable men of good will on both sides could work the matter out. And at the conference of OPEC three weeks later the principle of 'reasonable participation for host governments in the major old-established concessions on the grounds of changing circumstances' was formally endorsed.

Indeed, that the governments of the oil exporting countries should be able to become partners in the exploitation of their own oil if they are willing to pay for an ownership share seems, on the face of it, to be eminently reasonable. Because of this, it is possible, though I think unlikely, that the international companies might propose some

* Published in *Middle East Economic Survey*, Vol. XI, No. 44, 30 August 1968. Supplement.

arrangement which provided the *appearance* of partnership by enabling the governments to acquire an equity interest in the producing company. But it is highly unlikely that they would accept any arrangement which contained the reality of partnership, that is, one which gave any effective influence to the government partner over operations, and especially over offtake, prices and even investment, in spite of the fact that a government partner would have to put up a partner's share of new capital investment. The manifest instability of an appearance of partnership devoid of the reality makes it unlikely that any such compromise arrangement will be made unless the companies merely see it as a means of postponing the day of judgement and always providing it does not cost them too much money on current operations.

There are three broad grounds on which major international companies may fear partnership with governments. The first is political—an objection in principle to association with governments in this kind of business, or a fear that a government partner will interfere with operations on political grounds. The second relates to costs—a fear that payments to governments will exceed what the companies feel they can afford. The third relates to the effect of taking government partners on the industry generally—a fear that the companies' control over output would be so further weakened that prices would be severely affected and that the flexibility of the companies in managing their international operations would be seriously impaired.

In this discussion we shall be concerned primarily with the third fear, for this is the only issue that raises fundamental problems for the structure of the industry. The political question is not amenable to economic analysis; the problem of costs is a matter for bargaining, for governments can always attempt to increase their receipts from the companies in a variety of ways, and the extent of their success is entirely a question of their strength in negotiations. But control over supply brings us directly to the role of integration in the industry and to the forces bringing about a change in its structure. The special difficulties that arise with respect to the price of crude oil, and to equity participation for governments in the crude oil producing companies are both traceable to the same source: the fact that the producing companies are not simple crude oil producers and sellers but are an integral part of, and are owned by, larger groups which are engaged in refining and the marketing of products all over the world. Most of the revenues of the integrated groups come from the sale of products, not from the sale of crude oil. These facts are well-known, but their full significance has not always been clearly understood. Indeed, both the legal organization of the producing companies—as separately incorporated entities—and the fictions that have in the course of time been accepted about the role of these companies—as sellers of crude on a 'commercial' basis—have obscured the real issues. The governments of the exporting countries have relied heavily in arguments with their concessionaires on the so-called 'doctrine of changed conditions,' but there have been great changes in the last fifteen years which affect the possibilities of partnership of which they have not taken adequate account.

*The Profitability of Integrated Operations—No Sales of Crude Oil*

Before discussing the partnership issue, we must clarify certain relationships affecting the profitability of operations for international companies when there is a high degree of integration between crude oil production and downstream operations. It should be self-evident that monopoly over crude oil production will yield monopoly profits on crude oil only to sellers of crude oil. Integrated companies producing crude for their own use make no profits on the crude they do not sell; rather they make their profits from the sale of products, since profits arise from the difference between receipts from sales and the sum of all of the costs incurred by the companies making the final sales. Thus the value of crude to integrated companies that do not make outside sales of crude depends entirely on the profitability of refining it and selling the products.

Imagine, for example, a situation in which all production of crude is in the hands of a few companies whose owners want it entirely for their own use. Integration between crude production and refining is then complete since there can be no refiners other than those owned by the producers of crude. Monopoly profits may be made on the integrated business but there is no non-arbitrary way of assigning a share of the profits to any stage of it; the entire integrated operation must be looked on as one business.[1] This becomes clear if we enquire what happens to the value of crude to its owners in these circumstances if they should engage in intense price competition in product markets. Plainly, the price of products could be forced down to the point at which the integrated firms earned only enough to cover all costs at all stages of production—including replacement costs—plus the profit necessary to keep them in business, or, if demand were rising, to attract the necessary investment funds into expansion. Monopoly over crude oil production would yield no monopoly profits anywhere unless the companies were able to restrict price competition in product markets.

Even in such conditions of complete integration, it is nevertheless possible for the governments of the countries in which either crude production or refining is carried on (we assume that these activities are conducted in different countries) to impose an 'income' tax on the relevant affiliates and to insist that a price for tax purposes be set on the transfers of crude, thus attributing 'profit' to crude production. The companies would in any case have transferred crude from their producing to their refining affiliates at some price, if for no reason other than their accounting requirements or the requirements of customs authorities. These transfers could take place at any price between one that reflected the estimated costs of production of crude plus whatever management fees or 'profit' that the companies thought appropriate to allow their producing affiliates, and one that attributed all of the integrated profit to the stage of crude oil production, leaving only the necessary managerial fees or the administratively necessary profit in downstream operations. (Such a situation could arise if there were, for example, subsidies, such as depletion allowances against taxable income attributed to crude production, which encouraged the attribution of a substantial proportion of integrated profits to crude oil production). In any case, the amount of the total profit attributed to one stage of the industry or the other would be administratively determined, and tax considerations might well be the most important factor in such a determination. On the other hand, if the governments of the crude producing countries were in a strong position, they might succeed in insisting on a price which attributed by far the greater part of revenues to crude oil production. This could, up to a point, also coincide with the interest of the companies, depending on its effect on their total tax burden.

*The Effect of Market Sales of Crude*

So far, we have been discussing a situation in which the entire industry was assumed to be in the hands of a few fully integrated concerns that neither bought nor sold crude oil. The purpose of the analysis was to demonstrate that in such circumstances the price of crude would be of little significance to the companies apart from taxation and difficulties that might arise with governments. The profitability of the integrated companies would depend entirely on the extent to which price competition could be prevented in product markets on the one hand, and their total costs on the other.

Let us now assume that the dominant group of integrated companies sells some crude to third parties—to 'outsiders.' This implies, of course, that the integrated majors—as we shall call this group—no longer have a complete monopoly in refining, since it may be assumed that the crude oil sold is used in refineries other than those of the majors. The integrated companies make profits on their outside sales of crude oil, but because both the quantity and price of outside sales are likely to have repercussions on product markets, the companies will have a strong incentive to limit such sales, or to confine them to buyers with established markets, and to charge prices that will maintain the cost of crude to such buyers as high as possible in order to make it unprofitable for sellers of the refined products to cut prices in product markets. So long as neither the price of outside sales nor the quantity sold is such as to impair the position of the majors in product markets (which may well be the case, for example, if total demand is growing more rapidly than the major companies can expand), then the profit obtained on outside sales of crude will be a net gain for them. In a rapidly growing market where all sellers generally refrain from price competition in both crude and product sales, a relatively small and reasonably stable proportion of total crude production may be steadily sold to third parties without serious effects on product markets. But once crude sales induce or facilitate price competition in product markets, then any increase in the direct profits made on such sales by the integrated companies may lead to a decline in the profitability of their integrated transactions through a decline in product prices as a result of outsiders' competition.

An oil world in which a small group of like-minded integrated companies not only maintains a monopoly over crude oil production and successfully regulates the quantities and prices of outside sales made by all of them, but also contains price competition in product sales among themselves, is a useful 'model' to construct for analytical purposes in order to understand the underlying economic problem. But it is, of course, far from the reality of the actual world of oil, although there are some rather striking similarities between it and the state of the industry outside North America and the Communist countries until the middle 1950's. For nearly 20 years now, however, the quantities of oil moving in crude oil markets have been growing, and have clearly supported and encouraged price competition in product markets. The integrated majors were responsible for a great deal of these crude oil sales while another important part, especially in recent years, has come from new discoveries by independent companies. Nevertheless, so long as the majors see their business as an internationally integrated one in which sales of oil products are the important end result, increasing sales of crude by others, or increasing pressures which force them to dispose of increasing quantities of their own production to third parties, are always potential threats to the profitability of their integrated business, especially as the tax-paid cost of their own crude rises relative to the price in the market. The effective market price of crude oil must fall if the rate of increase of supply exceeds the rate of increase of demand at existing prices, and as it falls, not only are profits on crude sales affected but, of even greater importance for integrated companies, the price of products comes under pressure and the shares of these companies in product markets generally are threatened.

## Government Partnership

It is in the light of these basic considerations characterizing the relationship between crude and product markets from the point of view of internationally integrated companies that we must examine the question of government equity participation in the major sources of crude oil production. The heart of the issue lies in the significance for integrated companies of the acceptance of non-integrated partners in crude oil production, especially in conditions where the income attributed to crude oil producing companies is calculated with reference to prices which bear no relation to the prices of market sales.

To make the problem clear, we should distinguish several possible situations: (a) when the government wants partnership in order to obtain crude oil for its own use; (b) when the government merely wants a partner's share of the profits in addition to income taxes; and (c) when the government wants not only a share of profits but also participation in management and control. Moreover, it makes a difference whether or not the international company concerned itself produces sufficient crude to supply its own operations or, if the company is in surplus, whether or not the affiliates in which partnership is demanded are major sources of the company's total supply.

*To Obtain Crude:* If a government oil company wants to become a partner in a foreign-owned producing company primarily because it wants to obtain 'cost crude' for its own integrated operations, then it will be in the same position as the other integrated owners from the point of view of the planning of supply, and would presumably be willing to enter into planning the expansion of output on the same basis as they, that is, in relation to estimated sales of products. In these circumstances, the fact that the partner is a government poses no special problems for the private owners apart from the political apprehensions that the government might use its superior sovereign power to interfere in the company's business and commercial affairs. There would, of course, be other problems, such as the reconciliation of conflicting views of different owners, with differing requirements for crude, with respect to investment in the expansion of capacity, the arrangements to be made for sharing offtake, payment for overlifting, etc. But these types of problems, arising from the need to reconcile the differing interests of partners, have to be solved regardless of the type of partner, and do not arise simply because a government is involved.

*To Share in Profits:* As yet, however, a desire for crude for their own use is not an important reason why governments demand equity participation; more important is a desire to obtain greater revenues through obtaining a partner's share of profits attributed to crude oil production. For many years now the governments of the crude producing countries have in one way or another been steadily increasing their share of the profits of the integrated companies. So far as the major producing affiliates are concerned, the nominal rate of 'income tax' has remained at 50%, but some items allowed as costs or deductions (including royalties) have been eliminated or the amounts reduced. Moreover, the price of crude for tax purposes has been maintained in spite of falling prices in the markets for both crude and products, with the result that the proportion of integrated profits received by the producing countries has risen. Thus, if partnership is conceived of solely as an additional means of increasing government revenues, it can be looked on as an extension of the continuous process of bargaining over the division of the industry's profits between companies and crude-producing countries.

In the last analysis, partnership can give the government an increased income per barrel of oil produced in only two ways: either (1) the government takes its share of the oil and sells it at a price which yields a profit that exceeds the rate of tax it would have obtained from the companies after making allowance for the capital investment of the government; or (2) the private companies in one way or another increase their per barrel payments to the government. If the government partner takes its share of the oil but at the same time all oil continues to be valued at tax prices for the calculation of producing income, then the cost of oil to the other partners is increased unless the government pays the producing company the tax price for it. If on the other hand the government does not want to take oil for its own use, or to dispose of on its own account, but instead wants to sell it to the other owners (which involves them in automatic 'overlifting'), the cost of oil to the others will be increased unless they act only as selling agents for the government and the oil is valued for tax purposes at the prices realized, or they obtain it at cost for their own use. Hence negotiations over financial arrangements with respect to the government's share of oil would have to start from some such bases. In any case, whenever the private companies purchase oil, they stand to lose the tax privileges (including depletion allowances) which they would have had if they had produced it and could have claimed the difference between costs and tax prices as income for purposes of calculating their tax liability to their home governments. This is, of course, equivalent to an increase in the tax-paid cost of their oil.

*To Participate in Control:* The limits to which the international companies can go in making increased payments per barrel depend on their position in product markets and on the state of competition there, which is in turn affected by competition in the market for crude oil. But the revenues of governments can be increased by an increase in offtake at existing tax prices, and there is no reason to believe that, as partners in crude oil production, the governments of those countries which today are convinced that they should in all fairness have a larger share in total output would relax their pressures to this end. Since a non-integrated government partner is not itself directly concerned with product markets, and may even consider 'downstream losses' a problem that the companies should deal with, the pressures from the individual governments to increase the rate of supply are likely to enhance the danger that the aggregate rate of supply of crude oil will grow without adequate consideration of its relation to the aggregate rate of demand.

### The Problem Posed for the International Companies

For individual companies short of crude for their own operations, increased supplies would not be unwelcome providing that the cost did not exceed the price at which they could buy, but for companies with adequate or excess supplies, pressures to increase offtake would in effect be pressures to increase market sales. If the producing company in which the government became a partner provided only a small proportion of the supplies of the integrated owners, production might be increased by a substantial percentage without creating a serious problem of disposal, but even a small percentage increase in the output of one of the great crude producing concessions would create very large amounts of oil to be disposed of.

Because the international companies have taken their supplies from a number of areas, most of which could easily produce more at existing prices, they have in effect engaged in a loose kind of pro-rationing, which included in its scope a very large proportion of the world supplies of oil outside North America and the Communist countries. For each of the international companies, the determination of the amount and distribution of its oil supplies was based partly on its estimated requirements and partly on its ownership share in the crude producing companies. Some of the majors have thus been short of owned crude, while others have had to purchase it, often from their co-owners of the producing companies. The reconciliation of numerous conflicting interests—including the desire of the producing countries for increased output and the differing requirements of the individual international companies—while at the same time attempting to keep the rate of aggregate supply in line with the rate of aggregate demand, has for a long time posed difficult and delicate problems for the international companies as a group.

For a few years in the early 1950's the companies were reasonably successful in solving these problems, but their effectiveness in holding the balance while keeping supply in line with demand at existing prices became increasingly weakened as time went on, partly because they pursued policies which made the industry extremely attractive to any newcomer who could break in, and partly because they could not act as a unified cartel. Instead, within the limits set by the varying combinations of companies which joined together in production, refining or distribution in some areas, each of the international companies planned its own operations in the light of what it knew about the activities of the others, always subject to circumstances none of them could control. The advent of governments as full partners in the production of crude oil would immensely increase the difficulties of the international companies in their attempts to reconcile conflicting interests and would make it almost impossible for them to enforce price and output policies which prevented continued erosion of the profitability of the integrated industry.

In principle, the governments of the crude producing countries recognize the underlying problem and are fully aware that an increase in free market sales carries with it the likelihood of a deterioration not only in crude prices but in product prices as well. In principle, therefore, a government should be prepared, in demanding partnership, to agree that it would not press for increased output which would force greater market sales. But no individual government is in a position to appraise the overall circumstances of its major concessionaires and thus to take an active part in the determination of the international distribution of their supplies of crude oil. In these circumstances, cooperation by an individual government partner in policies designed to regulate output of the producing company in accordance with the outlets provided by the integrated operations of the other owners, together with their established contracts and the market sales which were considered possible at ruling prices without intensifying competition in product markets, would in practice imply government consent to a kind of prorationing operated by the companies. But this is precisely what many of the governments object to in the present situation; one of the reasons behind the demand for equity participation is to enable governments to have some control over, or influence on, offtake.

Clearly, the only alternative would be for the governments as a group, perhaps through OPEC, to join with the companies as a group in an effective cartel to plan the amount and distribution of oil supplies. But this is where matters now rest and we have come a full circle. The companies could not join such a cartel; the governments show no indication that in any case they would be willing to accept the necessary restrictions on their individual freedom of action.

## Conclusion

The peculiar problems raised by government demands for partnership in the great crude oil producing affiliates of the major international companies in the Middle East would not arise if these affiliates were simple producers and sellers of crude oil, for in that case the government partner would be just as directly concerned with market prices as the other partners, and just as directly concerned with the relation of these prices to the rate of supply. But because these affiliates are an integral part of international integrated firms producing crude for their own use, the *direct* link between crude prices and the supply of crude is broken. The fact of integration makes necessary the whole system by which prices are imputed to crude oil in order to attribute taxable profits to the producing companies. Moreover, integrated partners in crude production are directly concerned with market prices of products; non-integrated partners are not, and the existence of 'tax prices' for crude oil helps to insulate the government from the impact of market realities. This relative insulation gives rise to a difference between the government's evaluation of the eventual effect of developments in product markets and that of the integrated companies, and consequently a difference in the policies considered appropriate, even when the government recognizes the nature of the problem in principle.

A 'sleeping partnership' is possible if agreement can be reached over the extent to which the costs of the international companies are thereby increased. Effective partnership is possible with independent companies short of crude, or in producing operations which form only a small part of the crude supply of the integrated owners, or in companies engaged only in the production and market sale of crude oil. But since sleeping partners are unlikely to lie quiet indefinitely, it is highly probable that if the major companies accept government equity participation, the acceptance will mark the advent of a policy of withdrawal from participation in the ownership of the companies providing their major sources of crude oil. Fifteen years or so ago, it would have been easier for the major companies to accept partnership (but more difficult for the governments to insist on it) than it is today, partly because they could more easily have afforded the increased cost, partly because the deterioration in crude and product prices had barely begun, and partly because the ability of the governments to exercise pressure for greater offtake would have been less. The companies have always been acutely aware that 'excessive' supplies of crude would affect product markets, but the extent of their control over crude production and of refining in the major industrial countries, especially in Europe, would have made them less nervous that government participation would have a far-reaching influence on the very structure of the industry. Today, conditions have indeed changed, and the 'changed conditions,' which have steadily eroded vertical integration in the industry and have brought in newcomers and new forms of contract, instead of being conducive to the evolution of partnership in the major concessions, are in fact making it less possible for the great integrated companies to accept it for their major sources of crude oil.

The profitability to the integrated companies of owning the largest sources of their crude oil supplies is becoming increasingly dependent on the subsidy afforded crude production under the United States tax laws. But there is a limit to which this can continue to offset both the rising payments to governments in the face of falling prices, and also the political problems that arise with the governments of importing countries resulting from the fact that the price of crude oil for integrated refineries is a transfer price. If the oil companies should decide to accept government partnership, which is, by definition, a partial repatriation of capital—a partial withdrawal—they are likely to be on their way to a complete withdrawal from the ownership of their major concessionary companies, becoming mere offtakers or perhaps also contractors to governments. The effect on the price of crude oil of such moves is predictable: unless governments can form the cartel that they recognize they need, but can see no way of effectively bringing about, prices are likely to fall. Nevertheless prices have a long way to go before the production and sale of crude oil becomes unprofitable to the producing countries, or even to foreign companies under carefully specified conditions.

## NOTES

1 Theoretically one could argue that even without market transactions the appropriate price to be attributed to crude oil would be the price at which potential buyers could be transformed into actual buyers, thus reflecting the alternative use of oil by its owners. But this price would depend on the assumptions made by the potential buyers about the prospects of breaking into refining and marketing and about the future price of products. If they assume that the monopoly price of products could be maintained by the existing sellers, then they would pay a price for crude that reflected the level of product prices, after making allowance for the cost of establishing themselves in refining and distribution, plus the profit they required to find the business worthwhile. Clearly this alternative use of crude to its integrated owners cannot be treated independently of its consequences for the prices of products and thus can hardly be considered an independent measure of the value of their crude oil to them.

## XIII

## OPEC AND THE CHANGING STRUCTURE OF THE INTERNATIONAL PETROLEUM INDUSTRY*

*Discusses the influence of the Organisation of Petroleum Exporting Countries since its formation in 1960 on the changes in the structure of the international industry.*

In the last ten to fifteen years important changes have taken place in what might be called the 'structure' of the international petroleum industry. Next year OPEC will have been in existence for ten years. In this paper I shall attempt to examine what role, if any, the activities of OPEC have played in bringing about such changes.

The concept of 'structure' is an elusive one; it has been used in many different connections and for many different purposes. What is included in the definition of structure depends upon the purposes for which it is used. For example, simple descriptions of the 'structural' characteristics of an industry can include almost everything relevant to its operation, ranging from the conditions of demand and the organization of the market to the location of its raw materials, the state of technology and the characteristics of the work force. Virtually nothing need be excluded. But if the notion of structure is used for analytical purposes it must be defined with reference to the nature of the analysis and the type of question to be tackled with its help, and it is necessary to be more selective and more precise.

Here I shall use the term 'structure' to refer to the characteristics of the international petroleum industry that are in some sense the underlying determinants of particular 'variables' which are of especial interest. Hence the characteristics I shall call 'structural' for the purposes of this paper obtain their significance only with reference to the variables I want to consider. These are supply and demand, including the geographical distribution of output and sales, prices, and the distribution of income between companies and governments. The importance of changes in these particular variables is obvious.

We shall look back over the recent past and ask what has happened

\* Published in *Middle East Economic Survey*, Vol. 12, No. 19, 7 March 1969. Supplement. Republished in Z. M. Mikdaski, S. Cleland and I. Seymour, eds., *Continuity and Change in the World Oil Industry* (Middle East Research and Publishing Center, Beirut. 1970).

to these quantities; then we shall ask what have been the underlying determinants of the changes that have taken place, and call these determinants the 'structural characteristics' of the industry. Having done this, we can ask how the changes have taken place and what role OPEC has played in the process of change. This is the question we propose to examine.

Broadly speaking, the changes that have taken place and that we shall be concerned to analyse can be summarized as follows: The quantities of oil and products consumed have continued to increase rapidly and the quantities supplied have increased accordingly. But the rate of increase of supply has exceeded the rate of increase of demand, where both are defined in the 'schedule' sense, that is, as curves showing the quantities that sellers would be willing to sell, or buyers willing to buy, at a range of prices. This is evident from the fact that increased quantities of oil have become available at the same or at steadily falling prices. In other words, at existing or lower prices, sellers have been trying to increase their sales, and these attempts have pushed prices down in spite of the increase in the quantities that the market could be expected to take at any price within a relevant range. At the same time there has been a shift in the geographical distribution of production, with new areas becoming increasingly important in the production of crude oil as well as in refining and in consumption. Finally, the governments of the countries producing crude oil have steadily increased their share of the industry's profits, not merely of the profits attributed to crude oil, but of the profits on integrated operations generally.

*Demand*

Several types of changes in the demand for oil and products have occurred in the period we are considering. First and foremost, of course, is the continued rise in total consumption. The demand for oil is highly income elastic, and with increasing income in the rapidly growing industrial countries and in the industrializing countries, the demand for oil products has risen rapidly. Moreover, oil consumption has risen more than that of most fuels, due partly to technological change and partly to a relative fall in the price of oil products for which there is some price elasticity of demand, and to decreasing protection for coal in many countries. Oil has therefore slightly increased its share of the total energy market, while new uses of oil and gas in industry have expanded rapidly, especially in the petrochemical industry.

In addition, there have been shifts in product mix with consequent effects on the implicit value of crudes of different qualities. The demand for fuel oil, especially in Europe has increased in proportion to other products, thus favoring the heavier types of crude; at the other end, demand for naphtha has also increased, especially for the petrochemical industry, which had the opposite effect, favoring the lighter crudes. It is possible that these two tendencies have to a considerable extent cancelled each other out from the point of view of their effect on the values of crudes of different gravities. Rising concern in consuming countries with atmospheric pollution has led to policies which increasingly favor crudes with low sulfur content and makes necessary increasing investment to purify high sulfur crudes.

Finally there have been relative shifts in the geographical distribution of consumption, with the European and Japanese markets growing more rapidly than the United States market, while the absolute importance of markets in the developing countries of Asia, Africa and Latin America has increased. With the imposition of import quotas in the United States, the share of that market available to the exporting countries has been severely restricted.

Briefly, therefore, the structural changes on the demand side have been due to rising incomes unevenly distributed among the consuming countries, technological change affecting consumption, changes in the position of oil in relation to competing products either as fuel or as chemical feedstock, and changes in government policy affecting imports or competing fuels. Although the level and composition of demand can be looked on as part of the structure of an industry taken as a whole, they are also determinants of changes on the supply side, to which we now turn.

*Supply*

The production of crude oil and its geographical distribution have changed rapidly for a number of reasons: expansion of the production of existing dominant companies, entry and expansion of companies new to or previously insignificant in the international industry, changes in concession terms and the introduction of new types of contract governing the terms on which oil is produced, the entry of new oil-producing countries, and technological change, especially new technology making offshore exploration and production increasingly more economic. Since crude oil is of little use without processing, there have been concomitant changes in refining capacity. For a variety of reasons also the distribution of refining capacity since the Second World War has shifted from the centers of production to the centers of consumption. Thus a great rise in refining in Europe and in Japan has taken place leading to a heavy fall in the international trade of products in comparison with that of crude oil. There has also been a significant rise in non-integrated refining capacity in the developing countries, largely as a result of the creation of state-owned refineries. Moreover, the period of time for which refinery contracts for crude are made seems to have shortened, making possible more intense competition for refinery contracts on the part of crude suppliers.

Expansion of existing integrated companies would not in itself be considered a structural change in the industry unless there were a change in their relative size or in the ratio between the quantities of oil sold to third parties and the quantities consumed in their own operations. Although individual companies among them have always produced large quantities of oil surplus to their own requirements, most of the surpluses before the war seemed to have been disposed of to other majors on long-term contracts at special prices or through 'overlifting' from jointly owned producing operations. Relatively little was actually sold in arm's length transactions to third parties. In spite of the fact that each of the majors attempted to adjust its output to expected requirements in its integrated operations, obligations under existing contracts, and expected market sales at existing prices, surpluses were difficult to avoid in view of the increasing pressures to produce greater quantities and the difficulties of accurately predicting demand. During the 1950s there is considerable evidence that the majors were themselves shading delivered prices in order to obtain markets for their production.

Many new companies have also entered the industry. At the end of the Second World War there were very few 'independents' (that is non-majors, including for this purpose CFP) of any consequence in the Middle East or North Africa. Only the names of Getty and Aminoil (which includes eleven American independents) spring immediately to mind. Now one can easily name numerous important companies, notably Continental, Marathon, Occidental, Phillips, Indiana Standard, Arabian Oil, ENI, ERAP, other French and American as well as German companies, and the national oil companies. The foreign companies have been attracted by the prospect of finding their own oil at a cost less than the cost of buying it, for crude prices were long maintained at very attractive levels. Indeed, the search for oil seems uninhibited by falling prices, so important to companies remains the prospect of having their own production in increasing quantities. Most of the newcomers are integrated companies in the sense that they have some downstream outlets, but the extent to which they can make use of their own production within their own companies varies widely. It is indeed probable that the oil companies, taken altogether and including the majors, have been selling an increasing proportion of their oil to third parties, although no direct evidence on this is, so far as I know, available.

Newcomers to the industry have not been confined to companies, for new countries have joined the ranks of major exporters: Libya Nigeria, and Abu Dhabi are established producers with Oman and the UAR lurking in the wings. And new types of operation have, through technological advances, become economic, in particular on structures lying offshore under the continental shelf. Clearly oil reserves are becoming increasingly abundant.[1]

Changes in the relinquishment provisions of concession terms have also probably facilitated an increased supply of crude oil, since they have eased the entry of new companies. The earlier concessions were very large covering the major part, if not all, of the area of a country, and they lasted for very long periods of time. The 1933 Aramco concession agreement in Saudi Arabia made provisions for relinquishment at the discretion of Aramco of areas it did not exploit, but it was not until 1948 that the company agreed on a program of relinquishment. In most of the agreements with newcomers taking up concessions since the late 1950s, however, provision has been made for relinquishment according to a predetermined timetable, and existing concessionaires have voluntarily accepted the principle. By 1963 Aramco, for example, had relinquished about 75% of its original concession area; the Qatar Petroleum Company's concession had been cut by about one-third; the Kuwait Oil Company had given up about half of its original area; and even in Abu Dhabi, the Ruler has already been able to offer relinquished territory for new bidding. The fact that the established oil companies can no longer keep unexploited areas out of the hands of possible rivals for very long periods of time has greatly weakened their control over the exploitation of oil resources. In the longer run, financial terms such as those found in the ERAP agreements, as well as the establishment of joint ventures between national oil companies and foreign companies are also likely to increase the overall pressures of supply.

*Prices*

Although the quantities of oil supplied must equal the quantities demanded after taking account of changes in stocks (which in this industry are not normally a very great proportion of total consumption in any one period in view of the cost of storage), the amount that sellers are willing to supply at existing prices need not equal the amount that people are willing to buy at those prices. All of the considerations we have been discussing which have resulted in increased supply, together with the greater quantities of oil in the hands of producing companies without the downstream facilities to deal with their production, have increased competition in crude oil markets and have necessarily weakened the control over these markets previously exercised by the major companies, even if we accept that this control has never been as strong as is sometimes alleged. At the same time, difficulties in Iraq and Nigeria and the closure of the Suez Canal have undoubtedly eased the supply problem in recent years, just as the Iranian crisis of the early 1950s and the Suez crisis of 1956 eased it in the first half of the last decade.

Increased competition to sell crude oil, together with increased competition in product markets, which would normally be expected to arise partly because of it, has thus kept up the pressure on crude and product prices which first became painfully evident around 1958. And if we look to the future, the new types of contracts, such as those with ERAP, which weaken the hoped-for retarding effect of tax prices on the fall of market prices, the emergence of national oil companies into international market (after all even sales to Eastern bloc countries are likely to increase the pressure of total supply on the world market), and, possibly, the successful demands for government partnership in the companies operating the major concessions in the Middle East, will all work in the same direction.

*Division of Revenues Between Companies and Governments*

The last 'structural change' that we shall consider relates to the increased bargaining power of the governments of the countries producing crude oil. Although the greater influence of governments has been felt in a number of directions, I shall consider only the increased share of revenues that they have been able to exact from the companies. This has been effected in three major ways: the maintenance of posted prices in the form of tax reference prices; the expensing of royalties; and the elimination of marketing allowances. It is not necessary to discuss any of these at length here. Ever since 1960 the oil companies have not been able to reduce posted prices with reference to which government revenues are calculated. The effect of this in the face of falling market prices for both crude and products is to increase the percentage of their total profits that the integrated companies must attribute to their crude oil production for tax purposes, and to attribute taxable profits to the sales of crude oil which are greater than the profits actually made on such transactions. In both cases, a nominal 50% share of these 'profits' gives the governments a much greater actual percentage.

The arrangement under which royalties are to be treated as a cost instead of as part of income tax receipts again automatically increases government revenues. The expensing of royalties based on posted prices is not scheduled to be fully effective until 1972 (or 1975 if the so-called 'gravity allowance' is taken into account), but the arrangements have already led to a substantial increase in government receipts per barrel of oil produced. Allowances off posted prices to cover marketing expenses have been drastically reduced, which again improves the government position.

*Role of OPEC*

We can now ask, what has been the role of OPEC in bringing about these underlying 'structural' changes: the increased number of companies in the industry, their more heterogeneous character, the lessened significance of integration as a factor dominating prices as well as the flows of oil, the emergence of new types of contract and conditions of ownership, the emergence of new sources of production, the changing distribution of demand, and the increased bargaining power and intervention of governments in the producing countries. The answer is relatively simple: except for the last, OPEC has had either no role at all or a negligible one. OPEC, in spite of its studies, its public relations, its conferences and international ubiquity in the oil world, has not, with one exception, been an important factor with respect to the underlying changes that have shaped the industry in the last ten years and are likely to continue to shape it in the future.

OPEC has had no influence on the changes in demand nor on the distribution of refining; it has not been responsible for the rapid expansion of existing companies nor for the emergence of new oil producing countries; it has not brought about the increased supplies of oil moving in non-integrated channels.

It is, however, true that, through the confidence and encouragement that its very existence has imparted to governments desiring to weaken the bargaining position of their great major concessionaires and to obtain the short-term benefits of enhanced competition for new concessions, OPEC has probably been marginally influential in speeding up relinquishment, and in general increasing the rate at which new entrants emerged, and in undermining the older types of contract which were helpful in maintaining a firmer world price structure. But such changes would surely have come about fairly rapidly in any case —and indeed did come about in some degree in some newer areas before OPEC was even established. Governments would not have been willing much longer to accept the role of tax collector only, and were clearly showing signs of this before the creation of OPEC. It is also possible that through the maintenance of tax prices, for which credit must largely go to OPEC, the rate of fall of market prices has been retarded. But I do not think that this is more than a retardation (assisted at present by fortuitous conditions); it is unlikely to be of long-run significance.

The really important contributions of OPEC, for which it is justly famous, are the maintenance of tax revenues even though market prices have fallen, the expensing of royalties, and an acceleration of the generally enhanced position of influence that the governments of the producing countries have obtained. But the other 'structural changes' that I have outlined have occurred and are continuing quite independently of OPEC.

In my view, OPEC has never been as powerful in the international industry generally as were the major companies in their heyday. But this heyday did not last long; it began to fade in the early fifties, although the fading did not become obvious until the end of that decade. The companies failed effectively to contain the rate of supply in line with the rate of demand at existing prices for very long; OPEC has so far failed even more obviously in that task. But since it has at the same time succeeded brilliantly in its other task, the contradiction likely before very long to bring an even more far-reaching change in the structure of the industry is evident: increasing monopoly revenues are not consistent with decreasing monopoly power. Either the governments or the companies will give way, and if OPEC succeeds in maintaining the position of the governments, they may well find themselves with a greater responsibility for their own industry than they either expect or want. Oil has been leaking, not only into the sea, but into the innumerable interstices of the markets, and it will take more than a great expert on flaming or leaking oil wells to bring the total supply under as effective control as it was for the few years after the Second World War.

In saying all this, I would not wish to imply that OPEC's relative lack of influence on the structural changes in the oil industry over the past ten years has been due to any default by that Organization. Within its own sphere and objectives, OPEC has amply demonstrated its value to its member countries—not only through its achievements in maintaining tax prices and revenues, but also through the significant role played by its competent Secretariat in spreading greater awareness of the economics of the oil business.

## NOTES

1 It should be noted that this was written before it became clear that the rate of increase of demand had been grossly underestimated by the companies.

XIV

THE ROLE OF OPEC IN CHANGING CIRCUMSTANCES*

*Discusses the past and future role of the Organisation of Petroleum Exporting Countries as the general environment and operations of the industry change.*

VERY great changes indeed have taken place in the international oil industry in the last ten years; even the briefest comparison of the structure of the industry, of the patterns of oil production and trade, and of the types of companies involved in it in 1958 and today, would make very clear the extent of the changes that have occurred. Last March at the Petroleum Seminar in Beirut I gave a paper on OPEC and the changing structure of the industry in which I attempted to examine what role, if any, OPEC had played in bringing about these changes.[1] I examined the changes in demand and supply, in the significance of integration, in the companies active in the industry, in the terms of both old and new concession agreements, in prices and effective taxation, and in the role of governments. I came to the conclusion that, except for the increased importance of taxation and of the role of government, OPEC had *not* been a significant factor in the processes by which the changes in the industry over the last decade, had come about.

This conclusion does not imply (as I am afraid some thought it did) that I regarded OPEC as an insignificant or unimportant organization in the industry generally, apart from its effectiveness in increasing the revenues of its members. In my opinion, OPEC has been extraordinarily important—much more than one would ever have thought when the organization was established. It has helped to clarify the nature of the issues facing companies and governments, to spread understanding of the industry in the oil-producing countries, and to bring the companies to a fuller appreciation of the limitations on their position vis-à-vis the governments of the producing countries. These things should not be underrated, but nevertheless the success of OPEC in protecting its members from the effects of change and in improving the position of the governments of the producing countries

---

* Published in *International Oil and the Energy Policies of the Producing and Consuming Countries (A Collection of papers presented at OPEC's Seminar held in Vienna in July 1969).* Organisation of Petroleum Exporting Countries. Vienna 1970. Also published in *Middle East Economic Survey*, Vol. XII, No. 49, 3 October 1969. Supplement.

must not be permitted to obscure the fact that OPEC has, essentially, *reacted* to the changes that have taken place; it has had little role in bringing them about. This, of course, is not surprising: the creation of OPEC was a defensive act, and it is still quite young. The question now is, however, what changes can we expect in the future? Will OPEC be important in bringing them about, or will the primary role of the organization be, again, to protect its members so far as possible from changes it cannot influence?

No one associated with the international oil industry looks to the future with much complacency, for its underlying problems are by no means solved. I have no intention of attempting here to 'predict' the future in any precise sense, but I do want to analyse the significance of some of the fundamental and pervasive trends to which all who are involved in the industry are attempting to adjust. Essentially, the international petroleum industry is caught up in the convergence of three distinct movements, one of which is peculiar to the industry, while the other two are more generally characteristic of the developing world as a whole.

The first is, quite simply, the disequilibrium between supply and demand which has been facing the industry for nearly 15 years and which shows no signs of disappearing. I do not need here to explain this in any detail, for it is common knowledge that the rate of increase of supply has, for some time, been tending to exceed the rate of increase of demand, with a resulting pressure on prices. Indeed, OPEC was established precisely because the producing countries wanted to protect their revenues so far as possible from the effects of this disequilibrium.

The second 'trend' is of a different order and appears in many different circumstances and guises. It is political in nature and, again, quite simply, is a growing movement in the developing countries to reduce, so far as possible, the influence of the Western industrialized countries on their economies and societies. This can be looked at as a specific manifestation of nationalism.

The third 'trend,' if it can be called that, is less important, but does seem to influence attitudes, although it might be looked on as not entirely consistent, in some of its manifestations in any event, with the second. I refer to the apparently growing belief that the developed (and therefore presumably richer) countries of the world are duty bound to take measures at their own expense to help the less-developed countries.

The first two of these are particularly powerful forces, and it seems to me unlikely that any institutional arrangements can defeat them. I do not think that the conditions of either demand or supply that exist in the international petroleum industry can be controlled by arrangements emanating either from governments of producing countries or from oil companies. In order, effectively, to contain supply it would be necessary to protect the higher cost producing areas in the exporting countries at the expense of the lower cost producing areas. This kind of thing has been done with reasonable success in the United States, but it has required the direct enforcement of regulations by a powerful central government and the (at least tacit) consent of consumers who have accepted the argument that otherwise the military security of their country would be endangered. In addition, through their foreign operations, which have expanded faster than their domestic ones, the oil companies have been able to escape in some degree the restraints on their expansion inherent in US policies.

But why should consumers elsewhere accept price-maintenance arrangements? I do not think they will, and both reasoning and experience demonstrate the difficulty of enforcing international schemes that attempt to hold up prices in the face of persisting excess capacity unless consuming countries are prepared to cooperate. In a recent speech, the Minister of Petroleum and Minerals of Saudi Arabia, a most effective and knowledgeable man in petroleum affairs, has argued that consumers would, in the longer run, be hurt if oil prices fell because producers would be forced to form a cartel, which would be to the consumers' disadvantage.[2] With the greatest respect, I do not understand this argument. The purpose of a cartel is, of course, to maintain prices, and this can only be done by restricting increases in output in line with increases in demand. Shaikh Yamani seems to be saying essentially that consumers would gain if arrangements could be made to maintain prices now, because, if they are not maintained, then producers would have to adopt other means of holding them up—and this would be through the formation of cartel. Shaikh Yamani evidently believes that consumers are less at a disadvantage if prices are maintained by an alliance between the companies and the governments of the producing countries than they would be if the same result were accomplished by a producers' cartel. If by 'consumers' we mean consuming (or importing) countries, this argument only holds for those importing countries which also gain from the profitable operations of international oil companies owned by their own nationals. Otherwise, the importing countries are surely not so much concerned with *how* prices are maintained as they are with how *high* they are held. Hence, Shaikh Yamani must implicitly assume that a producers' cartel would be more effective in raising prices than the type of alliance between governments and companies that he suggests. This also seems unlikely, for, if it were true, I would certainly expect producers to prefer their own cartel and proceed to establish it, rather than to rely on an association with the companies.

It is possibly true that the producing countries, through OPEC, have in recent years succeeded in retarding the slide of market prices through their tax policy, which undoubtedly makes price competition more expensive for the companies. At the same time, the higher cost for the companies of lifting oil from OPEC countries due to taxation could be expected to stimulate exploration outside OPEC countries by adding a stronger commercial incentive to the existing incentive based on the search for security through diversification of supplies. There is no way of judging the relative importance of the various factors that determine whether companies will put greater exploration efforts in one area as contrasted to another, but there can be little doubt that OPEC countries have to consider this kind of problem in formulating their policies, even though at present they control the overwhelming proportion of the oil supplies on which the importing countries now draw. These factors are, of course, all well known and can as easily be over-emphasized as under-emphasized. But the implications of very large discoveries in Alaska (as yet unproved, to be sure), are surely being very carefully studied by OPEC with precisely this problem in mind.

In any case, it is not the consumers, but oil companies that search for, and develop, new supplies, and it is they who decide the distribution of offtake and development among their existing fields. In this respect Shaikh Yamani's argument is, from the point of view of the OPEC countries, surely based on correct assumptions, when he urged what is, in effect, an alliance between companies and governments through the device of government participation at all stages of the companies' integrated operations. If the incentive to develop OPEC oil is weakened for just those companies who are in the best position to market it without initiating price competition, then the interests of the companies and of consuming countries come closer together, while those of the companies and of the existing major producing countries draw further apart. But it is precisely at this point where the two powerful forces I have referred to converge to create conflict. The obvious commercial and financial interest of the producing countries, given the underlying economic trends in the industry, conflicts with their political interest, given the second underlying trend—the increasing pressures to reduce the influence of the industrialized West in order to promote their economic 'independence' and to extend their control over their own affairs.

The monopolistic position of the major companies in the great oil-producing areas of the world outside the United States and the Communist countries has long been a focal point of agitation. There has been and still is, much confusion of thought in the arguments advanced to attack the position of the companies, for there has been insufficient analysis of the various aspects of the companies' monopoly power, in order to distinguish those monopolistic elements that were advantageous to the producing countries from those that were disadvantageous. The countries gained from the ability of the companies to restrain competition and hold up prices; they lost to the extent that the monopolistic position of the companies in crude oil markets made it more difficult for the governments of the producing countries to change to their advantage the terms on which the companies operated. Similarly, the fact that the concessions of the major companies covered very large areas gave them a strong monopolistic position at the producing end of the industry, and left little scope for the type of competition for concessions of which the governments could take advantage.

It was, I think, insufficiently realized, that the power of the companies in product markets was an *advantage* to the producing countries, but only so long as there was little competition in the supply of crude oil, and that an optimum strategy would be one which would put pressure on the companies to improve concession terms, but which, at the same time, did not significantly increase competition in crude oil markets. To be sure, the success of such a strategy might have required some entry of newcomers, but probably not nearly as much as has, in fact, taken place. In practice, however, concession terms, including signature payments, etc., and oil revenues per barrel, have risen markedly, primarily, I think, because of the improved bargaining position of governments, although it may also be true that falling costs have made it easier for the companies to increase their payments. But clearly, the longer-run effects of the types of changes that have led to this situation have not yet worked themselves out, and it is this with which we are concerned here.

It seems likely that any 'excess monopoly power,' if I may use such a term, of the companies vis-à-vis the governments of the producing countries, as a group, has virtually disappeared. The companies are no longer in a particularly strong bargaining position because of their monopolistic position either in crude-oil producing areas or in product markets. It is true that the position of the major companies in product markets is still such that no *single* producing country could risk severing its ties with its major concessionnaires without serious damage; but the OPEC countries together, or even two or three of the major producers among them, could do so without risking an excessively great diminution of outlets for their oil, since the companies would have little choice than to continue to buy. Of course, prices might fall, but this could just as easily, and perhaps more easily, happen if competition in product markets were intense instead of muted.

At the same time, the 'bargaining power' vis-à-vis consumers, of the sellers of oil and products has diminished because of the increased competition among suppliers, and this must in the end undermine the position of companies and governments alike, unless the competition can be checked. The appeal of 'participation' seems to be gaining ground at the expense of 'nationalization,' and this switch in emphasis may well be due to an increasing realization that the oil producing countries are no longer in a weak position vis-à-vis the companies, while the position of the industry vis-à-vis consumers is weakening.

There is an old slogan 'if you can't beat them, join them,' but perhaps the governments of the producing countries are implicitly accepting a different notion: 'after you beat them, join them—if they are still strong.' This is essentially what Shaikh Yamani has argued, for he has pointed out that it is the strength of the companies in marketing which makes participation attractive, and has flatly stated that '. . . if they lose their power in the market, they will lose their attraction for us; they will mean nothing to us. If prices start on a downward slide and the majors prove to be powerless to arrest it, then we shall have nothing to do with them.'[3]

But suppose the majors are 'beaten,' (or in the process of being 'beaten') if I can continue the analogy, *both* in their power to 'exploit' the producing countries *and* in their power to maintain market prices. Then clearly the only argument for participation on economic grounds is that the alliance between the companies and the producing countries might improve the chances of maintaining prices (that is, let us be frank about it, of exploiting consumers). Again Yamani, with considerable perspicacity recognizes that this is the implicit assumption of his argument: 'Participation . . . is the only way to ensure stability in the market. Again, I repeat, it is not for financial reasons. In fact, downstream integration will require a substantial investment on our part. But this investment should be regarded as an essential insurance premium to safeguard the security of our future income from oil and to ensure that we do not find ourselves at the mercy of adverse market forces which we would be powerless to influence or control.'[4] But nowhere does he suggest in his speech just *how* 'participation'—downstream, upstream, or both—will stem the tendency for supplies to press on demand and will ensure the control of market forces. I see no reason to think it will, and, indeed, Yamani to the contrary, the pressures for participation come *not* because it is felt that market can be stabilized thereby (*au fond* this *is* a 'financial reason'), but because of the desire of the producing countries to play a more active part in the management and operation of their own industry. It is for this reason, as I have argued elsewhere,[5] that the major companies will find grave difficulty in accepting participation in their major sources of oil, and even more in their downstream operations.

Let us, for the moment accept what I might call the 'optimistic hypothesis'—that the companies do decide that their profitability will be enhanced by genuine partnership with producing countries at all stages of the industry. (I call this an 'optimistic hypothesis' because I do not think it likely, but it is always useful to examine both optimistic and pessimistic outcomes, given the uncertainty of the future.) In this case there are obvious difficulties for an internationally integrated company, using crude oil of many different varieties from many parts of the world, endeavoring to minimize its refinery costs and under the necessity of meeting the particular demands of its different markets, to accept widespread participation in its downstream facilities with an individual government which supplies particular types of oil from particular locations. The disruptions of its international 'logistics' could be very great, and the advantages of its international integration seriously eroded, although for large refineries running consistently on a particular crude oil (which is the usual case) participation in the operation in that refinery should not raise so much difficulty from this point of view. For refineries blending crudes, or where changes are likely to be made, partnership with a *consortium* of producers, might be less likely to interfere with commercial logistics because of the need for them to reconcile their differing interests before they could accept any particular policy. In these circumstances, OPEC might have an important role to play in reconciling its members' interests and ensuring that the producing countries take sufficient account of the commercial and logistic aspects of the flows of oil, as contrasted to the particular interest of one country in increasing offtake from its own fields.

But suppose we take the 'pessimistic hypothesis'—that neither the companies nor countries can stem the pressure of supply on demand, and that the companies would prefer to become contractors or offtakers of crude oil, rather than accept effective government participation in their major concessions and, even more, effective participation in their downstream activities. (I am not concerned here with the possibilities of token participation, accepted either as a political gesture or as a delaying tactic.) In these circumstances, what is the role of OPEC?

F. R. Parra, also in a paper given to this year's Beirut seminar, analyses the future role of OPEC with reference to the Declaratory Statement of Petroleum Policy as expressed in Resolution XVI. 90 of the Sixteenth Conference held in June 1968.[6] The Statement deals with control by governments over the exploitation of their petroleum resources, with participation, relinquishment, prices, and a variety of less far-reaching matters. Mr. Parra, after having analysed the several issues, concludes that apart from the protection of posted prices, the role of OPEC is to 'collectivize' the principles of the resolution, that is, to 'detect those areas on which a consensus is possible' and then to bring the 'consensus to the surface.' One cannot fault Mr. Parra's argument: clearly to be effective OPEC must speak with a reasonably unanimous voice and its members must act with reasonable consistency. On a number of issues both of these are still to be attained.

OPEC has what may be called an important 'technical' role in any circumstances—the improvement of legislation, the improvement of procedures for settling disputes, the standardization of accounting, etc. It has, also, a defensive role: the protection of its members so far as possible from the consequences of adverse changes in the industry as a whole. But I am raising a more difficult, and therefore a rather unpopular, question. What is its role, on the pessimistic hypothesis that the market will make impossible the indefinite maintenance of tax prices, and that the companies will not accept the reality of far-reaching participation? These are restraints on the effectiveness of OPEC in some of the most important directions envisaged by the resolution, restraints which OPEC may be powerless to remove. What then?

Clearly, the pessimistic hypothesis requires substantial readjustment in OPEC's view of itself, at least as this view is expressed by Resolution XVI. 90. It requires that OPEC take a long hard look at the effects of its own policies and the policies of individual countries on the rate of exploration and development in non-OPEC countries, on the incentive for substitution of other fuels for oil, on the attitudes of the poorer importing countries, and similar matters. Many producers' organizations for other commodities are busy along these lines, and, of course, the more the countries become independent of the companies, the greater becomes their direct interest in market prices and in demand. The companies have, after all, extensive opportunities for investment in other lines of business; most of the oil-exporting countries will not for a long time to come be able extensively to diversify their exports. Participation would, of course, increase this direct interest of the producing countries, but, at present, except for the proposed possibilities of guaranteeing fiscal stability for operating companies, OPEC has not seriously turned its attention to these aspects of the oil business. As the national oil companies develop, also, this kind of service could become important for them.

I should, perhaps, say a few words before I close about the third 'trend' I referred to earlier, which the OPEC countries share with other developing countries. The increasing insistence that the developed countries are morally obligated to take measures to facilitate the development of the less-developed countries of the world. I am, myself, sympathetic with this view, not only on humanitarian grounds, but, also because of the dangers inherent for the future of all of us in an increasing divergence of living standards among the countries of the world. But with one or two exceptions, the oil producing countries are not in a strong position to demand that they be given special consideration on the ground that money is needed for their own development. They are, for the most part, among the favored countries of the 'third world,' and, for the most part, they are not short of capital except so far as they may misuse what they have. As you all know, some of them have among the highest per capita incomes of the world. To be sure, none of them can as yet be referred to as 'developed' countries in any meaningful sense, but the reasons for this are not ones that more money can quickly overcome. It is true, that oil is often their only resource but this can, and surely will, change if the revenues they already receive are used with reasonable care. And finally, oil is an increasingly important item in the import bill of the poorer developing countries, who cannot be expected to view other than with misgivings the notion that they should pay high prices for it.

Whether or not the OPEC countries succeed in making progress along the road to participation, OPEC will have to begin to take a broader view of the position of the international industry than has so far been implicit in its operations. It was natural—and indeed correct —that, up to the present, OPEC should have viewed its functions in the way it has. But if present economic trends continue, the several possible responses to them—participation, further expansion of national oil companies or negotiations with present concessionnaires— will all require that OPEC expand its role in a constructive search for solutions for the industry generally; it can no longer concentrate on relationships with the companies to the extent that it has done up to now.

## NOTES

1 Essay XIII above.
2 In a paper presented at the Beirut Petroleum Seminar on May 30, 1969, the approved text of which was published in *Middle East Economic Survey*, 15 June 1969, (Vol. XII, No. 35).
3 *Ibid.*, p. 5.
4 *Ibid.*
5 See Essay XII above.
6 See *Middle East Economic Survey*, 16 May 1969, Vol. No. 29.

# SECTION III

## MIDDLE EAST ESSAYS

This section contains five essays on different topics, all relating to the Middle East. It begins with an article reviewing a book on the economic achievements (or lack of them) of Mohammed 'Ali of Egypt in the middle of the 18th century and concludes with a discussion of some aspects of Egyptian economic policy in recent years. In between are two articles on oil, but these are concerned more with the countries—Arabia and Iran—than with the role of the international firms, the subject of the previous section. Essay XVIII on Iran is published here for the first time. Essay XVI attempts to look at the monetary history of a selected group of Middle Eastern countries from 1952–1960.

XV

## ECONOMIC DEVELOPMENT AND THE STATE: AN OBJECT LESSON FROM THE PAST?[*][1]

*Discusses Dr. Rivlin's treatment of Mohammed 'Ali and suggests that a balanced view of his economic policy and achievements is still lacking.*

In the first decade of the nineteenth century a brilliant, vigorous, and ruthless individual gained power in Egypt. During the course of the next forty years he effected a thorough-going reorganization of government administration, made extensive capital investments in agriculture, greatly expanded both production and trade, and attempted to introduce a large measure of industrialization. Muhammad 'Ali Pasha was no twentieth century planner of economic development concerned to advance the welfare of the common man, but rather an ambitious officer in the old Turkish army, and he evinced no hesitation in using the full coercive powers of the state to achieve his ends. His exacting agricultural policies caused much suffering to the Egyptian *fallahin* and were partially self-defeating; his administrative bureaucracy was corrupt and inefficient; his capital investments were often wasteful; his industrialization program failed. Yet he is widely hailed as the 'Founder of Modern Egypt.' He is credited with effecting a transition from an agricultural subsistence economy to a modern market-oriented economy, with widening the economic horizons of the country, and with drawing it within the orbit of European trade and civilization. It is not surprising, therefore, that he has been the object of widely differing interpretations, and that nearly every discussion of him tends to choose sides.

From the point of view of economic policy and practice, we have the choice of appraising Muhammad 'Ali's work in the light of modern notions of economic development and the proper responsibilities of government, or in the light of the ideas current in his own times and of the restraints and possibilities imposed by time and place—in a word, in its historical context. Much of what Muhammad 'Ali wanted to do in the nineteenth century is not far removed from what modern planners in underdeveloped countries want to do today; so, not unnaturally perhaps, the publishers of the latest book dealing with

[*] Published in *Economic Development and Cultural Change*, Vol. XI, No. 2, January 1963. Copyright © 1963 University of Chicago.

him tell us on the jacket that the study 'should serve as an object lesson for present planners in underdeveloped countries.' Fortunately, the author has 'resisted the temptation' to present her work in this manner. Nevertheless, her appraisal of Muhammad 'Ali's economic policies is essentially from a 'modern' point of view, and a certain ahistorical treatment is evident in her judgments. Her concluding paragraph, for example, begins as follows:

> In short, Muhammad 'Ali's contribution to Egypt would undoubtedly have been more impressive had he curbed his private ambitions and had he, instead, devoted his energies to the improvement of the conditions of the people over whom he rules and to the development of sound institutions to replace those which he so ruthlessly destroyed' (p. 254).

In other words, if only he had acted like a modern planning official—in which case, of course, he might well have found himself speedily overthrown in the ruthless society in which he lived!

In spite of its title, the book deals with all aspects of Muhammad 'Ali's economic and administrative policy, including not only agricultural and fiscal policy, but also the organization of trade, industry, and manpower. The author's intention was to confine her discussion to matters of relevance to agriculture; in fact, both her descriptions and her judgments go far beyond this, and she expresses decisive opinions on such diverse questions as educational policy and the significance of Muhammad 'Ali's reign for the dissolution of the Ottoman Empire and the political future of Egypt. Apparently in the hope that each of the broad subjects, if dealt with in separate chapters, could be read independently of the book as a whole, the author abandoned a straightforward chronological treatment in favor of a series of chapters each devoted to a special topic. Thus, after a general introduction describing the events immediately preceding the rise of Muhammad 'Ali, we have three chapters on land tenure, which describe the conditions before the French occupation in 1798 and the changes that took place under Muhammad 'Ali, three on the administration system of the Pasha, the last two of which are entitled 'The Failure of the Egyptian Administrative System,' one on agricultural practices and output, and four dealing with trade, industry, military service, and irrigation, respectively. Some statistics and relevant documents are presented in a series of appendices.[2]

With this method of organization, the reader has to attempt for himself the integration of the discussion in the separate chapters, since data of crucial importance for an adequate understanding of a subject under discussion in one chapter can often only be found by searching in the index or thumbing through a number of other chapters.

ECONOMIC DEVELOPMENT AND THE STATE 249

For example, in Chapter VI, on the 'Failure of the Administrative System,' we are told that the government's 'administrative failure' resulted partly from:

> . . . its practice of controlling and supervising the economic activities of the agricultural population. . . . The introduction of a government monopoly of crops began in 1812 in Upper Egypt where the entire grain crop was seized for the account of the government by the *kashifs* who were instructed to prohibit the direct sale of grain by the cultivators to the merchants and to collect all cereal, including the amount reserved by the peasants for their own consumption . . . this seizure producing a serious shortage of cereal for local consumption (pp. 112–13).

Now, the annual Nile flood decisively affects the economic position of Egypt from year to year; if we do not know whether the flood was favorable or unfavorable we can make serious errors of judgment, but it is not until 50 pages later, in the chapter on 'Trade and Agriculture,' that we learn that the flood of 1811 was poor, and that in 1812 there was 'barely enough wheat in Egypt to feed the native population' (p. 173). These facts set the government's action in a very different context; in particular, we can no longer accept the statement that it 'produced' the shortage of cereals. There are numerous examples of this type of failure to present all the essential information at the same time.

Dr. Rivlin has taken a very narrow view of what constitutes an agricultural policy, and this has led to a serious imbalance, and even contradiction, in her treatment. Because the importance of raising revenue dominated Muhammad 'Ali's policy, she concludes that if he possessed an agricultural policy at all, 'it was merely a one aspect of his fiscal policy and inseparable from it' (p. 252); she therefore puts the emphasis of her discussion on the techniques for collecting revenue and on the effects of heavy taxation and discriminatory price policies on the agricultural population. In consequence, the administrative procedures and economic consequences of *raising* revenue are described in detail, whereas nothing like the same attention is given to the *expenditure* of the revenues in agriculture, and no attempt is made to balance the one against the other. Only 33 pages are devoted to the positive agricultural achievements of Muhammad 'Ali, while the chapter on irrigation emphasizes his failures and inefficiency: 'Care must be exercised in assessing Muhammad 'Ali's contribution to Egyptian irrigation. . . . His judgment was . . . often faulty, his actions hasty, the execution of his commands unsatisfactory. In fact, it even appeared that work had either been confined to people who were blind or had no knowledge of the science of hydraulics. . . .' (p. 249).

As a result, the two aspects of his agricultural policy are never sensibly integrated, and no consistent picture emerges; the author presents apparently contradictory impressions and makes no attempt to reconcile them. Thus, in the discussion of Muhammad 'Ali's military policy, we are told that in the 1820's and 1830's 'It appeared that anyone who could walk was recruited into the forces' (p. 209), and

> ... when conscription was combined with the other burdens imposed upon the peasants, its effects were disastrous. The land was drained of cultivators by conscription, the *corvée*, and disease; the remaining agricultural population, which had become less competent as a result of voluntary mutilation, could not meet Muhammad 'Ali's unreasonable demands for more crops and higher taxes. The farmers either abandoned the land to escape the government's heavy exactions or were forced into bankruptcy (p. 211).

Indeed, all the detailed evidence presented supports this grim picture of crippled, diseased, and virtually worthless workers, deserted villages, and hopelessly recalcitrant peasants, the whole becoming progressively more miserable and unproductive throughout the entire reign of Muhammad 'Ali. Hence, the reader will be amazed to find in the 33 pages directly dealing with agricultural output in reasonably quantitative, instead of chiefly impressionistic terms, that agricultural production increased very substantially during the reign of Muhammad 'Ali, foodstuffs alone nearly tripling between 1821 and 1844, while at the same time cotton was introduced and expanded rapidly (see tables on pp. 158 and 337). If we search among the appendices, we will even find that the rate of increase of population also rose, even if the accepted figures are exaggerated, as Dr. Rivlin holds. I find it quite impossible to reconcile the two pictures. Dr. Rivlin denies that the area under cultivation increased (p. 270), although she stresses the importance of the Pasha's investments of capital, especially in the construction of *sayfi* canals and other forms of irrigation, permitting multiple cropping of the same land, and of the introduction of cotton and cash crops. Nevertheless, all of these could hardly have produced the results recorded on a land 'drained of cultivators;' we can only accept the statistics of production if we assume that her picture of 'disaster' in the countryside is grossly overdrawn. The Egyptian *fallah* was in a miserable state before the advent of Muhammad 'Ali; the conflicting evidence as to whether or not he was substantially worse off under Muhammad 'Ali does not yet permit a final, conclusive answer.

This failure to reconcile apparently contradictory conclusions runs through the entire book. But apart from the absence of any attempt to ensure consistency between the judgments in different chapters

ECONOMIC DEVELOPMENT AND THE STATE   251

and to weigh and evaluate contradictory evidence, the real problem arises from a failure to make the analytical distinctions relevant for an appraisal of Muhammad 'Ali's policies taken all together. Nowhere are we presented with a clear statement of what he was trying to do, a statement in which the relevant aspects of his problem are separated in such a way that we can form a balanced judgment of the forces with which he was dealing and of the degree of success with which he dealt with each.

Muhammad 'Ali identified himself with the Egyptian state; he was ambitious to make it economically and politically powerful. He needed revenue, partly to maintain his own position, but largely for the economic and military expenditures he thought necessary to achieve his ambitions for Egypt. Agriculture was the chief source of revenue, and his problem therefore was to increase *both* the productivity of agriculture and the 'surplus' that could be extracted from it to meet expenditures for administration, capital investment, and the military. There may be serious conflict between these two aims, for excessive exactions from the cultivators will reduce productivity, but Dr. Rivlin puts so much emphasis on the latter that, as noted above, we cannot understand how output could have increased as it did. Some attempt to show, with similar detailed evidence, how and in what way Muhammad 'Ali's productive investments, together with his more sensible price policies, affected the *fallahin*, and to present both in an integrated discussion, would have caused her at least to recognize the apparent inconsistency of saying in one chapter that in addition to ruining the peasants by 'excessive taxation,' Muhammad 'Ali 'excluded them *entirely* from the benefits which they might have derived from the commercial opportunities that had developed during his reign. . .' (p. 118, italics added), while pointing out in another chapter that he did offer inducements to the peasants to obtain the output he wanted. For example, in the early years cotton prices fixed by the Pasha 'were high to serve as an inducement to the *fallahin* to undertake cotton cultivation;' moreover, 'The Pasha made generous advances to the *fallahin* of seed, oxen, and water-raising devices. . .' (p. 140); and after 1834 a 'favorable export market made it good policy to stimulate greater productivity by increasing the purchase price' of rice (p. 145).

The Pasha's attempts to raise revenue took two basic forms: taxation and trading profits. Both were associated with serious administrative problems, injustice, and oppression; but the latter form, which gave rise to his notorious monopoly policy, brought him up against the powerful vested interests of the foreign merchants. Dr. Rivlin discusses the oppression of the *fallahin* and the administrative problems in great and interesting detail, making much new material available in the process. The difficulty is that she cannot understand how such an inefficient and corrupt system as she describes could work at all—an incomprehension shared by many other European observers; hence, as her chapter headings already tell us, she concludes that it was a 'failure.' The trouble, as before, is that the evidence is all negative and the discussion unbalanced; comprehensible or not, the system did in fact 'work,' and surely there must be some evidence to explain why and how it worked. We can agree that an inefficient, corrupt, and disgruntled bureaucracy, and a hostile agricultural population, created great difficulties, and that Muhammad 'Ali's reforms never achieved as much as he had hoped. But how are we to reconcile the detailed story of unrelieved inefficiency and failure and the conclusion that 'disorder reigned in every part of the government with the situation becoming progressively worse' (p. 105) in the chapters on administration, with the statement at the end of the book that Muhammad 'Ali 'organized a civil administration capable of executing his orders effectively'? (p. 252). Once more the evidence presented contradicts the final appraisal.

In dealing with the monopoly policy of the Pasha, it is essential to distinguish the use of monopoly as a device for squeezing revenue out of agriculture—a form of taxation of ancient and modern popularity with governments—and its use as a means of encroaching on the profits of merchants. Neither administrative inefficiency nor oppression of the fallahin was the reason for the tremendous opposition of the Europeans—it was the effect on their trading opportunities. The chapter on 'Trade and Agriculture' is largely devoted to Muhammad 'Ali's struggle with the foreign merchants, but a balanced appraisal of this struggle cannot be made in the absence of a clear picture of the position of the merchants in the economic life of Egypt in the period. It is unfortunate that Dr. Rivlin could give us no considered discussion of this problem, but only fragmentary, and to some extent conflicting, information. We are told that in 1819 'the native merchants were forced out of business completely' (p. 176), and later, that Muhammad 'Ali 'destroyed the native merchant class' (p. 254); yet, from other evidence presented, it is clear that these merchants were active all during the period. For example, in the chapter on 'Military Service and Agriculture,' Dr. Rivlin describes an incident in 1831 in which 'merchants who had come from all parts of Egypt to attend the *annual* fairs held in Cairo and Alexandria' were conscripted into the army (p. 203, italics added).

There is considerable confusion over the relative importance for the formation of agricultural prices of crop yields, government policy, and merchant speculation. It is evident that speculation was important, at least at times, in raising prices when crops were poor (cf. p. 153).

At one point, when in 1833 Muhammad 'Ali finally agreed to the merchants' demand that he should sell his crops at public auction, prices were driven up by the speculators 'who were usually the representatives of second-rate business firms . . . willing to earn smaller profits. . . . The larger firms, unwilling to operate on narrow margins or to take unnecessary risks, soon withdrew from the market' (p. 181). We are told that foreign merchants 'without capital or credit had swarmed to Egypt to make their fortune' (p. 176), but Muhammad 'Ali was handicapped in curbing their activities by the Capitulations. As early as 1819 foreign merchants were 'outraged at not obtaining their due share of the trade' (p. 176); it would have certainly been interesting to have been given some idea of the profit margin they insisted on.

The discussion of Muhammad 'Ali's industrial policy need not detain us. In nine pages Dr. Rivlin describes his industrial projects, advances the usual reasons for his failure, and concludes that Egypt would have been better off if the resources used had been left in agriculture. On the controversial question of whether the Convention of 1838 forcing free trade on Muhammad 'Ali was an important element in the failure of his industries, she has only this to say: 'It has been said that the Anglo-Turkish Convention of 1838 was the most significant cause for the failure of Egypt's infant industries, but the evidence does not justify such a conclusion' (p. 199). No effort at all is made to examine the tariff provisions of the Convention with respect to changes in imports of goods that competed with Egypt's industries, although we do find that by 1844 the 'import of British cotton goods was particularly heavy. . .' (p. 197). Let it be said at once that a judgment on the Pasha's industrial policy is not necessary to a discussion of his agricultural policy, although an analysis of the effect on agriculture of the diversion of productive activities, and of the industrial program generally, would have been. Unfortunately, even the latter is dealt with only superficially.

\* \* \*

This review has been both long and critical. Its length is justified by the fact that Dr. Rivlin's book is important; her study presents much new information, and no scholar of the period can ignore it. The criticism is necessary because in an attempt to 'rectify the distortions foisted upon a credulous public by official historiographers and other persons infatuated with the Muhammad 'Ali legend' (p. vii), Dr. Rivlin has swung much too far in the opposite direction. It is true that there has been a tendency on the part of historians to dwell at length on the achievements of Muhammad 'Ali and, although recognizing the oppressive side of his regime, to devote relatively little space to it. But this imbalance is not corrected by dwelling at length on the oppression and inefficiency of the regime and devoting relatively little space to its achievements. Moreover, since the achievements can hardly be denied in the final appraisal, an unclear and disjointed picture emerges. Nevertheless, Dr. Rivlin, in emphasizing the details of the dark side of Muhammad 'Ali's policies, has perhaps made it more likely that future historians will avoid the one-sided appraisal common in the past.

Finally, can it be said that this study yields an 'object lesson for present planners'? There are, of course, certain obvious and non-controversial 'lessons,' for example, the importance of information. As Dr. Rivlin makes clear, many of Muhammad 'Ali's mistakes were due to a simple lack of information about such crucial matters as the size of the population, the area of cultivation, and the output of crops; the same tendency to plan on excessively optimistic assumptions and to refuse to accept unpalatable facts is common today. Nevertheless, although we learn much from history, we are, for the most part, on treacherous ground if we try to draw general conclusions from a story confined to a specific time and place. From the misery of the agricultural population in the period, some might conclude that oppressive and monopolistic policies are self-defeating, whereas others would be impressed by the demonstrations that ruthless and totalitarian techniques can nevertheless powerfully advance economic development in the face of popular resistance. From the failure of the industrial projects, some might conclude that countries whose advantage lies in agriculture should not attempt industrialization, whereas others would see in it the importance of attaining political autonomy in order to prevent foreign interference.[3] From the fact that Muhammad 'Ali's trade policy 'increased the country's dependence upon European markets and made it susceptible to fluctuations in the European economy' (p. 253), some might decide that trade should be restricted because it leads to interdependence, whereas others would point to the economic advantages of a widening market. And so on. Indeed, the danger is plainly very great that one's own preconceptions, one's own definitions of 'success' or 'failure', will determine the 'lessons' one decides one has learned from a study of Muhammad 'Ali and his period.

## NOTES

1 Helen Anne B. Rivlin, *The Agricultural Policy of Muhammad 'Ali in Egypt*. Cambridge: Harvard University Press, 1961. Pp. xiv + 393.
2 There will not be space here to discuss these appendices, in some of which quite unnecessarily cumbersome arithmetical manipulations are set out,

the purpose of which is not at all clear. I shall cite only one example: on pp. 256–57 of Appendix I is a table of the classification of land in Egypt. In 1844 village land was divided into two categories, Ma'mur (taxed) and Ab'adiyah (uncultivated), i.e., ma'mur land plus ab'adiyah land equalled village land. In a footnote to this appendix (p. 359), Dr. Rivlin finds it an 'interesting coincidence' that if she takes the figure representing land taxed during the French occupation and subtracts it from total village land, then adds this difference to the difference obtained when she subtracts ma'mur land from the same figure, she gets an amount equal to ab'adiyah land. In other words, if $m + a = v$, then it is an 'interesting coincidence' that $(v - x) + (x - m) = a$! She would have got exactly the same result if she had taken the figure representing the population of whales in the North Sea; hence, it is quite illegitimate to find any significance in the fact that the figure chosen for the exercise represented the land taxed during the French occupation.

3 See, for example, Charles Issawi, 'Egypt since 1800: A Study in Lop-Sided Development,' *Journal of Economic History*, XXI, 1 (March 1961), 7.

# XVI

## MONEY, PRICES, AND ECONOMIC EXPANSION IN THE MIDDLE EAST, 1952–1960*

*Analyses monetary developments in the period 1952 to 1960 in Egypt, Syria, Lebanon, Iraq, Iran and Turkey, examining changes in the supply of money in relation to prices, national income and imports.*

PERHAPS the chief difficulty facing an economist attempting an empirical analysis of the process of change in underdeveloped countries is the scarcity of reliable statistics. Nevertheless, both the economists in the countries themselves, and those outside who want to increase their knowledge and understanding and improve their judgments, must make do with what is available. Monetary statistics, deficient though they may be, are among the more reliable of the available statistics, and monetary analysis is of central importance wherever development programmes financed by monetary expansion are actively pursued. Rightly or wrongly, investment in industry tends to take pride of place in these programmes. Investment requires finance, and governments are often in the dubiously happy position of being able to persuade or require the 'monetary authorities' to provide the money they want for this purpose as well as for their other needs. At the same time, the claims of agriculture, of private investment and especially of private consumption have a way of making themselves felt, and these too require money, which has been forthcoming in not inconsiderable amounts. Economic expansion has in fact been proceeding vigorously in the Middle East—but not without its financial troubles. The amount and character of the real expansion can best be seen in national income and product accounts where they exist; the financial expansion, and the characteristics thereof, are displayed in the monetary accounts.

These two sets of accounts are designed to set forth different aspects of the working of an economy. The national income and product accounts are concerned with the amount, origin, and alloca-

* Published in *Rivista Internazionale di Scienze Economiche e Commerciali*. Anno IX (1962), No. 5. I want to thank for critical advice and comment the students in my Seminar on Current Economic Problems of the Middle East in the London School of Economics, Professor Fritz Machlup of Princeton University, and my husband Professor E. F. Penrose. Mr. A. de Frias, research student in the London School of Economics, made the calculations.

tion of the final product and with the contribution of foreign transactions to national income. There is no place in them for financial transactions, nor any way of showing the borrowing and lending accompanying the production and consumption of goods and services. The accounts of the monetary system, on the other hand, are concerned with the sources and recipients of finance, with the volume of money and other highly liquid liabilities of the monetary system, and with the sectors of the economy that were chiefly responsible for the creation of these liabilities. In some ways, however, the monetary accounts can be used to help in the evaluation of the national income accounts of countries whose national income statistics are unusually unreliable, since a separate analysis of the different uses to which new money was (or must have been) put, and of price movements, can throw some light on the likely movements of real income. For example, if the money supply had been increasing rapidly in a given country and prices had been rising in the same proportion while there was no evidence that substantial changes in velocity or increases in the proportion of imports had taken place, it will become difficult to accept, without very close examination, national income estimates which showed a large rise in national income in the period.

In nearly all underdeveloped countries the money supply has been expanded at a rapid rate, and consequently the relationships between increases in the money supply and other functionally related quantities are worth investigating. In this paper I propose to make such an investigation for six countries of the Middle East: Iran, Iraq, Lebanon, Turkey, Egypt and Syria. These countries form an interesting group for such an analysis, both because of their similarities and because of their differences. In all of them the bulk of the population is engaged in agriculture; they are all to be classed as 'underdeveloped,' and their development is hindered by similar cultural, institutional, and economic considerations. Yet they are very different. In addition to agriculture, Iran and Iraq produce oil; they are both among the world's great oil producers and receive oil revenues, which largely finance their development programmes. Lebanon is a commercial and financial center of great importance in the Middle East; transit and entrepôt trade provide by far the largest part of its foreign earnings; it is the only country of the six whose government has no large-scale development programme. Turkey, on the other hand, has for many years been engaged in extensive efforts to develop domestic industry with inadequate foreign earnings. Egypt and Syria have both begun an accelerated development programme within the period we shall consider. All of the countries have undergone severe political disturbances since 1952, when we begin our analysis.[1]

For these six countries we shall first examine the increases in the quantity of money between 1952 and 1960, and the sectors of the economy responsible therefor; we shall then look at the relationship between money and prices; and finally we shall investigate some of the variables other than money that are relevant for an analysis of the impact of an increased money supply on prices.

## THE QUANTITY OF MONEY

There has been much debate over the question what liquid assets in the hands of the public, besides currency, should be considered as money; there is no unambiguous answer, and the definition of money must depend on the problem in hand. Since demand deposits in commercial banks are not only immediately convertible into currency without loss, but are also often directly used as a means of payment, they are usually classed as part of the money supply, whereas time deposits, savings deposits, and other slightly less liquid assets are classed as 'quasi-' or 'near-' money. Furthermore there is a presumption that if demand deposits earn no interest they will be held for the same reasons people hold currency, that is, to provide immediate liquidity largely for transactions purposes, whereas assets held in the form of near-money are looked upon more as liquid savings than as a means of payment. In other words, one of the criteria used to distinguish money from other liquid assets involves an assumption about the attitudes and behaviour of people.

This assumption may not be so applicable in the less developed countries as it is in countries with a more highly organized banking system, and in many ways demand deposits in underdeveloped countries are probably closer to quasi-money than they are to currency. I shall take account of this where it seems relevant to do so, but in general I shall follow the classifications used in the *International Financial Statistics* published by the International Monetary Fund, where money is defined to include currency and demand deposits in the hands of the public.[2] Government deposits and currency in the banking system are excluded from the money supply.

The changes in the supply of money (M) and of quasi-money (QM) in each of the six countries over the period 1952–60 are presented in Table I. It will be seen that there was everywhere a very substantial increase in liquidity (M+QM); in all countries except Egypt the money supply alone more than doubled, the increase exceeding 200% in Turkey and Iran. In addition, the proportion of currency in the total money supply fell in all countries except Egypt and Turkey, while the ratio of quasi-money to money rose substantially in all countries. In other words, demand deposits grew faster than currency, and quasi-money grew faster than money.

Before we examine the significance of these changes, let us analyse the changes in the assets and liabilities of the monetary institutions of these countries in order to see what sectors of the economy were chiefly responsible for the changes in the volume of liquidity.

DETERMINANTS OF THE CHANGE IN THE MONEY SUPPLY

In Table II is presented a summary of the changes in the assets and liabilities of the monetary system (which together determined the net change in the supply of money) for each of the six countries between 1952 and 1960, as published by the International Monetary Fund in *International Financial Statistics*. Since the nature of the monetary surveys published by the IMF may be unfamiliar to a number of readers, and therefore the meaning of the table not clear, I will digress briefly to explain.

The monetary surveys are essentially summary balance sheets of the monetary institutions of each country, usually the central bank, the commercial banks, Treasury currency, and postal savings. The chief liabilities of a monetary system are money (currency and demand deposits), quasi-money (largely savings and time deposits of various kinds), government deposits, and the capital and surplus of banks. All of these are claims of the public or the government on the monetary institutions. The liabilities are created by the acquisition of assets by the monetary institutions for which they pay out currency or create bank deposits. They may acquire foreign exchange or other claims on foreigners, and they may acquire domestic government and private securities of all kinds. Hence, at their creation all liabilities are matched somewhere by assets of equivalent value.

For the countries with which we are here concerned three main categories of assets are distinguished by the IFS: claims on foreigners, claims on governments (and official entities where appropriate), and claims on the private sector of the economy. Liabilities are classified as money, quasi-money, government deposits, capital accounts of the banking system itself, and a number of miscellaneous items. For several reasons it is not possible always to classify appropriately all assets and liabilities, and therefore some of them are simply listed as 'unclassified'; often this item is distressingly large from our point of view. An increase in any asset automatically brings about an increase in one or more of the liability items (either money or a non-money liability), or a decrease of another type of asset. From Table II the effect on the money supply of the net change in each category of asset and liability can be seen for each country. As with all our other statistics there are grave deficiencies in many of the figures, especially for Iran and Turkey and, to a lesser extent, for Lebanon. Table II should be examined in conjunction with Chart I, which shows the changes in the assets of the monetary system of each country over the entire period.

Apart from the substantial increase in the money supply, the most striking characteristic of the monetary situation that is brought out in the table and chart is the importance of the private sector in spite of the official emphasis on government investment and long-range development plans in most of the countries. In all countries except Iraq claims on the private sector accounted for more than 40% of the total assets of the monetary system in 1960, and in all countries there was an almost continuous expansion over the whole period. In Turkey, Syria, and Lebanon, claims on the private sector dwarfed all other influences on the money supply, and even in Egypt and Iran, where government investment programmes were extensive, expansion of claims on the private sector usually kept pace with, and sometimes exceeded, those on the government.

There was a rapid expansion of government borrowing in all countries except Iraq and Lebanon, where claims on the government had virtually no influence on the money supply. Even in Iran with its extensive oil revenues, government borrowing rose steadily over the whole period, but the government's attempt to brake expansion a little after 1957 is shown up by some falling off in the rate of increase; the acceleration of government programmes in Egypt after 1954 and in Syria after 1955 shows up clearly.

The role of foreign assets varied substantially from country to country. In Iraq, the effect on the money supply of foreign assets (which were acquired primarily on government account as a result of the receipt of oil revenues) was to a great extent offset by the increase in government deposits up to 1956.[3] This, of course, reflects the fact that the government was not spending its revenues as fast as they accrued. In Iran the data on foreign assets are difficult to interpret because a revaluation took place in 1957; much of the effect of this is hidden in the 'unclassified liabilities' item,[4] and comparability in the series is completely broken at this point. Furthermore, the period 1951–54 was the period of crisis following the nationalization of the oil industry in 1951; after this Iran received substantial quantities of United States emergency aid. Nevertheless, over the period foreign assets had an expansionary effect on the money supply in Iran as well as in Iraq. In Lebanon, too, there was a steady increase in the monetary system's official holdings of foreign assets, which, together with the expansion of the private sector, accounts for about the entire increase in the money supply. In Egypt and in Syria, on the other hand, the fall in holdings of foreign assets was an important offset to the other expansionary factors, being largely responsible for

## SIX COUNTRIES OF THE MIDDLE EAST, 1952-60

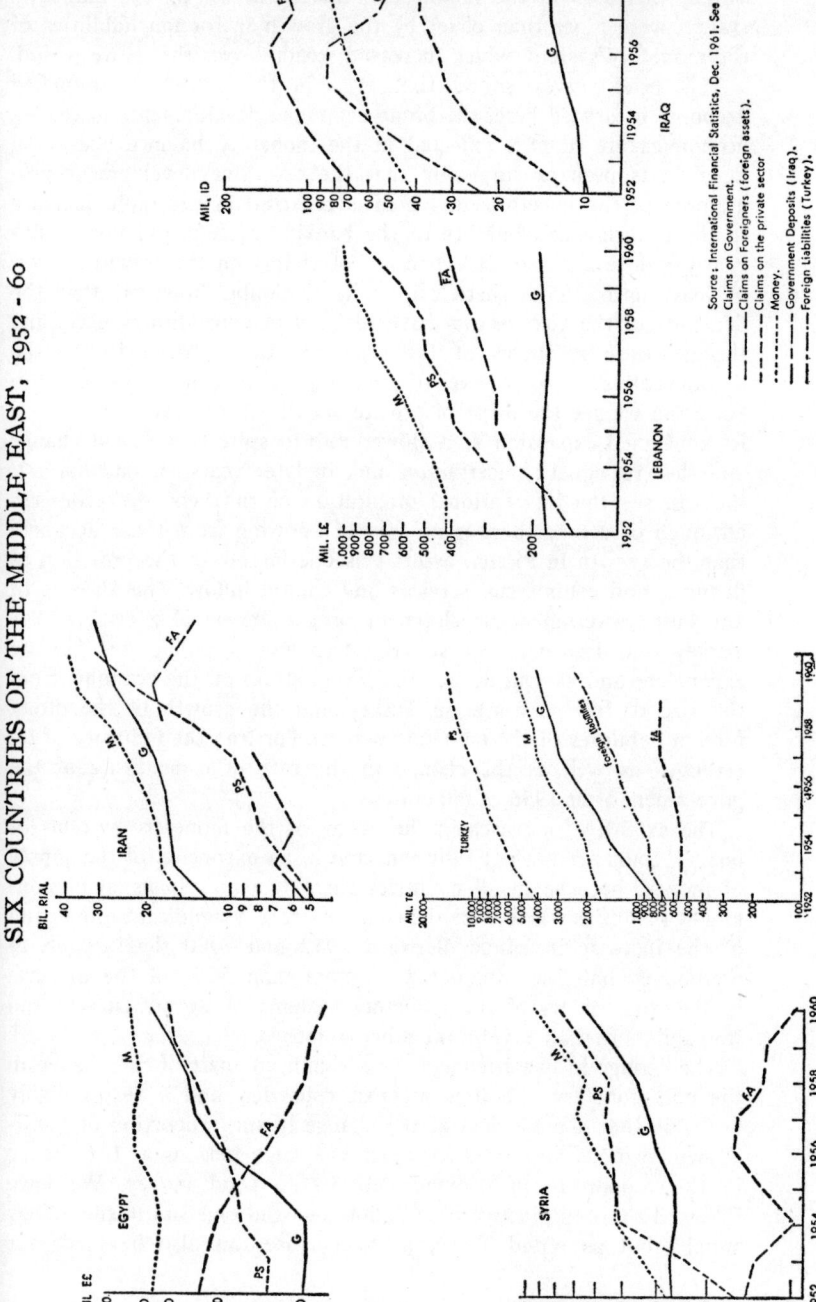

the relatively small increase in the money supply in the former. In Turkey increases in the holdings of foreign assets by the monetary system were more than offset by the growth in foreign liabilities of the monetary system, which increased steadily over the entire period.

This brief survey shows that even in the extremely simplified accounts presented here, the broad economic developments in the six economies are in part reflected in the monetary balance sheets. In Egypt it is perhaps surprising that the extensive development programme of the government, which is reflected in the rapid increase in the government's liability to the banking system, did not significantly reduce the rate of expansion of claims on the private sector, at least until 1959. There can be little doubt, however, that the demands of the two sectors have resulted in expenditures exceeding the domestic resources of the economy, thus giving rise to the substantial use of its reserves of foreign assets as displayed in Chart I. For Syria we see the burst of private activity in the early 1950's and its continued expansion at a slower rate in spite of political change and the consequent uncertainty, and, in later years, of bad harvest. We can see the international orientation of the Lebanese economy, although of course there is no way of knowing from these accounts that the growth in foreign assets was due largely to the provision of financial and commercial services and capital inflow. The absence of any large government development programme is also evident. For Turkey and Iran we can see the tremendous public and private expansion; and we can deduce the severe strain on the economy from the size of foreign assets in Turkey and the growth of the direct foreign liabilities of the monetary system. For Iraq the influence of oil revenues as well as the change in the rate of expenditure of the government after 1956, stand out clearly.

The extensive increases in the assets of the monetary systems of our six countries are not fully reflected in an expansion of the supply of money, because in all countries there was an expansion of non-money liabilities which were the counterpart of a significant proportion of the increase in assets. Between 1952 and 1960 the increase in non-money liabilities amounted to more than 50% of the increase in the total assets of the monetary systems of Egypt, Turkey, and Iraq and more than 25% in the other countries.

The change in quasi-money was a significant part of the change in the non-monetary liabilities in most countries, and it seems highly probable that a great deal of the change in the proportion of quasi-money to money occurred for much the same reasons as the change in the proportion of demand deposits to total money. We have followed the IMF practice of including demand deposits in the money supply, but as noted above, a strong case can also be made for interpreting a significant part of the rise in demand deposits in a short period as an increase in personal savings or hoards, rather than as an increase in the circulating monetary media in countries where the use of the check as a means of payment is not widespread. If we do this, we find that 'money' defined as currency alone increased somewhat less than currency plus deposits, especially in Iran and Lebanon, (See Table I), but in general the difference is not very great in view of the very substantial increases in both.

## MONEY AND PRICES

A reasonably accurate measure of changes in prices is more difficult to obtain than a similar measure of changes in the money supply, deficient as the latter may be. Ideally, we should like an index number which would represent unsubsidized, untaxed market prices of final goods for sale against money. In practice we have to choose between a cost of living index and a wholesale price index, neither of which has a high degree of reliability and neither of which covers anywhere near the range of commodities relevant for our problem. For the most part the wholesale price indexes include only raw materials, foodstuffs, and simply processed commodities. They thus omit the manufactured consumer goods, which form such an important part of the imports of these countries and an increasing proportion of the domestic production in many of them, as well as capital goods. The cost of living indexes are based on the consumption of a selected type of urban family, but the weights used rarely reflect at all reasonably the present pattern of consumption. It will be seen that in all countries except Egypt and Turkey the cost of living index has risen more than the wholesale price index. Since there are good reasons for suspecting that both price indexes understate the rise in prices that has actually occurred, I shall use in each case the index that has risen most. This is an extremely rough-and-ready procedure, but under the circumstances it seems unavoidable.

The movements of the indexes of money, wholesale prices, and the cost of living in each country are presented in Chart II.[5] The most important phenomenon brought out in this chart is the remarkably small increases in prices in Iraq, Syria, Lebanon, and even Iran, in relation to the increases in the quantity of money. In Turkey and Egypt, on the other hand, the postulates of the 'crude' quantity theory of money—that prices increase in the same proportion as the quantity of money—would seem almost to hold if we compare money with the wholesale price indexes. In Iran there was a substantial increase in prices, especially in the cost of living, but neither the increases in money nor the inflation of prices were nearly as severe as they were

s

TABLE I

MONEY SUPPLY—SIX COUNTRIES OF THE MIDDLE EAST 1952–60

| Year | EGYPT (mil. E£) | | | | SYRIA (mil. S£) | | | | LEBANON (mil. L£) | | | | |
|---|---|---|---|---|---|---|---|---|---|---|---|---|---|
| | Money (M) | Currency/ (M) | Quasi-money (QM) | QM/M | Money (M) | Currency/ (M) | Quasi-money (QM)[1] | QM/M | Money (M)[2] | Currency/ (M) | Quasi-money (QM)[1] | | QM/M |
| | | | | | | | | | | | Total | Foreign currency deposits | |
| 1952 | 355·4 | ·52 | 62 | ·18 | 298 | ·79 | 12 | ·04 | 424 | ·45 | — | — | — |
| 1953 | 343·9 | ·49 | 70 | ·20 | 347 | ·78 | 16 | ·05 | 414 | ·47 | — | — | — |
| 1954 | 334·8 | ·50 | 75 | ·22 | 435 | ·77 | 24 | ·06 | 478 | ·49 | 81 | 54 | ·16 |
| 1955 | 347·7 | ·48 | 79 | ·23 | 429 | ·74 | 29 | ·07 | 555 | ·45 | 111 | 72 | ·20 |
| 1956 | 401·2 | ·52 | 78 | ·19 | 518 | ·79 | 32 | ·06 | 627 | ·51 | 124 | 82 | ·20 |
| 1957 | 415·0 | ·51 | 86 | ·21 | 593 | ·78 | 36 | ·06 | 790 | ·43 | 165 | 115 | ·21 |
| 1958 | 391·4 | ·53 | 105 | ·27 | 559 | ·75 | 43 | ·08 | 827 | ·46 | 190 | 136 | ·23 |
| 1959 | 413·8 | ·48 | 123 | ·30 | 671 | ·73 | 51 | ·08 | 997 | ·38 | 243 | 165 | ·24 |
| 1960 | 424·2 | ·52 | 120 | ·28 | 754 | ·68 | 74 | ·10 | 1,115 | ·36 | 482 | 369 | ·43 |
| % incr. 1952–60 | M=19% | C=19% | QM=93% | | M=153% | C=119% | QM=517% | | M=163% | C=108% | QM=495% (1954–60) | | |

| Year | IRAQ (mil. ID) | | | IRAN (bil. rial.) | | | TURKEY (mil. T£) | | |
|---|---|---|---|---|---|---|---|---|---|
| | Money (M) | Currency/ (M) | Quasi-money (QM) | QM/M | Money (M) | Currency/ (M) | Quasi-money (QM) | QM/M | Money (M) | Currency/ (M) | Quasi-money (QM) | QM/M |
| 1952 | 43 | ·69 | 10 | ·23 | 14·16 | ·54 | 2·45 | ·17 | 1,604 | ·71 | 988 | ·62 |
| 1953 | 52 | ·66 | 11 | ·20 | 18·17 | ·52 | 3·26 | ·18 | 1,895 | ·70 | 1,417 | ·75 |
| 1954 | 62 | ·66 | 12 | ·19 | 18·51 | ·52 | 3·68 | ·20 | 2,074 | ·66 | 1,616 | ·78 |
| 1955 | 66 | ·65 | 14 | ·21 | 20·22 | ·48 | 4·12 | ·20 | 2,646 | ·68 | 1,912 | ·72 |
| 1956 | 76 | ·64 | 17 | ·23 | 23·59 | ·45 | 5·31 | ·23 | 3,317 | ·70 | 2,504 | ·75 |
| 1957 | 82 | ·61 | 22 | ·27 | 27·69 | ·43 | 5·79 | ·21 | 4,073 | ·72 | 3,327 | ·82 |
| 1958 | 99 | ·64 | 23 | ·23 | 36·33 | ·37 | 7·30 | ·20 | 4,348 | ·70 | 3,546 | ·82 |
| 1959 | 114 | ·67 | 28 | ·25 | 39·96 | ·36 | 9·80 | ·25 | 4,964 | ·69 | 4,227 | ·85 |
| 1960 | 108 | ·68 | 33 | ·30 | 44·37[a] | ·32 | 10·93[a] | ·25 | 5,427 | ·71 | 4,478[a] | ·83 |
| % incr. 1952-60 | M=148% | C=145% | QM=232% | | M=213% | C=87% | QM=346% | | M=238% | C=234% | QM=353% | |

[1] Includes foreign currency deposits held by private sector with commercial banks.
[2] Includes some interbank deposits.
[a] Third quarter only.

For description of items see source.

Source: IMF, *International Financial Statistics*, Dec., 1961.

in Turkey. The wholesale price indexes of Syria, Lebanon, and Iraq, by contrast, were actually lower at the end of the period than they were at the beginning, in spite of the fact that in all three countries the money supply increased by around 150% (and currency alone by more than 100%).

In view of the statistical deficiencies of the indexes, one cannot take very seriously small differences between the movements in the different countries but even if the indexes should understate the rise in prices by as much as 50% the broad outlines would remain unchanged: inflation of prices closely related to the quantity of money in Turkey and Egypt, a rise in prices in Iran which was disproportionately large in relation to the increase in the quantity of money as compared to Syria, Lebanon, and Iraq, where price increases were much less than the quantity of money. Taking into consideration our general knowledge of these economies in the last decade, I think one can safely say that in none of the four countries have inflationary pressures on prices yet been severe, whereas they have been severe in Iran and Turkey. At the same time, it is clear that with the possible exception of Egypt increases in the money supply must have significantly outstripped increases in real income even in countries where prices were not rising rapidly.

Of course, the 'crude' quantity theory would be applicable only if both the 'real' demand for money to hold and the supply of goods were constant, so that every increase in the supply of money was used to purchase an unchanged supply of goods. In fact an increase in the supply of money can be used quite differently. In general, people can (1) purchase an increased supply of goods for sale from increased domestic production; (2) purchase imports; (3) purchase things not entering into the national income accounts or into the price indexes, such as securities, real estate, and other goods not currently produced, the money thus remaining in the 'financial circulation'; and (4) hold money to increase liquidity reserves. Only to the extent that new money is not used for any of these purposes will it constitute effective demand for an unchanged supply of goods and thus pull up prices in proportion to the increase in the money supply.

Although the data available are insufficient to enable us to make an adequate analysis of the use of money in these different ways, some useful indications can be worked out. What we really need in order to estimate the first is a measure of the increase in real domestic output available for domestic sale, whereas what we have for all countries (except Iran) for most of the period are national income estimates at current prices.[6] Even if we deflate the national income estimates for price changes and deduct exports, we should still have to take account of the fact that the monetization of transactions is

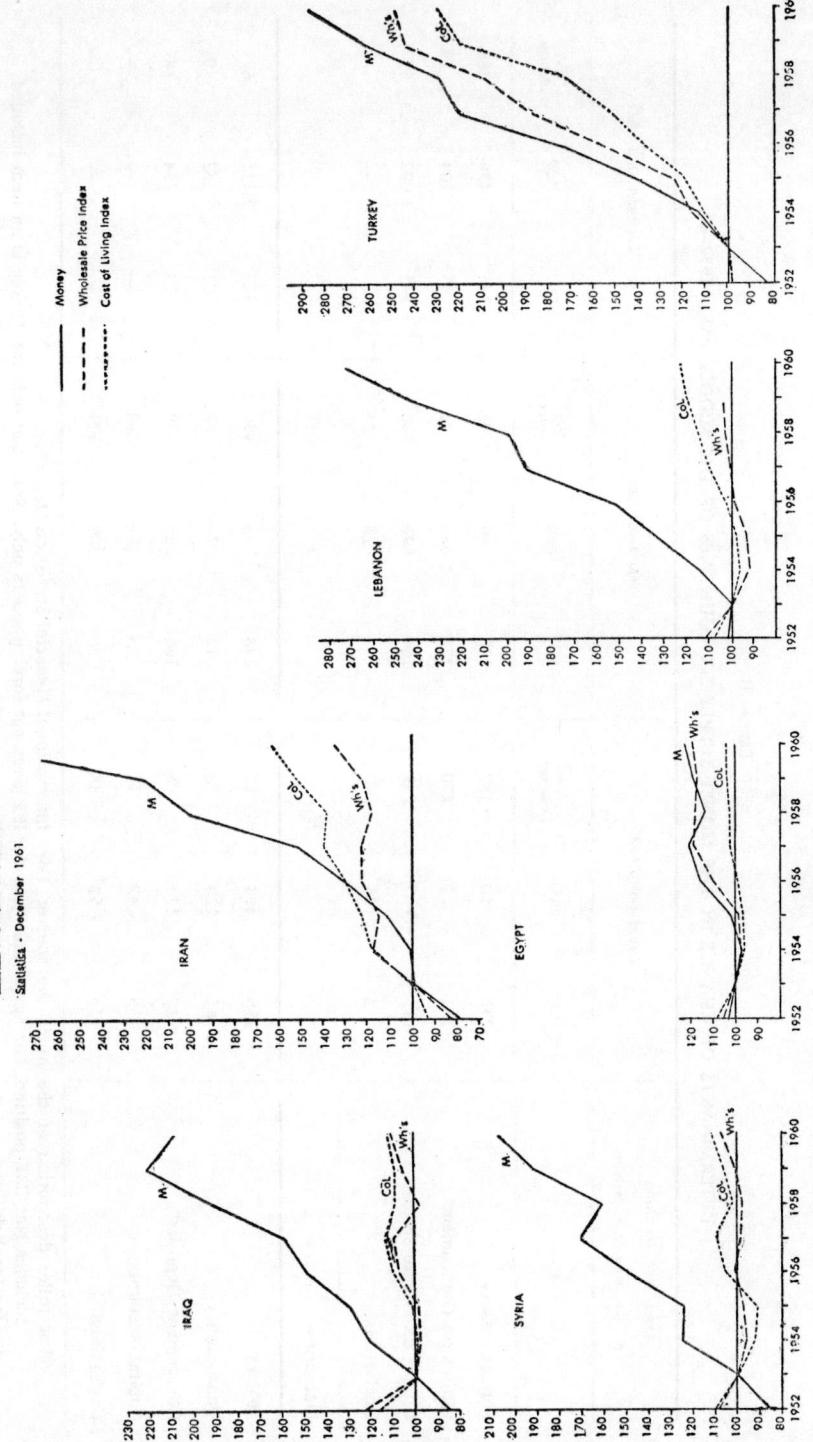

## TABLE II

### DETERMINANTS OF CHANGE IN THE MONEY SUPPLY SIX COUNTRIES OF THE MIDDLE EAST. 1952–60

| Assets and liabilities of the Monetary System | Egypt (mil. E£) | | | Syria (mil. S£) | | | Lebanon (mil. L£) | | |
|---|---|---|---|---|---|---|---|---|---|
| Assets | 1952 | 1960 | Net change | 1952 | 1960 | Net change | 1952 | 1960 | Net change |
| Foreign Assets[1] | 237 | 50 | −187 | 148 | 46 | −102 | 148[6] | 436 | 288 |
| Claims on Government[2] | 101 | 371 | 270 | 271 | 566 | 295 | 85 | 104 | 19 |
| Claims on Private Sector | 137 | 310 | 173 | 166 | 696 | 530 | 345[7] | 1,002 | 657 |
| Unclassified | 8 | 25 | 17 | 3 | 26 | 23 | — | — | — |
| **Liabilities** | | | | | | | | | |
| MONEY | 355 | 424 | 67 | 298 | 754 | 456 | 424[8] | 1,115 | 691 |
| Quasi-money[3] | 62 | 120 | 58 | 12 | 74 | 62 | 81[7] | 482 | 401 |
| Government Deposits[4] | 17 | 115 | 98 | 166 | 205 | 39 | 73 | 254 | 181 |
| Capital Accounts | 36 | 47 | 11 | 26 | 94 | 68 | — | — | — |
| Unclassified[5] | 14 | 52 | 38 | 86 | 209 | 123 | — | — | — |

For fuller description of the items see source: IMF, *International Financial Statistics*, Dec. 1961.

[1] Includes Net IMF position. For Egypt and Syria the IFS gives net foreign assets only. Note that only for Turkey is an item included for foreign liabilities.

[2] Includes claims on official entities.   [3] Includes foreign currency deposits of private sector.

[4] Includes counterpart funds, and deposits and prepayments for imports of official entities.

| Assets and liabilities of the Monetary System | Iraq (mil. ID) | | | Iran (bil. rial) | | | Turkey (mil. T£) | | |
|---|---|---|---|---|---|---|---|---|---|
| Assets | 1952 | 1960 | Net change | 1952 | 1960 | Net change | 1952 | 1960 | Net change |
| Foreign Assets[1] | 66 | 108 | 42 | [9] | 11·40 | — | 533 | 704[12] | 151 |
| Claims on Government[2] | 8 | 31 | 23 | 12·24 | 32·35 | 20·11 | 774 | 4,334 | 3,560 |
| Claims on Private Sector | 11 | 53 | 42 | 5·35 | 39·54[10] | 34·19 | 4,213[11] | 14,290[11] | 10,077 |
| Unclassified | 2 | 29 | 27 | ·34 | 1·01 | ·67 | 625 | 6,223 | 5,598 |
| Liabilities | | | | | | | | | |
| MONEY | 43 | 108 | 65 | 14·16 | 44·37[10] | 30·21 | 1,604 | 5,427 | 3,823 |
| Quasi-money[3] | 10 | 33 | 23 | 2·45 | 10·93[10] | 8·48 | 988 | 4,478 | 3,490 |
| Government Deposits[4] | 24 | 34 | 10 | 2·40 | 7·79 | 5·39 | 840[13] | 4,808[13] | 3,968 |
| Capital Accounts | 6 | 21 | 15 | — | — | — | 1,251[14] | 3,759[14] | 2,544 |
| Unclassified[5] | 3 | 26 | 23 | [9] | 19·36[10] | — | 893 | 6,411 | 5,518 |
| (Foreign Liabilities) | | | | | | | 670 | 2,552 | 1,882 |

[6] Official holdings only. [8] Includes some interbank deposits.
[7] 1954. Earlier figures not comparable. [9] Due to revaluation 1952 figures not comparable with 1960. Some of the effects of revaluation are hidden in Unclassified Liabilities. Iranian statistics are difficult to interpret in this respect. For further information see Source.
[10] 3rd quarter 1960. [11] Includes credits to state economic enterprises.
[12] 1959–1960 figure not comparable due to revaluation.
[13] Includes government lending funds. [14] Includes bonds.

probably increasing in many of these countries and that this represents an increase in domestic goods for sale but not necessarily an increase in national income.[7] For the second, we have statistics of imports, and we can get some idea of the changing significance of imports from the point of view of the relation between the money supply and prices from two statistics: the ratio of imports to income and the ratio of money to imports. The effects of an increase in the financial circulation and of an increase in a willingness to hold cash balances will show up in the income velocity of circulation of money, or the ratio of income to money.

In order to analyse the reasons for the differences among the different countries in the relation between the movement of prices and of the quantity of money I have examined the relevant ratios for all countries except Iran: the ratio of money to income (the inverse of the income velocity of circulation), the ratio of imports to income (the ex-post average propensity to import), and the ratio of money to imports. The movements in these ratios are shown in Chart III. The basic statistics are for all countries extremely crude and unreliable, although some are better than others. The monetary data and the import data are probably the best, the weakness of the price indexes have already been noted, and the income estimates could hardly be weaker.[8] In spite of the crudity of the data, however, they do demonstrate certain economic relationships which can be expected to grow in importance for most of these countries as they strive to accelerate their economic development; and at least the direction of change confirms to a surprising extent what we might have expected on the basis of general reasoning.

Let us look first at the ratio of money to income (M/Y).

## The Velocity of Circulation

In Lebanon, Syria, and in Iraq (until 1959) the ratio of money to income rose substantially; in Egypt there was a tendency for the ratio to fall, and in Turkey there was no significant trend. Thus the available data indicate a fall in the income velocity of circulation or a rise in the propensity to hold money with rising money income, and consequently a partial offset to the impact on prices of the increases in the supply of money, in precisely those countries and those periods where the rise in prices was least in relation to the money supply. This moderating influence was absent in Turkey, where prices and money rose roughly at the same rate, and in Egypt. The movements of the ratios are thus consistent with the observed differences in the relation between money and prices of the different countries. The conclusions with respect to velocity that we might draw from the movement of this ratio are for the most part even more marked if

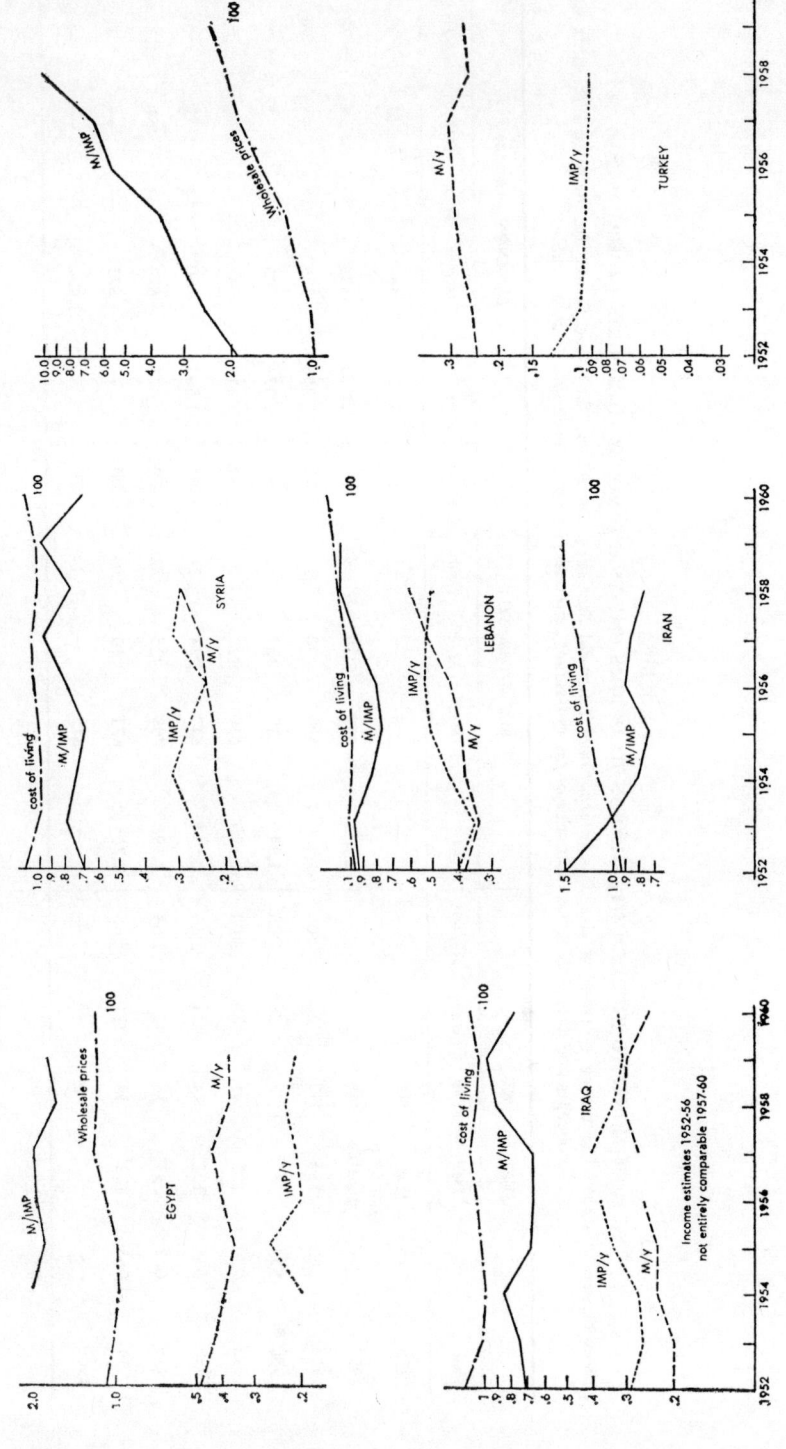

RATIO OF MONEY TO INCOME (M/Y), IMPORTS TO INCOME (IMP/Y), AND MONEY TO IMPORTS (M/IMP) IN RELATION TO PRICES

## TABLE III

RATIOS OF MONEY TO INCOME (M/Y), IMPORTS TO INCOME (IMP/Y) AND MONEY TO IMPORTS (M/IMP.) 1952–60

Unless otherwise stated the source of these statistics is *International Financial Statistics*, December, 1961. See Source for detailed discussion of concepts and data. – N.B. *The statistics of the individual countries are not internationally comparable.*

| Year | EGYPT (mil. E£) | | | | | SYRIA (mil. S£) | | | | | LEBANON (mil. L£) | | | | |
|------|------|------|------|------|------|------|------|------|------|------|------|------|------|------|------|
|      | Y[1] | IMP | M/Y | IMP/Y | M/IMP | Y[2] | IMP | M/Y | IMP/Y | M/IMP | Y | IMP | M/Y | IMP/Y | M/IMP |
| 1952 | 748 | Data not comparable | ·48 | — | — | — | 434 | — | — | ·69 | 1,115 | 444 | ·38 | ·40 | ·95 |
| 1953 | 780 | | ·44 | — | — | 1,881 | 440 | ·18 | ·23 | ·79 | 1,168 | 424 | ·35 | ·36 | ·98 |
| 1954 | 868 | 164 | ·39 | ·20 | 2·04 | 2,180 | 593 | ·20 | ·27 | ·73 | 1,256 | 559 | ·38 | ·45 | ·86 |
| 1955 | 931 | 187 | ·37 | ·26 | 1·86 | 1,946 | 617 | ·22 | ·32 | ·69 | 1,374 | 708 | ·40 | ·52 | ·78 |
| 1956 | 970 | 201 | ·41 | ·20 | 2·00 | 2,330 | 630 | ·22 | ·27 | ·82 | 1,417 | 763 | ·44 | ·54 | ·82 |
| 1957 | 935 | 192 | ·44 | ·21 | 2·16 | 2,514 | 612 | ·24 | ·24 | ·97 | 1,503 | 800 | ·53 | ·53 | ·99 |
| 1958 | 1,027 | 233[3] | ·38 | ·23 | 1·68 | 2,198 | 709[3] | ·25 | ·32 | ·79 | 1,325 | 676 | ·62 | ·51 | 1·22 |
| 1959 | 1,083 | 222[3] | ·38 | ·21 | 1·86 | 2,266 | 671 | ·30 | ·30 | 1·00 | — | 821 | — | — | 1·21 |
| 1960 | — | — | — | — | — | — | 819 | — | — | ·70 | — | 1,051 | — | — | — |

|  | IRAQ (mil. ID) | | | | IRAN (bil. rial.) | | | | TURKEY (mil. T£) | | | |
| --- | --- | --- | --- | --- | --- | --- | --- | --- | --- | --- | --- | --- |
| Year | Y | IMP | M/Y | IMP/Y | M/IMP | Y | IMP | M/Y | IMP/Y | M/IMP | Y | IMP | M/Y | IMP/Y | M/IMP |

| Year | Y | IMP | M/Y | IMP/Y | M/IMP | Y | IMP | M/Y | IMP/Y | M/IMP | Y | IMP | M/Y | IMP/Y | M/IMP |
| --- | --- | --- | --- | --- | --- | --- | --- | --- | --- | --- | --- | --- | --- | --- | --- |
| 1952 | 217 | 62 | ·20 | ·29 | ·70 |  | 9·38 |  |  | 1·51 | 12,424 | 1,557 | ·13 | ·13 | 1·03 |
| 1953 | 259 | 68 | ·20 | ·26 | ·75 | Not | 16·88 | Not | | 1·08 | 14,696 | 1,491 | ·13 | ·10 | 1·27 |
| 1954 | 268 | 73 | ·23 | ·27 | ·85 | avail- | 22·51 | avail- | | ·82 | 14,785 | 1,339 | ·14 | ·09 | 1·55 |
| 1955 | 289 | 97 | ·23 | ·34 | ·68 | able | 27·00 | able | | ·75 | 18,220 | 1,393 | ·15 | ·08 | 1·90 |
| 1956 | 303 | 113 | ·25 | ·37 | ·67 |  | 25·59 |  |  | ·92 | 21,197 | 1,141 | ·16 | ·05 | 2·91 |
| 1957 | 299[5] | 122 | ·27 | ·41 | ·67 |  | 31·98 |  |  | ·87 | 26,623 | 1,112 | ·15 | ·04 | 3·66 |
| 1958 | 319 | 110 | ·31 | ·34 | ·90 |  | 46·59 |  |  | ·78 | 33,873 | 882 | ·13 | ·03 | 4·93 |
| 1959 | 380 | 116 | ·30 | ·31 | ·98 |  | — |  |  | — | 40,095 | [4] | ·12 | — | — |
| 1960 | 432 | 139 | ·25 | ·32 | ·77 |  | — |  |  | — | — | [4] | — | — | — |

For statistics of money see Table I.

[1] Income estimates 1954–59 for Egypt are the estimates of A. R. Abdel Meguid in *Egypte Contemporaine*, April 1961.
[2] Income data are taken from U. N., *Monthly Bulletin of Statistics* and represent net domestic product at factor cost in 1956 prices.
[3] Includes trade between Egypt and Syria from official sources.
[4] Turkish import values published only in US $—see source.
[5] Income estimates 1952–56 not entirely comparable with 1957–60. Source of estimates 1957–60, Ministry of Planning, Government of Iraq.

we examine the ratio of currency to income in the different countries, since in general the proportion of currency in the total money supply moved inversely to the ratio of money to income except in Egypt and Turkey.

The fall in the velocity of circulation of money, if we accept the evidence that it occurred in Syria, Lebanon, and Iraq, could, of course, be the result of a number of separate influences. In the first place, it could reflect a rise in personal savings of the public held in the form of liquid balances. This would be consistent with the argument that in underdeveloped countries part of the difficulty of development lies in a shortage of opportunities for profitable investment. Similarly it could reflect, as indeed it does in Iraq to a considerable extent, a rise in government savings in the form of foreign exchange revenues. National income estimates include net increases in a country's claims on foreigners over the period to which the estimates refer. If these increases in foreign assets are matched, not by increases in domestic money but by increases in government deposits (which are not part of the money supply), then national income is larger but not the money supply, and M/Y is therefore correspondingly less.

On the other hand, unstable political conditions or the economic measures of governments may have encouraged the holding of money by the private sector for speculative or precautionary reasons, or may have caused a rise in intermediate and financial transactions in relation to final transactions (for example, a rise in speculation in real-estate or securities), or may have induced a fall in business credit in relation to final sales. It has been alleged that in most of these countries there has been an increase in financial, speculative, and real-estate transactions, which do not add to the national income, but do tend to increase the demand for money. We have no way of measuring this, or of determining whether or not such activities are more important in some countries than in others, but if such an increase has occurred, the effect would be to raise the ratio of money to income, that is, to reduce the income velocity of circulation.

Finally, some of the fall in income velocity could be due to increasing monetization of income transactions as the economies have developed. There are a number of ways in which the monetization of transactions has probably been increasing in the Middle East. Payments that used to be made in kind (rents, wages, taxes) may be increasingly made in cash, and cash sales of surplus by farmers may be rising as well as cash purchases. Since the imputed values of non-monetary transactions are included in estimates of national income, the monetization of these transactions might not affect the national-income estimates, but would affect the demand for money and hence the income velocity. More money would be required in relation to income, but a correspondingly increased supply of money would be matched, so to speak, by increases in output for sale. This process would show up in the statistics as a fall in the income velocity of circulation of money.

## Imports and Prices

In any event, and no matter what the cause, a rise in cash balances per unit of income will reduce not only the impact on prices of an increase in the supply of money, but also its impact on the demand for imports. Unless we assume a negative marginal propensity to import, an increase in income in an open economy will lead to an increased demand for imports, which, although subtracting from the quantity of money, will also put pressure on the balance of payments, unless exports or capital inflow increase at the same time, and on the country's foreign reserves.

In all of the countries considered here it is reasonable to assume that the marginal and average propensity to import would tend to rise as money income rose, partly because an increased demand for imported consumer goods can be expected in spite of the development of domestic output that has undoubtedly taken place, and partly because of an intensified demand for capital goods for public and private investment. Indeed, the increased total demand is to some extent reflected in the changes in the foreign assets of the monetary systems of the different countries that we have described in Chart I, where we saw the steep decline that took place in Egypt throughout the whole period, the decline in Syria since 1956, and in Iran since 1957, and the 'debtor' position of Turkey. In Iraq, foreign assets began to decline with the acceleration of government expenditures in 1956, but in Lebanon, which has no government development programme to speak of, they continued to increase throughout the period, in spite of a near doubling of the value of imports.[9]

Nevertheless, when income also increases a rise in imports does not in itself indicate a rise in the *propensity* to import, which can only be shown by a rise in the *ratio* of imports to income. This ratio, however, will reflect a 'behavioural' propensity only if imports are freely permitted; if imports are restricted, the ratio becomes a mere ex-post statistical calculation showing what happened, and it is misleading, though common practice, to call it a 'propensity'. Since in fact import restrictions, and especially exchange restrictions, have been imposed with varying degrees of severity in most of these countries, and in addition political events such as the Suez crisis and the Lebanese civil war, have impeded free entry of imports, it would not be surprising if IMP/Y failed to rise significantly. If, in spite of such restrictions, imports did rise faster than income, they would have tended to moderate the impact of the increased money supply on domesic prices.

Such a moderating influence was completely absent in Turkey, where import restrictions were more severe than in the other countries, and IMP/Y actually fell while prices soared. In Egypt, where imports were also extensively restricted, IMP/Y rose little, if at all, except for 1955. On the other hand, in Iraq and Lebanon, and in Syria (except for 1956), where imports were more freely permitted, the increase in the ratio was significant.

Clearly imports can rise for a number of reasons other than an increase in domestic money income, notably as a result of a fall in the prices of foreign goods relative to those of domestic goods, and as a result of increased foreign grants, loans, or investments. Moreover, the movements in the ratio IMP/Y can be influenced by changes in income which do not give rise to increased demand for goods and services. In Iraq, for example, part of the oil revenues received were held as government deposits or investments abroad; these revenues, though unspent, contributed to the increase of national income as measured by the statistics but had no effect on demand for domestic or foreign goods. In these circumstances, imports could be affected only when the government spent its revenues, either for foreign goods imported directly on government account, or for domestic goods and services, thus raising domestic income, indirectly inducing an increased demand for foreign goods. Thus, for Iraq the rise in IMP/Y, if calculated with reference to the relevant aspects of national income, would be substantially greater than shown here.

In general, a rise in IMP/Y will help counteract inflationary tendencies due to an expansion of the money supply regardless of the reason for the rise in imports, for, as we have shown earlier, an increased supply of money, after adjustments have been made to take account of changes which show up in the velocity of circulation, must have been used either to purchase imports or to purchase domestic output. If we take the statistics presented here at their face value, it is clear that the chief reason for the failure of prices to rise faster than they did in Syria in view of the rapid expansion of the money supply must have been increased domestic output, whereas in Iraq and Lebanon, imports will have been an equally, if not a more, important factor. In Egypt, on the other hand, the ratio of imports to income, like the ratio of money to income, did little to temper the increased supply of money, and in Turkey the restriction of imports is clearly of decisive importance in explaining the inflation of prices.

In view of the limited extent to which national income estimates are available, and of their extreme unreliability, I have examined another ratio, the ratio of money to imports, in order to see if it helps in our analysis. In comparing the movements of this ratio with movements in prices, we ignore all uses of money other than the purchase of imports or of an unchanged domestic output. In other words, the effect on prices of changes in the relation of income to both money and imports is left out of account.[10] In a sense we can look at imports as an offset to the impact of increases in the supply of money on prices, or we can look at changes in the income variables (M/Y and IMP/Y) as an offset to the impact of increases in the supply of money on imports and the balance of payments. So far as I am aware the ratio of money to imports has been used for theoretical analysis only by Polak in an extremely interesting discussion of credit expansion, imports, exports, and monetary reserves.[11] Plainly a country in which imports expand very rapidly with increases in the money supply could afford very little credit expansion without running into trouble with its international reserves. The ratio of money to imports may be of some help in the absence of income data, but its usefulness is very limited since any explanation of the connections between movements in M/IMP and movements in prices must go back to the other variables, M/Y and IMP/Y.

For Turkey we find an extremely steep rise in M/IMP; so steep that changes in the income variables would have had to be very great indeed to prevent a similarly steep rise in prices. For Lebanon, too, we find a substantial rise which corresponds fairly closely in direction, but not in magnitude, to the year-to-year movements in prices. In Iraq and Syria, on the other hand, the rise over the period was relatively moderate, and in Egypt there was no significant trend. Again, in both Syria and Egypt the annual movements of the ratio corresponds closely to the direction of changes in prices from year to year. The movement of the ratio in Iran perhaps helps to explain the failure of the wholesale price index to rise as rapidly as the cost of living index after 1954, for imports are undoubtedly more important in the former, whereas the latter will reflect rising prices in the cities of some commodities and services which are not sensitive to increasing imports. We know, however, that the Iranian statistics are very much less reliable than those of the other countries and we cannot safely draw even rough conclusions from the existing data.

To sum up, the movements of the three ratios that we have been discussing are, when taken together, consistent to a surprising degree, given the extremely uncertain quality of the basic data, with the observed relationship between money and prices in the different countries. To be sure, in view of the rapid increase of the money supply, we might have expected a somewhat greater degree of inflation in Syria, Iraq, and Lebanon than the indexes show. But apart from the fact that the price indexes probably understate the actual increase that took place, the check provided by a fall in the velocity of circulation and the rise in imports was substantial. There seems to be some tendency, however, for an acceleration of the rate of increase in prices in all three of these countries in the last year or so along with a continued increase of the money supply. In Turkey, by contrast, there was no check and uninterrupted inflation proceeded *pari passu* with the increase in the money supply. Iranian statistics are so bad that not much of an analysis can be made; nevertheless, the fall in M/IMP was substantial and clearly only extensive oil revenues have prevented price rises similar to those in Turkey. In Iraq and Lebanon, foreign-exchange earnings, together with accumulated foreign assets could for a time permit increases in imports sufficient to prevent serious inflationary pressures even if the money supply continues to increase at a rapid rate. The prospects of substantially increasing foreign-exchange earnings of Egypt and Syria are less bright, and in the light of the steep decline in foreign assets in recent years, these countries may well face disturbing inflationary pressures as the import restrictions that will be necessary if expenditures continue at the present rate are applied.

In this paper I have dealt only with financial aspects of the money supply in relation to imports, income, and prices. I have not dealt directly with other significant influences, such as civil wars, revolutions, and other political developments, although some of their economic effects will be reflected in the variables analysed; although I have been dealing with agricultural countries, I have even largely ignored variations in crop yields from year to year. I have done this deliberately, partly because a detailed examination of the relevant factors for each country would take us far beyond the scope of a single article, but primarily because I wanted to make a comparative analysis of the broader more pervasive influences that persist underneath the short-term political and even economic disturbances. The period is short; the data are inadequate, but in spite of this, the analysis of the monetary statistics of the six countries not only helps us to understand the course of events in the past, but also points out the problems of financial policy that will affect many of them as they push their programmes of economic development.

## NOTES

1 1952 is the earliest year for which the statistical series I want to use are available on a reasonably comparable basis for each country.
2 The National Bank of Egypt takes a different view and its definition of money differs from that of the International Monetary Fund in that it includes time and savings deposits in the money supply. The bank argues that this 'is justified by the practical experience acquired in the Egyptian

money market, which reveals that the cost of emergency withdrawals from time and notice accounts is extremely low. Commercial banks in Egypt usually allow immediate withdrawals from notice accounts without charging their client with any expenses other than the interest on the notice account for the few days that ought to pass between notifying the bank of the depositor's desire to withdraw and the actual withdrawal. This practice renders the value of time deposits very close to that of demand deposits'. National Bank of Egypt, *Economic Bulletin*, Vol. XIV, No. 1 (1961). Although it may well be true that from the point of view of easy availability time deposits are not easily distinguishable from demand deposits, it may also be true that they are in general less active than demand deposits, and that from this point of view it may be useful to separate them from the money supply. On the other hand demand deposits are probably less active than currency and the same argument could be used to restrict money to currency alone. I repeat, there is no generally applicable distinction.

3 For this reason I have shown the movement of government deposits in Iraq in Chart II.

4 The unclassified liabilities for Iran include liabilities of certain official institutions as well as 7 bil. rials revaluation proceeds. See notes in sources for Table 1.

5 There are always difficulties of interpretation when an analysis is started in a particular year. As explained earlier, I have started with 1952 because it is the earliest year for which the statistical series I want to use are reasonably comparable. It must be remembered, however, that in 1952 most countries were still undergoing readjustments following on the post Korean-war boom. Except in Iran and Turkey, where inflation was persistent, prices fell until 1954, after which they tended to rise again. The attached chart of wholesale price movements will help to put the period 1952–60 into some perspective.

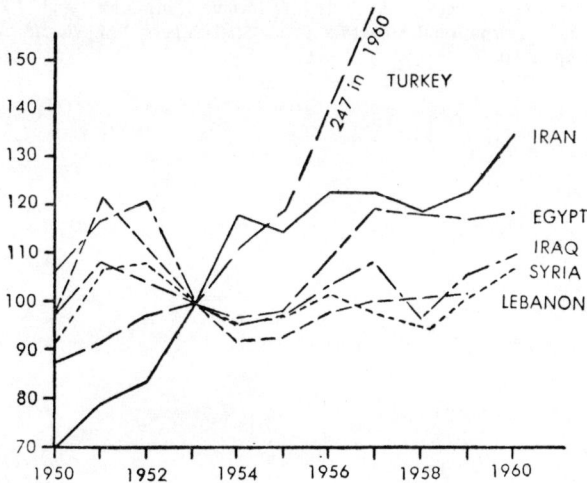

6 The Syrian national income statistics are in constant prices. I did not re-inflate them because I do not know what price index was used to deflate them in the first place. However, the price movements were slight and consequently the direction of change is probably little affected, which is

the important thing for my purposes. In any event the national income estimates of the different countries are *not* comparable.
7  See below.
8  As an illustration of the difficulty of drawing conclusions from such statistics compare, for example, the difference between my calculations of M/Y and those of Anwar Ali for 1954, in 'Banking in the Middle East', (IMF *Staff Papers*, Vol. VI, No. 1, p. 52). Although my calculations and those of Ali rank the countries in the same order, there are in at least two cases very significant differences in the size of the ratios. The differences are, of course, due to revised estimates of either the money supply or of the national income. We can only assume that later statistics are an improvement on earlier ones and that in time the figures for the series I am using will be improved. Whether or not this problem is considered serious enough to destroy the usefulness of the statistics for analytical purposes is a matter of opinion. I think they are of some value for analysing change when the direction of change is clear and for comparing different countries when the differences are large and the definitions reasonably comparable.
9  Of course large increases in imports can occur without a decline in foreign assets if they are paid for with current foreign acquisitions; conversely, not all of a decline in foreign assets can be attributed to imports. Egypt for example, had to pay out in the period substantial sums to the Sudan pursuant to the agreement with respect to the High Dam and on account of the change in currency arrangements. She also paid out sums as compensation for the nationalization of the Suez Canal and certain other foreign properties. Revaluation of holdings of foreign securities is also at times an important factor.
10 Statistically M/IMP can be derived by the simple, though meaningless, process of multiplying M/Y and Y/IMP, and thus eliminating income.
12 J. J. Polak, 'Monetary Analysis of Income Formation and Payments Problems', International Monetary Fund, *Staff Papers*, Vol. VI, No. 1 (Nov. 1957), pp. 1–50.

## XVII

## OIL AND STATE IN ARABIA*

*Examines some important ways in which the discovery and development of oil has affected economic and political change in the states of the Arabian peninsula.*

THERE are undoubtedly few, if any, aspects of the political, social or economic life in Arabia that are today entirely untouched directly or indirectly by the development of the region's oil resources. And since the organization and activities of governments are themselves responsive in some degree to changes taking place among their peoples, a full examination of the relation between oil and state in Arabia would not only require an extensive historical treatment of topics about which much is already well-known but would also inevitably lead us into a discussion of most aspects of Arabian life. Education, including study abroad, greatly improved communications, including radio and even television, travel and wider contacts with foreigners, have all been instrumental in changing attitudes and aspirations among the population, and all have directly or indirectly been made possible on an increasing scale through the availability of oil revenues. The organization of the state in the established oil exporting countries has been radically changed, the position of the religious authorities significantly undermined, and the personal power of ruling families eroded. Doubtless such changes would have come about eventually in any case, but they have certainly been hastened by the development of oil, which necessarily opened the countries with accelerating rapidity to influences from elsewhere, and especially from the industrialized western world.

Oil was discovered in Bahrain in 1932 and exports began in 1934. Revenues, which were around $9000 in 1933, exceeded $1m by 1940 and $10m by 1954. In 1968 they were estimated at around $20m. The next discoveries in Arabia were made in 1938 in Kuwait and Saudi Arabia, although their rapid development, and especially that of Kuwait, had to await the end of the war. Kuwait's revenues in 1948 were only $12m; by 1958 they exceeded $350m and in 1967 they reached $700m. Between 1950 and 1960 Saudi Arabia's revenues rose

---

\* Presented at a Conference on the Arabian Peninsula at the School of Oriental and African Studies of the University of London in March 1969 and to be published in *The Arabian Peninsula* under the auspices of S.O.A.S. Allen & Unwin. 1971.

from around $50m to over $330m, and by 1968 approached $1000m. Part of Kuwaiti and Saudi revenues come from the Neutral Zone, where oil was discovered in 1953. Oil was found in Qatar in 1939 but exports did not start until ten years later, the wells having been plugged and the installations stripped during the war. Revenues rose from only $1m in 1950 to over $100m in 1967. Oil exports started from Abu Dhabi in 1962 and revenues exceeded $100m by 1967. The country is now the world's twelfth largest oil producer and there is more to come, for new fields are still being discovered and new concessions granted. The Fateh field in Dubai was discovered in 1966 and exports are about to begin. Production also began in the Sultanate of Oman in 1967 and when exports are established payments are estimated to reach between $50m and $100m by the end of 1969.

The influence of oil makes itself felt in many ways: through the direct effect of the productive operations of the oil companies; through the direct receipt of foreign exchange revenues by governments or rulers who must needs deal with them; through the ways in which these receipts are spent; through the relations with oil companies and other foreign interests of countries producing (or hoping to produce) oil; through the changes in the international political relations of oil (or potential oil) countries both within the Arab world and with countries outside it, which follow from the enhanced economic significance with which oil endows its possessors; or through the entry into oil operations by the governments themselves. Because of the pervasive influence of oil directly on the role and structure of the state and indirectly through social and economic changes, any discussion in a paper of this length must be highly selective. I propose, therefore, to confine myself to an analysis of the significance for the role of government of a few of the peculiar characteristics of the organization of the oil industry in the Middle East. I shall conclude with a brief discussion of the entry into the oil industry of the State itself through national oil companies.

There are three important characteristics of the oil industry in the Middle East which are of especial importance for an understanding of the relation between oil and state: (a) Receipts from the export of oil do not naturally accrue to indigenous exporters, as is the case for most export commodities in the non-communist world, but come directly to the government; (b) the exploration for and the production of crude oil has been traditionally conducted under a system of concessions to foreign companies; (c) the most important companies producing crude oil are wholly-owned affiliates of internationally integrated firms and in consequence most crude oil is produced for use within the integrated framework of the major firms and is not sold in an open market.

Because of the first characteristic, the effect of export receipts on economic activity was in the first instance dependent on the way the government chose to spend the money, and this in turn necessarily gave the government a dominant role in the economy whether or not it was adequately prepared for such a role or even wished to assume it. The second necessitated a special type of negotiation between companies and governments, the outcome of which gave the companies an extraordinary degree of control over the countries' resources and was largely responsible for the prolonged nearly complete domination by a very few companies over Middle East oil resources. The third made the exporting countries dependent on the companies for markets and made necessary an arbitrary formula for the valuation of the oil produced.

*Dominance of Government Revenues*

Production for export, like any other productive activity within a country, uses domestic resources, including labour, and thus creates employment and domestic incomes. This is a primary, direct effect of production itself, the importance of which depends upon the amount of resources used. By and large the oil industry does not use a great deal of labour in relation to the value of its output, and it imports most of its other inputs. Of course, where the domestic population, and especially the supply of skilled labour, is small, even oil production may absorb a large proportion of the available workers. When the export industry is in foreign hands, the expenditures of the foreign companies will also bring foreign exchange to the country since the companies must acquire the domestic currency for their domestic expenditures by purchasing it with foreign exchange, or else expend foreign currency, gold, or other acceptable coins or metals directly. The sale of output abroad gives rise to foreign exchange receipts, but the effect of these receipts on the economy depends on how they are spent, and this in turn will differ according to who receives the sales proceeds.

If they are received by domestic exporters—which might be termed the 'normal case'—then it could be expected that much of the proceeds would be exchanged by them for domestic money which they would then spend on consumption or further investment. This would induce further employment and create more income to the extent that the money was not directly spent for additional imports. If the exporters are foreigners, they too may retain some of their receipts for further investment in the country, but a further part—if not all—of the money may be retained abroad and contribute nothing to the domestic economy. If the government taxes foreign exporters, then most of the proceeds from exports may accrue to the domestic economy only through the tax receipts of the governments. This would put the government in the position of having to spend money or to promote expansion of private credit if the export receipts were to have any effect on the economy at all.

In Arabia, governments and rulers have received revenues from oil companies in many forms—concession rents, fees, royalties and income taxes, but by far the most important form for those countries in which export production has been established is the income tax. The expenditures of the oil companies in the several countries are a further contribution to their foreign exchange receipts, but because of the amounts of money involved, the direct impact of oil on the economy comes primarily through government expenditures. Sooner or later the pressing need to spend its money reasonably (or to get it into the hands of private individuals other than royal relatives and court favourites) has forced the governments into extensive reorganization (sometimes after a change of rulers), and to create financial institutions, budgets and methods of administration which could deal with the many problems involved. Not the least of these have sometimes arisen because of the arrival on the scene of hordes of foreign businessmen, contractors, freelancers or even foreign government representatives, all pressing their own proposals for the expenditure of the local government's funds.

When oil was discovered in *Bahrain*, *Qatar* and *Kuwait* they were all British protectorates and the expenditure of the oil revenues, together with the changes in the organization of the government, were powerfully influenced by British guidance and advice. From the time when oil revenues started in *Bahrain* in the early 1930's, only one third of them went to the Shaikh for his personal and family affairs; half of the rest were devoted to meeting current government expenses and to the financing of development projects, and half were reserved for investment abroad against future needs. Government services of all kinds increased rapidly and the appropriate government agencies had to be created to provide them. Oil revenues levelled off in the late 1950s while population continued to increase. The Shaikh has apparently not been willing to respond so readily as he had earlier to British advice, with the result that appropriate adjustments have not been made to declining per capita oil revenues.

In *Qatar*, on the other hand, oil revenues have continued to increase. As in Bahrain, only a third of the revenues was allocated to the Shaikh, the rest being reserved for public expenditure of one kind or another and for investment abroad.[1]

In *Kuwait* the history of the administrative and governmental adaptation to the expenditure of the oil revenues is a history of the gradual transformation of a system of direct and personal family rule to a constitutional one in which administration is, in principle at least, in the hands of civil servants, properly organized in government departments headed by ministers responsible to an elected national assembly. Since the very idea of a civil service is completely alien to traditional culture, the transformation is by no means complete. Nevertheless, given the fact that less than a generation has passed since oil revenues have been available on a really large scale, Kuwait has moved with great speed in the acquisition of the attributes of a modern state, and the organization and administration of the Kuwait Fund for Arab Economic Development is almost a model of its kind.

On the other hand, very large revenues accruing to the Government gives it almost unlimited ability to employ people, and since employment in the 'civil service', unlike employment in industry or agriculture, requires almost no complementary productive inputs and men often can easily be employed even where there is no 'job' to fill, government employment tends to burgeon beyond all bounds, with personal influence the key to getting a post, especially in an economy so very heavily dependent on expenditures by government. The World Bank Mission to Kuwait reported that in 1953 nearly 10% of the *population* (not labour force) was in the civil service. Nearly a third of these were illiterate and many had private jobs as well, there being no clear distinction between public duty and private interest. There have, of course, been great advances in this respect in the past 15 years but the problem still remains.

When oil exports began, *Saudi Arabia* was neither a British protectorate with effective foreign advisers nor did she have a ruler who had much understanding, or even wish to understand, how oil revenues might be used to benefit his country. In the early years of oil exports, therefore, not only were the revenues received by the King to a very considerable extent spent in non-productive ways, but in spite of a flow of revenues reaching over $300 million in 1955—a thirty-fold increase in ten years—the King had actually to borrow money. The country got into severe financial difficulties; the free market rate of the riyal fell by about half, inflation set in and consumer goods became in short supply. An extensive governmental and financial reorganization was required and under the leadership of Prince Faisal as Prime Minister, was taken in hand. Although budgets of a kind were published as early as 1947/48, it took roughly ten years before a reasonably sound system of financial management could effectively be imposed. The government had been too unprepared, technically and psychologically, to cope with the expenditure of so much money so quickly, but by the 1960s the problem of using oil revenues had been largely responsible for the establishment of modern administrative and financial institutions and for the emergence of an outlook on the country which gave a high priority to economic development.

*Abu Dhabi*, coming much later into oil money, and with the benefit, again, of strong foreign advisers seems (again after a change of rulers) likely to be in a position to handle its funds with greater sophistication, as probably will *Dubai* whose people have had a longer contact with the outside world than have those from the more inward looking Arabian states, such as *Muscat* and *Oman*. Here there will apparently be a somewhat more severe struggle between a conservative ruler suspicious of all things new and the forces of modernity which will press with growing insidiousness on the society.

*The Concession System*

An oil concession is an agreement between one or more oil companies and the government within whose borders the relevant territories lie. Except in the United States, underground oil is everywhere the property of the State and can be exploited by private companies only with the express permission of the government. The early concession agreements in Arabia, as in the rest of the Middle East, gave a single concessionaire exclusive rights over the exploitation of oil resources in a very large proportion of the territory of the country for a very long period of time in return for a variety of stated payments to the government. The result was the establishment of a highly monopolistic control over the exploitation of the region's oil resources.

All of the production of Arabia came under the control of a very few international companies whose territorial rights covered much larger areas than they could effectively exploit.[2] This, combined with the apparent profitability of oil operations, led to demands for extensive changes in existing concession terms after the second world war. Such demands gained added strength from the much more favourable terms that the governments had been able to obtain for the new concessions granted in the 1950s and 1960s. The abundance of low cost oil, together with the price policies of the major companies, attracted many newcomers to the Middle East, and competition for concessions among companies other than the established 'majors' who had traditionally dominated the field, immensely enhanced the bargaining position of governments. Indeed, some of the new agreements are more in the nature of contractual than concessionary relationships and it is possible to see them as pointing the way to a move from 'concessions to contracts',[3] thus radically changing the nature of the relationship between oil companies and governments.

The major demands of governments have been for changes in financial terms in order to give them a greater share in revenues; for relinquishment by the companies of unexploited areas held under exclusive concession thus enabling governments to offer concessions to other companies; and for equity participation for the government in the companies producing oil. We shall discuss each of these in turn.

Originally the oil companies were exempt from all taxes under their concession agreements but paid to the government a royalty of four gold shillings per ton of oil produced. With the great expansion of production after the war the governments were pressing for an increased share of the profits attributed to crude oil, and in 1950 Aramco in Saudi Arabia led the way by agreeing to pay the Saudi King 50% of its profits in the form of an income tax.[4] Other governments soon obtained the same terms and oil revenues jumped, partly because of the tax itself, but partly also because the '50-50' arrangements made it necessary to determine and announce a price for crude oil, which was higher than the price that had previously been attributed to crude oil by some, if not all, of the companies. The governments of the producing countries granted the companies a substantial marketing allowance for tax purposes, but this was later gradually eliminated.

The next important financial changes occurred after the formation of the Organisation of Petroleum Exporting Countries in 1960, for OPEC was responsible for a virtual freezing of the 'tax prices' of crude oil—the so-called posted prices—which meant that the companies were no longer free to lower these prices at will.[5] For the first time the governments obtained partial control over the prices attributed to their crude oil. OPEC was also instrumental in forcing the companies to 'expense' royalties—that is, to treat the $12\frac{1}{2}\%$ royalty they paid on crude as a cost instead of as part of the tax on profits paid to the governments. Again oil revenues per ton of oil produced rose substantially, although the full effect of phasing in royalty expensing is being spread over a number of years in view of the weak position of crude and product markets since 1964 when the expensing agreement was made.

The original concession agreement in Saudi Arabia made provision for relinquishment at the discretion of Aramco of areas it did not exploit, but it was not until 1948 that the company agreed on a programme of relinquishment. In most of the agreements with the newcomers taking up concessions in the late 1950s and early 1960s, provision was made for relinquishment according to a predetermined timetable, and other concessionaires voluntarily accepted the principle. By 1963 Aramco had relinquished about 75% of its original concession area; the Qatar Petroleum Company's concession area has been cut by about one-third; KOC has given up about half of its original concession; and even in Abu Dhabi, the Ruler has already been able to offer relinquished territory for new bidding. The fact that major companies can no longer indefinitely keep unexploited areas out of the hands of possible rivals has further weakened their monopoly over oil resources.

A number of Middle Eastern governments, but in particular Saudi Arabia, have strongly pressed their major concessionaires to admit the government as a full partner in the companies producing crude oil. This has been consistently refused, but the new concession agreements with non-major companies provide for equity participation. The move toward partnership with the state oil companies was led by French, Italian and Japanese companies as a means of facilitating the expansion of their Middle East oil interests. Although the established major oil companies had in one way or another been forced to improve the terms on which they operated in the 1950s, they were adamant in maintaining the '50-50' principle in the division of profits as well as their exclusive ownership of their producing affiliates.[6] In bidding for new concessions, the French, Italian and Japanese showed little hesitation in breaching both principles. Moreover, the governments have also pressed for the right to participate in 'downstream' operations—that is, in refining and marketing—and to have a share of the profits attributed to these stages of the industry. Again, the major companies have, by and large, insisted that this was impossible.

The first concessions in the Arab Middle East providing for equity participation and violating the '50-50' principle were made by the Japanese in 1957 and 1958 in the agreements signed with Saudi Arabia and Kuwait for offshore rights in the Neutral Zone.[7] Subsequently Saudi Arabia led the way in forcing the pace of such agreements and continually improving their terms. In 1965, Petromin, the Saudi Government-owned company, and Auxirap, a company backed by the French government, formed a partnership for the exploitation of an area in the Red Sea zone of Saudi Arabia which had been relinquished by Aramco. The agreement provided for a 40% shareholding but equal voting rights for Petromin, for the extension of Petromin's activities into refining and marketing, and for taxation at existing rates. In effect, the government was to receive some 80% of the profits made on crude exports under this arrangement. Even more favourable agreements, providing for partnership, downstream integration, and a higher take for the government, were signed at the end of 1967, one between Petromin and ENI (the Italian Company) and one with American independents.

Although Saudi Arabia has pushed most vigorously in the new agreements, both Kuwait and Abu Dhabi have made similar arrangements with newcomers. In 1968 the Kuwait National Assembly ratified an oil agreement between the Kuwait National Oil Company and Hispanoil, a Spanish group sponsored by the Spanish government. Under the agreement the joint company was guaranteed a 25% share of the Spanish market for 15 years. Abu Dhabi also signed an agreement with Maruzen Oil Company of Japan in December 1967 with very favourable financial and participation terms.

As stated above, however, no progress whatsoever has been made in obtaining equity participation in the concessions of the major companies—Aramco, KOC, and the IPC group which also owns the Qatar Petroleum Company. The heart of this problem lies in the significance for integrated companies of the acceptance of non-integrated partners in the production of their raw material.

## The Problem of Partnership with Governments

The important companies that produce oil in Arabia are all wholly-owned affiliates of internationally integrated oil companies. Their output accrues to their parent companies, each of which is entitled to a share in output proportional to its equity ownership in the affiliate. Thus they are not independent producers themselves exporting to world markets at world prices and free to make independent contracts for their crude and to export to whatever markets would gain them the highest prices. They are subsidiaries, producing oil in accordance with the international plans of their owners. As already noted, the parent companies taken together dominate the international petroleum industry at all levels and for the most part the oil they produce in their Arabian affiliates is produced for use in their own refineries or is sold among themselves on special terms.

Although the governments want to engage in integrated oil operations through their national oil companies, they are not yet in a position to use a partner's share of crude oil from the major concessions. Nor do they seek partnership merely to obtain greater revenues. The oil minister of Saudi Arabia stated very firmly last year that Saudi Arabia's long-term plan was to *control* all oil operations. And it is the possibility that increased control by a non-integrated governmental partner will result in pressures to increase output that creates the biggest problem for the oil companies. Merely to permit the government partner to take a partner's share of crude oil would not be a serious obstacle; the possible greater cost to the companies could be looked upon as part of the continuous process of bargaining over revenues, which the companies do not expect to cease in any case; but control that enabled the governments to press for increased output which the companies would be expected to dispose of is another matter, for it would involve serious pressure to increase market sales and hence to cut prices. The problem is manageable for crude-producing affiliates that supply only a small part of the parent company's crude requirements, but even a small percentage increase in the output of one of the major affiliates would create very large amounts of oil to be disposed of.

In principle, the governments of the crude-producing countries recognize the underlying problem, and are fully aware that an increase in free market sales carries with it the likelihood of a deterioration not only in crude prices but in product prices as well. In principle, therefore, a government should be prepared, in demanding partnership, to agree that it would not press for increased output which would force greater market sales. The Saudi Oil Minister has, for example, made it clear that the Saudi government, which has been opposing OPEC's attempts to establish quotas to control the output of each country's oil, would change its policy and support such measures if product prices fell to the point at which the maintenance of posted prices for crude oil became untenable.[8]

The difficulty is, however, that no individual government is in a position to appraise the overall circumstances of the parents of its major concessionaires and thus to take an active part in the determination of the international distribution of their supplies of crude oil. In these circumstances, cooperation by an individual government partner in policies designed to regulate output of the producing company in accordance with the outlets provided by the integrated operations of the other owners, and by their established contracts and the market sales which were considered possible at ruling prices without intensifying competition in product markets, would in practice imply government consent to a kind of prorationing operated by the companies.[9] But this is precisely what many of the governments object to in the present situation; as we have seen, one of the reasons behind the demand for equity participation is the desire of governments to have some control over, or influence on, offtake. Clearly, the only alternative would be for the governments as a group, perhaps through OPEC to join with the companies as a group in an effective cartel to plan the amount and distribution of oil supplies. There is as yet no evidence that this would work.

Although it seems, on the face of it, to be eminently reasonable that the governments of the oil exporting countries should be able to become partners in the exploitation of their own oil if they are willing to pay for an ownership share, it can now be seen that such a change could have far-reaching consequences for the organization of the international companies, who might well prefer to withdraw from ownership of their major sources of crude oil and become contractors to governments. This, the Saudi oil minister has set himself firmly against for fear that it might cause a drastic reduction in oil prices. The oil companies may offer a compromise in the form of arrangements which give the appearance of partnership without granting the government partner any share of control over offtake. In my view, this would only be a stop-gap measure and merely postpone the fundamental issue; it would be unlikely to last until the end of the concession period, which is not until 1999 for Aramco and 2026 for the Kuwait Oil Company.

The governments are not putting all their hopes of entering the oil industry on the prospect of obtaining an equity interest in their major concessions. In addition to the partnership arrangements made with their new concessionaires, most of them have established national oil companies, which we shall now briefly discuss.

*National Oil Companies*

The two important national oil companies in Arabia are the Kuwait National Oil Company and the General Petroleum and Minerals Organisation (Petromin) of Saudi Arabia. KNPC was established in 1960 with a 60% government and a 40% local private ownership. Petromin was created in 1962 as a public corporation wholly owned by the Saudi government.

Both companies intend to build up an international oil business, which will be in some degree integrated, in cooperation with foreign enterprises as well as on their own initiative. KNPC has completed the construction of an advanced export refinery and has set up an international marketing department with headquarters in London. A Danish subsidiary of the company has completed a terminal in Copenhagen and is already storing oil products. Supply contracts have been accepted in several parts of the world. As noted above KNPC has joined with Hispanoil in a joint enterprise (Kuwait Spanish Petroleum Company) owned 51% by KNPC, to operate the concession obtained from the government. KNPC also does all the local marketing of products in Kuwait.

Similarly, Petromin has lost no time in entering the petroleum industry. In addition to local marketing, which is the first activity usually reserved to a national company, Petromin has a 75% interest in a refinery in Jiddah and a 71% interest in a lube oil blending plant to be set up also in Jiddah. The company is gaining experience in exploration through its joint ventures with the French, Italians and others under the new agreements discussed above. It has also set up service companies for drilling and geophysical surveys, which work on contract for oil companies. In the international field, Petromin is planning to undertake refining and distribution enterprises outside Saudi Arabia, has sold oil to Rumania under a barter agreement, and is setting up a tanker company.

Petromin has not confined itself to petroleum operations proper but is attempting to advance the industrialization of Saudi Arabia, building on her mineral resources. A large urea plant owned 51% by Petromin is under construction using natural gas as the raw material; sulphur production is another major project of the company; a steel rolling mill in Jiddah—the first stage in an iron and steel manufacturing programme—went into operation in 1967, and further stages are being planned. Provision is made for a joint petrochemical venture in the new oil agreement with the Italian companies. Most of these ventures depend in some degree on foreign cooperation and usually involve joint companies with a 51% ownership for Petromin.

In addition to the national oil companies, a new Arab organisation has been created which is in some respects a kind of international Arab oil company. The Organisation of Arab Petroleum Exporting Countries was established in 1967 on the initiative of Saudi Arabia, with Kuwait, Libya and Saudi Arabia as founders. It has two types of function. One is to establish cooperation among the members and promote their interests. The other is more interesting. It is to 'utilize the common resources and potentialities of members in establishing joint ventures in various phases of the oil industry such as may be undertaken by all the members or those of them that may be interested in such ventures' (Art. 2. e). The organization is a juridical entity and can operate as a commercial organization. It could therefore make direct arrangements with consuming agencies on behalf of its members if this were desired. It is not designed to displace OPEC, but to supplement it for the member Arab states. Only an Arab state for which oil constitutes 'the principal and basic source of its national income', is eligible for membership. There are eight members but Egypt is excluded, since oil is not the principal source of national income. Whether OAPEC will become an important organisation in the Arab oil world remains to be seen.

*Conclusion*

We have discussed the more important changes during the past twenty years in the direct relations between the governments of the oil-exporting countries of Arabia and their foreign concessionaires, and we have examined the evolution of the role of governments in the oil business itself. Starting as little more than a collector of royalties, the governments of the larger exporters are now actively engaged in almost all aspects of the industry and in addition bargain with their concessionaires on equal terms. Great changes have taken place in the general political circumstances of the world as a whole, including the emergence in Arabian oil of important countries, such as Japan, France and Italy, who were willing to take an independent line in oil matters, as well as changes within the Middle East itself, including changes in the competence and aspirations of governments. These changes have placed the governments in a very different position today than they were when the original concessions were negotiated. The very need to spend oil revenues with a reasonable degree of efficiency, forced extensive administrative reorganisations in the structure of the state and sometimes extensive political changes as well.

There are many aspects of oil and state in Arabia that we have not discussed. To mention only a few: the role of oil in boundary disputes and the progress made in the fixing of boundaries that would have been unimportant in the absence of the possibilities of oil; the settling of offshore rights, in particular the disputes over the Persian Gulf between Iran and Saudi Arabia; the interstate economic cooperation that has taken place, including the Kuwait Fund for Arab Economic Development.

The most important issue in the relation between companies and governments in the coming few years will center around the question of equity partnership for the governments of the oil-producing countries. In Arabia this will involve primarily Kuwait and Saudi Arabia, for the other governments are not in a position to enter the industry. Saudi Arabia has taken the lead in pressing demands for partnership. What type of compromise, if any, can be achieved on this issue will depend partly on what happens to the markets for oil and products. If, for some reason or other, the rate of supply is kept under reasonable control (recently the dispute of the companies with Iraq as well as other political events in the area have helped to ease the so-called oil surplus) and prices do not come under severe pressure, the companies may be able to work out acceptable arrangements which would last until the governments began to feel the desire for a more effective control than such arrangements would be likely to give. If, however, the prices of crude oil and products resume their downward trend, the companies will feel less able to offer more to the governments and may prefer to withdraw from the ownership of their crude-oil production. Such a move would fundamentally change the role of the governments of the producing countries in the international oil industry, and would certainly raise more urgently than ever the question of the establishment of a governmental producers' cartel.

While these more sophisticated disputes go on, the face of the Trucial Coast will rapidly change as the expenditure of oil revenues accelerates in Abu Dhabi, Dubai and Oman and as exploration (with its attendant fees to governments) continues in the lesser Shaikhdoms.

## NOTES

1 It should be recognized, however, that much of the money expended for 'public purposes' will inevitably become diverted to private hands without any apparent productive *quid pro quo*. Moreover, a third of oil revenues easily made multimillionaires of the rulers and their families.

2 The early history of the concessions has been set forth in several authoritative publications which discuss the influence of political factors, including the rivalry between Britain and the United States, as well as of economic factors, such as the inability or unwillingness of any but the large and well-financed companies to undertake the risk and expense of exploration and development. The ownership of oil in Arabia is closely bound up with the economic position of the parent companies, including their access to markets. See especially George Lenczowski, *Oil and State in the Middle East* (Cornell Univ. Press, 1960; S. H. Longrigg, *Oil in the Middle East* (Oxford Univ. Press, 2nd ed., 1961); the United States Federal Trade Commission, *The International Petroleum Cartel* (Washington, D.C., 1951). All of the oil of Saudi Arabia, Qatar, Bahrain and Kuwait (with the exception of the Neutral Zone) was owned in unequal shares by eight international companies, seven of which produced 85% of the output of the world outside North America, China and the Soviet bloc in 1950 and also controlled nearly 75% of refinery throughput and probably a similar proportion of marketing.

3 This is the title of a paper presented to the Vth Arab Petroleum Congress (Cairo, 1965) by the Organisation of Petroleum Exporting Countries.

4 It should be noted that the tax cost Aramco nothing since the sum paid the Saudi Government could be deducted from the company's liability under the U.S. income tax laws. In other words, the payment was at the expense of the U.S. Treasury and the arrangement was accepted by the Treasury. At this time the United States government was particularly concerned to strengthen the position of the U.S. oil companies in the Middle East. In 1948 also there had been a substantial increase in royalty payments in terms of dollars when Saudi Arabia and Iraq insisted that payments should be based on the free market price of gold in the Middle East rather than the price of $35 an ounce fixed by the U.S. and British governments.

5 When product prices fall, the value of crude oil is reduced, and in such circumstances the companies would normally reduce posted prices for crude. In fact oil prices had been falling for ten years up to June 1967 but the companies have been required to pay 'income taxes' to the governments of the producing countries on a posted price unchanged since 1960.

6 Under the San Remo Agreement of 1920 Iraq had been promised a 20% equity share in the oil company to be formed in Mesopotamia, but she was later persuaded to give it up in return for a royalty. The problem posed for the major integrated companies by the demand of governments for partnership is discussed below.

7 The first such agreement in the Middle East was made in Iran in 1957 with the Italian state company.

8 Referring to Saudi Arabia's latest oil agreements, in a seminar in Beirut in June 1967, the Minister pointed out that taxes and royalties in these agreements were based on posted prices, but that there was also a provision requiring the foreign partner to market Petromin's share of crude oil production at an agreed realized price, for the determination of which a formula had been set out. He added, 'However, if it turns out that the realized price for sales to third parties is below what we are prepared to accept, we are entitled to institute a reduction in the overall production from the venture as

a whole. In other words we are prepared to cut back output rather than allow the price to fall below a certain level. And the motive for this lies in our recognition of the fact that any cut-price marginal sales can have a damaging effect on the whole price structure.' Quoted in *Middle East Economic Survey*, Vol. XI, No. 32 (June 7, 1968), p. 3.

9 Although the level of offtake in any one country is under the control of the oil companies, they are subject to political pressures from the governments. Iran, for example, has used its political bargaining power very effectively to gain increases in production. One of the problems in establishing production quotas arises from the fact that a number of countries are dissatisfied with their *share* of the market, and since market shares can only add up to 100%, an increase in the share of some (including new producers such as Abu Dhabi) must reduce the share of others. In spite of the Saudi Minister's statement above, he has also made it clear that his government would act to safeguard its own interest if the oil companies succumbed to the pressure of other governments at the expense of Saudi production.

## XVIII

## IRAN AS A PACEMAKER IN MIDDLE EAST OIL

*Discusses the leading role that Iran has traditionally played in the development of the oil policies of governments in the Middle East.*

IRAN has always played a central role in the development of the oil industry in the Middle East. In Iran the first oil discoveries in the area were made; Iran made the first attempt to nationalize foreign companies in the area and was first to establish a national oil company; and Iran is now taking the lead in the development of new types of oil contracts and relationships with foreign companies. The changes pioneered by Iran are of great importance, not only for Iran itself, but also for the industry generally. In this most international of industries developments in one country can never be isolated from developments elsewhere. This article will examine some aspects of Iranian oil policy and their significance for the Middle Eastern industry.

The policy and actions of the Iranian Government with respect to oil are marked by three outstanding characteristics: a deeply felt commitment and clear intention that Iranians shall increasingly participate in and eventually control their own oil industry; a high degree of skill, imagination and subtlety in approaching the objective; and a conviction that Iran is a natural leader for the Middle Eastern industry. The interplay of these three characteristics of Iranian policy in practice can be seen in almost every important move the Government has made. The first characteristic provides the motivation for action and sets the direction of policy; the second shapes the means adopted and the third explains many aspects of Iran's attitudes in OPEC and *vis-à-vis* her sister oil producers in the Middle East.

The strength of Iran's determination to bring its oil industry under national control was clear even in the early 1950s, for although Mossadeq's hope of eliminating the foreign capitalists was not in the end practicable, the broad basis of his policy—that Iranian oil should be run by the Iranians—was almost universally approved in the country and 'nationalization' in some form became a political necessity. However, the complicated provisions of the Consortium agreement barely concealed the fact that, in the circumstances, nationalization alone made little difference in practice. The more significant result of the conflict with the Anglo-Iranian Oil Company was the replacement of a single monopoly concessionaire by a consortium of companies, thus putting the control of production in Iran on much the same basis as that in the rest of the Middle East.

On the other hand, the establishment of the National Iranian Oil Company (NIOC) in 1951, the very first of the national oil companies in the major crude-oil exporting countries, and anticipating by nearly 10 years any other such company in the Middle East, was a significant step both nationally and internationally. Nationally, it took the local industry out of foreign hands and, even more important, it gave institutional form to Iran's aspirations. In doing this it put in motion an expansion of national activity which has steadily increased in scope. Internationally, NIOC not only provided a model for other oil-producing countries, but it led the way in negotiating new forms of oil agreements and in bringing the oil exporting countries into direct contact with international markets. In addition, the petroleum legislation of 1957, which established procedures for joint arrangements between NIOC and foreign companies, contained many provisions which made clear the Government's determination that the conditions under which Iranian oil would in future be exploited by new companies were to be laid down by Iran. Participation, with the foreigner acting primarily as the operating partner, was to be the order of the future.

Iran has approached this objective in a manner both realistic and imaginative, and, as a result, has achieved original and important successes in three distinguishable, though interrelated, fields—the terms of oil agreements, the direct role of the national oil company, and the methods of putting pressure on the Consortium to secure the levels of offtake and revenues desired by the Government. In all of these areas Iran's lead has had an important influence on the evolution of policies elsewhere in the Middle East, and in a variety of ways has set the pace of development. Moreover, the gains have been achieved without serious conflict with the oil companies, although there has been a certain amount of 'brinkmanship' in Iran's dealings with the Consortium.

In the new oil agreements the assertion of a national right to 'participation' has taken a variety of forms, as is evident from any summary of their salient provisions. The first agreements after the legislation of 1957 provided for a government share in the equity capital of companies formed with foreign partners to exploit oil resources. This led to an effective 75/25 division of total profits, since the Government not only imposed a 50% tax on profits but also received 50% of the remaining half as an owner. Iranians also had an equal share in management. In the more recent agreements, provision is made for half of any oil discovered to be set aside as a national reserve, and the foreign company's right to crude from the other half consists of an entitlement to purchase at special prices. The position of the foreign partner in these agreements has often been described as that of a contractor to the NIOC, but the foreign contractor also undertakes to market some of NIOC's share of the output.

These agreements are a far cry from the concession arrangements between the Government and the Consortium, which, for all practical purposes, approximate to the old-style agreements prevailing for the major concessions elsewhere in the Middle East. In spite of the advantages of the new agreements to the country and their much heralded novelty, however, the oil produced under them and the revenues received are negligible in comparison with the operations of the Consortium. Of a total of $938 million in oil revenues received by Iran in 1969, $908 million came from the Consortium companies. In 1969, of a total output of 166 million tons, 152·5 million came from the Consortium. Hence, in by far the greater part of its industry, Iran has no equity share, no real managerial control and no direct voice in the determination of output, investment or liftings by the companies, although the Government is always consulted and must be listened to. Nor is there any indication that any change is on the way. To be sure, the members of the Consortium discuss their production schedule with NIOC and the final programme is usually the result of a compromise, but the needs of the companies in the light of their commitments to lift crude oil elsewhere must be the overriding factors, since the companies have to dispose of the oil they take. Iran is thus subject to the same constraints on its ability to 'run' its industry as are the other countries in the Middle East. It has reacted in two ways, both of which have important implications.

The first is to extend as rapidly as possible the direct activities of the national oil company in the international arena and thus into downstream operations in world markets. The Government is serious in its attempt to make the NIOC into an integrated international company. NIOC has its own sources of crude oil, largely made available by the Consortium, and has secured a number of crude-oil contracts abroad, mostly in Eastern Europe; it is joining other companies in the ownership of refineries to obtain outlets for its crude oil; and in November, 1969, the U.S. subsidiary of NIOC even filed an application with the U.S. Oil Import Administration for a quota to import oil into the United States. Iran is clearly blazing the path that other national companies are only beginning to follow, for none of the others is as yet nearly so far advanced as the NIOC.

Iran, like the other oil exporting countries, does not want to remain primarily a seller of crude oil; downstream integration—active participation in refining and marketing abroad on much more than a token scale—is the objective, but this objective will be very difficult to realize. Certainly NIOC will be able to sell crude in increasing quantities as time goes on and will find it possible to participate in the ownership of refineries here and there, and perhaps even in the occasional small distribution network. But in comparison with the very large quantities of crude oil flowing out of the country every year, the international activities of NIOC are likely to be very small for a very long time to come. It does not seem likely that the position of the great international oil companies in the physical handling of oil in world markets will be significantly impaired by the expansion of national oil companies in the forseeable future. To some extent the oil companies may 'move over' to make room for these small rivals, especially if signs appear that the national companies are prepared to engage in price competition to find markets, but they will not move very far; nor will they join with them extensively as partners in downstream operations. On the other hand, if the national companies do aggressively compete for markets, the pressures on the world price structure, which are already serious, will be intensified. Of this the Government is well aware, and the steps it has taken have been taken very cautiously indeed.

Secondly, realizing that in any circumstances it will take a long time to build up its own internationally integrated oil company, Iran is determined that in the meantime the oil industry shall make the maximum possible contribution to its economic development programmes. In this connection it has engaged in astute and successful 'brinkmanship' in pressing its demands on the Consortium. The Shah has plainly stated that a situation in which foreigners can control the rate of development of his country's resources is intolerable. Nevertheless the fact is that foreign oil companies do have this control and there is nothing much that can be done about it except to persuade them in some way or other to raise offtake and revenues.

It was Iran's bitter experience in the 1950s that taught the oil-exporting countries the danger of pushing the oil companies too far and the advantage the companies possessed in their ability to shift their source of supply from one country to another. Iran has no desire at present to get too near the fire, but at the same time it cannot afford to allow the companies to feel sure about how far it will go. Iran has put pressure on the Consortium with originality and imagination, basing its claims for increased offtake, increased revenues, and special treatment generally, on the needs of its development programme, the size of the population and of the country, and the advantages that its political stability and international reliability offer the companies.[1]

The argument Iran presents has been outlined and published at length in the world outside Iran while simultaneously strongly worded demands on the companies are expressed by the Shah himself. Such tactics have never been used in this way and on this scale to wring concessions from the oil companies, but Iran's astute combination of threats and persuasive public argument has been remarkably successful.

The Iranians are frank about it—they insist openly on preferential treatment over the other oil-exporting countries on the grounds that they deserve it. They know that the dispute in the early 1950s cost them their place as the leading oil producer, and they believe that they should regain it. They know that the companies can take a greater proportion of their supplies from Iran, although admittedly at the expense of other producers. They know also that the companies must be nervously watching the deteriorating political situation in the Middle East. The Government is prepared to take full advantage of present circumstances in order to ensure the financing of the development programme and to assert Iran's right to leadership of the industry in the Middle East, not only in the formation of policy, which it has in practice assumed, but also in the amount of oil produced. In the Middle East Iran's only serious rival in both respects at the present time is Saudi Arabia.

Iran's case may be a strong one, but it has not always commended itself to her Arab neighbours, and a certain amount of friction has arisen in OPEC and elsewhere. This may become an important weakness in Iran's policy, depending upon the evolution of events in the oil industry and in the Middle East generally. The bargaining power of the oil-producing countries *vis-à-vis* the companies is probably stronger than ever before, and in recognition of this OPEC has proposed cartel arrangements which would require the cooperation of the producing countries in programmes to allocate and regulate production. Iran, like the other countries in OPEC, is willing to cooperate with other producers, but not at the expense of the expansion of its own industry, and her demands have at times provoked a reaction from some of the Arab producers, notably Kuwait and Saudi Arabia, who have warned that they will not accept a less than proportionate expansion of their own output. The oil companies, with interests in many countries will clearly be in an impossible position if together their host's insist on a percentage increase in output greater than the average percentage increase in consumption. In 1971 the emphasis was still on prices and a high degree of unity emerged between Iran and the other producers in the Gulf in the negotiation of an extremely favourable five-year price agreement with the companies.

The Consortium agreement provides for extensions after 1979, and the first five-year extension will be due then. This extension is not supposed to give rise to new negotiations, but it is widely assumed that there will be some renegotiation. Iran may then attempt to introduce into the Consortium agreement some of the provisions of the new-style contracts, such as, in particular, those which provide for government participation and guaranteed offtake as well as greater government control over the rate of development and production. Iran is undoubtedly technically capable of managing its industry, but is not much nearer being able to sell the output without the cooperation of the international companies than it was in 1951. The companies may be faced with a choice of accepting the Government as a full partner in both upstream and downstream operations or of becoming contractors to the Government with offtake agreements. They will probably prefer the latter, but if Iran continues to pursue its objectives with the caution and realism which have characterized its policies so far, the change, which ever way it goes, should come peacefully with no disruption of oil supplies—although there may be substantial effects on prices.

## NOTES

1 Iran also stresses that it is not directly involved in the dispute between Israel and the Arabs. If the situation in the Middle East deteriorates much further, Iranian oil will obviously be of crucial importance for the companies.

## XIX

## A NOTE ON DEVELOPMENT PLANNING AND THE ROLE OF THE ENTERPRISE WITH SPECIAL REFERENCE TO EGYPT*

*Examines some of the ways in which central planning may inhibit the dynamic expansion of enterprises and discusses the problem with reference to Egypt and to the importance of a diversified range of manufacturing exports for the economic future of the country.*

IT is becoming increasingly realized that one of the most difficult problems for state policy with respect to the formation of government plans to promote economic development and to control an economy relates to the role of the enterprise (the 'firm') in economic life. Apart from agriculture, it is in this area where the failures of central planning in the socialist economies have perhaps become most evident in recent years, and the changes that have been taking place in the economic management of some of these economies are to a large extent the direct result of the failure of earlier conceptions of the relation between the enterprise and 'the plan'.

But the problem is important also in so-called 'mixed' economies where there is a significant private sector in industry, although here it appears in a different form and is usually posed as the problem of the relationship between the 'private sector' and the 'public sector'. Any general economic plan must be concerned with the relation between these two sectors and it is usually recognized that the firms in the private sector will only fulfil a constructive role under certain conditions of freedom and security. But the plan must also be concerned with the proper role of firms in the public sector, where the general question of the relation between the growth of the firm and plans for the economy is much the same as in a centrally planned economy.

In a socialist economy the firm or enterprise is publicly owned, and in the socialist countries of Eastern Europe the problem of the role of the enterprise has received much attention, notably in Czechoslovakia where far-reaching changes in the economic system

---

\* The substance of this paper was first delivered as a lecture before the Société d'Economie Politique de Statistique et de Législation d'Egypt in Cairo in February 1968 and subsequently published by the National Institute of Planning of the United Arab Republic, Memo. No. 827, February, 1968.

were being introduced before the Soviet invasion. In discussing these changes, one Czech economist, for example, explained that an

'... important goal of the changes in the system of management is to make the economy more dynamic. The previous concept understood economic development only as a collection of quantitative relationships, so that it was essentially static. With this conception, the most important task of planning to assure economic growth was to attain as high a rate of growth in savings as possible . . . there is now a new concept of relations between enterprises and society. Formerly this relation was founded on the principle that society decides for the enterprise how much it should save and the enterprise was then obliged to pay into the State Budget the planned profit. Applying this principle led, of course, to a number of anti-social tendencies. . . . If a complex and integrated management of the enterprise is to be made possible . . . it is essential for the enterprise to have considerable freedom in managing its own finances, for it to decide itself how it can use its money, for instance, on wages or on investments, etc.'[1]

Central planning can and does take a variety of forms with varying degrees of direct control over enterprises, but the notion that central planning of some kind is desirable for an economy rests primarily on the assumption that the 'market' left to itself will not bring acceptable results in resource allocation, economic stability, income distribution and various other economic and social objectives. In particular, private profit and private control of industry are deemed inadequate as the guiding principle and the primary means for the regulation of economic activity. Probably nowadays most people would not quarrel with this assumption (although the further assumption of some that profit is not only inadequate but also immoral is, of course, another matter). Most economists recognize that some form of government economic management is desirable and the area of controversy is largely limited to debate over the scope and form of such management.

## The 'Static' Nature of Plans

Many development 'plans' are little more than a series of hoped-for projects; others involve the determination of a desired pattern of output in more or less detail for the economy as a whole including the division of the national income between consumption and investment, which is supposed to determine the rate of growth of the national income, the structure of industry, again in more or less detail, the allocation of manpower, etc. Available resources are appraised in both real and financial terms, and the planning process essentially involves the allocation of resources and flows of finance in accordance with a desired pattern and growth of output. In the more technically sophisticated planning, macroeconomic models of the economy are created, input-output tables constructed, and all the technical apparatus of modern 'quantitative analysis' is brought into play.

As Mr. Sokol pointed out in the passage cited above, a plan conceived of as a series of projects or of macro-economic quantitative relationships is essentially static. It is static in the sense that there is no built-in mechanism to induce changes in unforseen directions in response to changing circumstances. An economy where the activities of enterprises are governed entirely in accordance with 'targets' determined in advance by a central plan will not itself be characterized by an endogenous or organic process of growth in the sense that movement from one position to another is a result of an interaction between the producing enterprises and their environment. There is, of course, nothing to prevent the growth of such an economy as investment takes place according to the plan, but in a very real sense the growth is a growth imposed from above, its efficiency depending on the 'preferences', that is, the theories, of the planners. These preferences may be appropriate, and a satisfactory rate of growth may be achieved—at least for a while—but it can hardly be said to result from an inherent 'dynamism' in the economy itself in the sense that is usually implied by the term 'self-sustaining' growth.

There has been a great deal of discussion in economic literature about the pre-conditions for 'self-sustained growth', and the importance of raising the rate of saving to some allegedly critical percentage of income is given pride of place. But if this term means anything at all it surely means that there is somewhere built into the economy a dynamic relationship which by its very nature both creates investment opportunities and promotes further investment. This is not characteristic of a plan, nor will the mere raising of saving and investment in some way or other produce such a result. It is only achieved when individual productive units—the firms—are continually and restlessly searching for ways of expanding. In this process lies the primary source of economic dynamism.

In theory, a plan can be made dynamic through the introduction of appropriate procedures for revisions in the light of events, but this is more like a model of 'comparative statics' than a model of dynamic interrelationships. In principle such interrelationships can also be incorporated into planning by the introduction of 'feedback' and response systems. This involves an elaborate 'computerization' of the economy and, so far at least, the fruitfulness of models of this sort as a means of centrally controlling economic activity remains to be demonstrated. They are essentially mechanical models and whether they can take adequate account of, or give adequate scope to, the contribution that can be made by the judgement and genius of a myriad of individuals in the society needs much more investigation than it seems to have had.

In any event, as noted above, the experience of many of the Eastern European countries has not so far been satisfactory in this respect; their difficulties have, to a large extent, stemmed directly from the inability of the planners to take adequate account of the role of the enterprise and the nature of its growth. Not only have the 'preferences' of planners often proved grossly inadequate (the theories of development held by the planners are crucial in a planned economy) but the reactions of the enterprises in the economy to the framework imposed on them by plans have also often been perverse. Moreover, for countries that must depend heavily on a diversified variety of manufactured exports, freedom of enterprises to respond directly to the market may be of decisive importance. As is suggested below, this may be vital for the future of Egypt.

*The Role of the 'Firm'*

In any economy, the basic unit of industrial production is the firm or enterprise, for it is the firm that acquires the factors of production, organizes them in the productive process, designs the methods of production and the products, and usually surveys markets and arranges sales. In particular, the firm is the organization through which innovations are put into practice. The enterprise I am calling the 'firm' is more than a factory or plant, that is, more than a physical producing unit operated by workers and managers to produce the types and quantities of goods for which the plant is designed. A firm is an organization which, in addition to operating one or more factories, has the broader task of selling goods, searching for markets, raising finance, including finance for expansion, determining its pattern of output and the use of its resources in new ways. In a 'market economy' it is the primary dynamic unit, acting as an intermediary between final purchases of products and the sellers of productive services.

An established firm is a continuing organization of people knit together in a defined institutional framework working to obtain an income which is in part paid out to the suppliers of productive services (including management), in part to owners (who may be private individuals or governments), and in part retained for further investment. To this end, an enterprising firm often undertakes research of various kinds, promotes inventive activity among its people where practical, and experiments with new ways of producing and selling its existing products or with the introduction of new products, and expands in new directions. Many of the activities that firms in a private-enterprise economy engage in may be considered socially undesirable or even economically disadvantageous to the economy as a whole, but this is another question. The point I am trying to make here is that the firm is the elemental focal point of economic dynamism. If for any reason firms are unwilling or unable to act in an enterprising manner which furthers the industrial development of the economy there will be no other source of dynamic industrial growth to call on.

## Planning and the Role of the Firm

It is widely alleged that in underdeveloped countries private firms and individuals have in the past, and still are, unwilling or unable to take on the task of industrial development, lacking either the ability to overcome the peculiarly difficult obstacles in a backward economy, the enterprise to attempt to do so, or the incentive to take up industrial pursuits in contrast to commercial or financial activities or speculation. In consequence, government action is held to be required not only to create the necessary infrastructure and accept the necessary risk involved in the establishment of industrial plants, but also to control and direct the activities of the 'private sector'. Again, there are probably few who would challenge the proposition that considerable government action is required to speed up and direct development and that for such action governments should plan carefully what they want to do. In most countries nowadays some sort of macroeconomic plan exists but in many countries, including Egypt, planners have taken inadequate account of the dynamic role of the enterprise in the process of growth, and, in consequence, of the importance of giving sufficient scope for its expansion as the initial conditions which may have inhibited its growth change. It is obvious that the mere freedom to expand is not enough to ensure dynamic behaviour on the part of firms, but I shall argue that a development policy is not likely to produce self-sustaining growth until it first induces and then gives scope for such behaviour.

I should like to make clear that the problem discussed here is not, in principle at least, related to the question of whether or not enterprises are publicly owned. It has long been standard in the analysis of business enterprises to distinguish between ownership and management, and it is widely recognized that for large firms in any event, the effective control of the firm, including the making of fundamental policy, lies primarily with management. Indeed, one could make a strong case for the proposition that the chief significance of private ownership is that it permits the existence of capital and stock markets, and that the chief essential difference between a privately-owned large firm and a public enterprise is that the former must pay attention to capital markets and share prices. In addition, in the private sector new firms come into existence in a different way. In this paper, however, I am not concerned with the problems of public ownership, but with the relation between a central government plan and the freedom of action required by enterprises if they are to become the source of an organic and dynamic growth in the economy. If we leave aside the problem that arises for private enterprise in a mixed economy when government policy regarding private firms is either frankly antagonistic or simply vacillating and unclear thus creating such an atmosphere of uncertainty that firms refuse to take risks or make heavy commitments, there are broadly three ways in which the scope for enterprises which would permit them to fulfil a dynamic function, tends to be inhibited by government planning:

1. By the government telling firms what to do and prohibiting independent action, and especially investments, without prior approval.

2. By the absorption of scarce resources in new projects or in projects imposed by the 'plan' on such a scale that existing enterprises are denied adequate resources.

3. By the imposition of regulations which destroy or pervert the incentive of the firm itself, or of its managers, to take special risks or make special efforts.

The first occurs in the so-called 'command economy', which has been characteristic of a number of Eastern European countries and is apparently what some Egyptian economists had in mind when they argued for a widespread nationalization of industry on the ground that this would enable the state to control and direct the flow of investment in the economy. A hierarchy of government agencies of one kind or another has been set up in Egypt to control the investment of enterprises in accordance, in principle at least, with the plan.

The second occurs most commonly when the plan envisages the construction of new projects, and especially large ambitiously designed projects, which absorb the greater part of the foreign exchange available, thus creating a 'foreign exchange shortage' which compels the government to refuse permission for the expansion of existing enterprises in many projects even universally accepted as desirable. It should be remembered that any government can in this way create a 'foreign exchange shortage'; as will be argued below, such a course may be particularly dangerous for an economy which must depend on the development of a diversified range of manufactured exports. Again, there is reason to think that this kind of policy has inhibited expansion in Egypt.

The third way in which the dynamism of enterprises may be

restricted tends to occur in economies where certain types of social policies, in particular policies with respect to income distribution are carried to great lengths without regard to their economic consequences, or where the enterprise is prevented from establishing a direct relation with the market, as is the case when it is required to sell through an intermediate selling organization which has no connection with it.

Let us now consider in more detail and in the light of the nature of the firm and of its growth, the way in which these three types of restraint on enterprise affect the firm.

*Direct Control of Activities of Firms*

Like other human organizations, an enterprise involves the life of the men who work in it and, if it is to operate effectively, must call forth some sort of commitment from them, especially from those whose function is to undertake responsibility and devote what talents they have to creative, non-routine activities. Men seem to try to find a satisfaction or meaning in their working lives by identifying themselves with the institutions in which they work and finding in the success of the institution a purpose worthy of the commitment they make.[2] Cooperation to this end leads to the development of the firm as an organism in its own right which establishes its own goals and reacts aggressively or defensively to the outside world to advance its interests or to protect itself from the adverse consequences of external events. The psychology of social organizations is not within the special expertise of economists and I have not space to analyse this aspect of the problem here. But since economics is a social science, such psychology cannot be entirely ignored in a discussion of the role of the firm in economic activity, and the general point I want to make is that firms cannot be treated as passive agents executing other people's decisions; they will inevitably react creatively or perversely according to the circumstances.

The central planning of a 'command economy', in principle at least, tells the firm what it ought to do both in terms of targets to be fulfilled and in terms of the general policies governing its operations. It is true, that even in such an economy, planning starts to a considerable extent from the bottom up, so to speak. Firms are consulted about their requirements and objectives and they put forth their own plans for consideration. They are not left out of the planning process. Nevertheless, the 'coordination' of the plans of firms in the light of the overall 'strategy' adopted for the economy is in the hands of a variety of superior organizations, which make the final decisions, hand down the targets and instructions, and lay down the criteria according to which the success of firms will be judged. The scope for enterprising activity on the part of firms is thus confined to their participation in plan-making and in putting forth suggestions for revisions, but always subject to the higher authority of the controlling organizations. A variety of techniques for decentralizing decision-making can be built into the structure of control, but this is merely another way of saying that the firm may be given a certain amount of freedom or autonomy in particular areas. But at some point such freedom for the enterprises becomes incompatible with central planning of this kind, which rests on the assumption that a coordinated plan governing the actions of firms will give better results than would be achieved by allowing firms to react to opportunities without reference to the objectives of an overall plan for the economy.

With this approach to the relationship of the firm to the plan, several problems arise. In the first place the plan may be wrong in important respects, wrong in the sense that it requires actions of the firm which are patently unreasonable or it prevents the firm from taking action that is patently sensible. If one is to rely on the criticisms of economic policy that have been published in socialist countries, such results are common, ranging from the simple overproduction of goods the market will not absorb as a result of attempts by firms to fulfil the targets given to them, to the employment by the firm of workers it cannot use (and which may even be in demand elsewhere) in order to fulfil their employment targets. If the firm at the same time is required to make profits to fulfil financial targets, it often reacts to protect itself from the adverse consequences of failure, by manoeuvring to obtain especially low targets, by using its resources inefficiently in order to justify higher prices, by altering the quality of products, etc. Here the scope for 'enterprise' may be very great as the 'organism' engages in defensive manoeuvres in its own interest, but the 'enterprise' is of course perverse from the point of view of the growth of the economy.

These are negative results of enterprise. If we turn to the more positive side of the enterprising activities of a firm, a problem arises when a firm wants to grow more rapidly than the growth of the market for its existing products. Sometimes a firm finds fairly early in its career that it can profitably enter new lines of activity while continuing to expand in existing lines. This is commonly due to the fact that the rate of growth of the market for existing products is insufficient to absorb the firm's resources and energies, especially its managerial and entrepreneurial resources together with the specialized knowledge it may have obtained about the potentialities of its markets, the raw materials it produces, or of related products.[3] In underdeveloped countries, in particular, an individual firm may often find advantages in moving forwards or backwards in the industry in which it is engaged or in producing some of its own requirements.

In this respect the normal expansion of a firm pays no attention to the conventional categories in which industries are classified. The organization for the control of industry in accordance with a plan, however, is likely to be set up in accordance with such categories, and each firm will be classified as belonging to some particular industry. If a firm puts up a proposal for expansion which cuts across the organizational categories, the expansion may be natural, logical and efficient from the point of view of the firm but illogical to the tidy minds of the bureaucracy. It is not difficult to see why such proposals may cause much in-fighting among the control organizations concerned. If, for example, there is a controlling organization which includes the rubber industry and another which includes automobile manufacturing, any attempt of a firm classified in the latter to produce its own tyres is likely to meet opposition from the former, which may insist that there is excess capacity somewhere among its own factories which could be turned to tyres. The issue is likely to be decided according to the balance of bargaining power among the controlling organizations without reference to the nature of the intangible qualities of management or enterprise which may have given rise to the proposal in the first instance. I found a number of such cases in Egypt in 1968.

Each industrial firm is unique in the same sense that each human being is unique, for it consists primarily of a collection of particular individuals operating in a specific set of circumstances which are not duplicated elsewhere, and each has a unique history. This uniqueness gives rise to unique opportunities for each particular firm in addition to the opportunities that are open equally to all firms in particular lines of activity.[4] The natural course of expansion of a firm involves change and continual innovation regardless of the existence of industrial 'census categories', which are convenient (though often misleading) statistical boxes irrelevant to the course of change and the process of growth. To impose restraints for such reasons on the expansion of firms can well be fatal to the development of self-sustaining growth of the economy, for it may not only prevent firms from taking advantage of the unique opportunities that come their way, but even kill their incentive to do so. The restraints are imposed in the name of 'coordination', but coordination of industrial activity that takes the form of an allocation of activities and targets to firms in accordance with static statistical classifications designed for other purposes, takes no account of the nature of the growth of firms and gives rise to a severe conflict between the imperatives of economic growth and the provisions of the plan.

## Diversion of Resources from Existing Enterprises

Many of the things that firms want to do will be 'little things', in the sense that they appear insignificant to the planners against the background of the 'big projects' which seem to have preference in the formulation of a strategy of growth. And this leads us into a discussion of the second way in which planning may interfere with the dynamic function of the firm, especially from the point of view of the development of underdeveloped economies. The first point to note is that as yet economists know very little about the process of economic development. There are many (competing) theories, some stressing balanced and others stressing unbalanced growth, some emphasizing light and others heavy industry, or capital-intensive or labour-intensive techniques, some insisting that agriculture must have priority and others that this should be given to industry, etc. But a reasonable body of agreed and tested knowledge is still to come. In view of the insecure theoretical foundations for development planning, together with very imperfect ability to predict the future, it is highly probable that in very important respects any plan is likely to be seriously deficient, and if the plan is wrong, projects will be started that are inappropriate, activities will be forced on existing firms that are uneconomic while other activities are restricted. The result is demoralization of management, discouragement of enterprise, and inefficient investment. It is reported, for example, that in Czechoslovakia the capital/output ratio began to appear negative in the early 1960's, presumably largely as a result of malallocation of investment. The national income statistics of a country may show a rising income when investment expenditures take place regardless of whether or not the new plant when it comes into operation will be productive, for such expenditures are themselves counted as part of the national income. But if the investments turn out to be unproductive, no permanent increase in income occurs.

In underdeveloped countries there is reason to suspect that much inappropriate investment takes place. This is sometimes defended as an inevitable part of the trial-and-error process of development—as no doubt it is—and in any case as providing a useful means of shaking up a 'traditional' economy because of the general ferment consequent on the expenditures and the activities accompanying them. But whether or not the large investment projects of the state are appropriately chosen, there is a grave risk that such projects will pre-empt such a large proportion of the available foreign exchange that the expansion of existing enterprises is stifled. And if the appropriateness of the broad economic theories adopted by the planners is doubtful in view of the uncertainty surrounding both our understanding of the process of economic development, and our expectations of the future, there is no overwhelming presumption that the projects chosen by governments will in general be a better use of the resources available than the alternative projects which might arise if existing enterprises were encouraged and enabled to take advantage of the opportunities for expansion open to them.

Again, however, I should like to emphasize that this argument is not an argument against the government undertaking important projects deemed desirable for the development of the country, and especially those projects that create the necessary infrastructure. Nor can one assume that every proposal for expansion by existing enterprises should be sanctioned, but there is much to be said for giving priority *in principle* to projects that existing firms want to undertake, providing that they are not obviously inconsistent with the objectives of development. This is, of course, a vague proviso, but if the importance of promoting dynamic behaviour on the part of firms is accepted then a presumption that the proposals of existing enterprises should be given priority in principle would go a long way towards achieving this end. Many of these proposals are likely to involve relatively small investments, but in total the sum of them could well be quite large. Moreover, many of them in the nature of the case may be for the production of products that require no subsidy and little or no protection, thus laying the foundation for potentially competitive exports. For Egypt, this latter possibility may be of great importance.

It seems likely that Egypt's economic future will depend heavily on the possibility of developing a diversified range of manufactured exports as, for example, Japan did very early. Expansion of Egypt's existing standard exports is unlikely to be sufficient to support extensive further development and so far as the establishment of other obvious industries is concerned, such as petrochemicals, fertilizers, and iron and steel, etc., she is likely to find not only that it will be difficult to become competitive with exports from existing industrial countries, but that a number of other developing countries are thinking along the same lines. But given the level of industrial development already in the country and the supply of managerial and technical talent, which is by now very considerable, it seems reasonable to suppose that, with the application of more entrepreneurship, selling enterprise and energy, her existing firms could find numerous small opportunities for themselves. Indeed, even the few and brief discussions that I have had with those dealing with Egyptian industry convinced me that there was considerable scope for such expansion if the incentive and ability to undertake it were not almost deliberately impaired by existing policies, which are excused partly by the 'foreign exchange shortage'. But this 'shortage' is largely a reflection of the fact that the government has chosen to use the funds available for investment in other types of project.

This raises the question of whether Egyptian planners and government officials may not be underestimating the potential of the existing enterprises in the economy. There seems to be a widespread assumption that from the point of view of entrepreneurship, managerial ability, and technical competence Egypt has not advanced very much in the past 20 years except insofar as the planners and government officials themselves take action. In my own discussions, the above proposals with respect to granting more freedom to firms to exercise their own enterprise and judgement were frequently met with the reply that in the past such entrepreneurship had not been forthcoming, which had made it necessary for the government to take over, and that there is no reason to assume that things had changed much in this respect. I did not have the impression, however, that anyone had tried to test this assumption. Furthermore, there is a widespread reluctance to trust individual managers even with foreign exchange to go abroad to find markets in which to sell their goods. It is possible that the static hand of control, by stifling entrepreneurship, thus justifies its own existence, while ignoring the possibility, which seems to me to be equally probable at this point in Egypt, that many of the 'pre-conditions' for dynamic industrial growth have in fact been already created and that effective responses would emerge if permitted.

## The Structure of Incentives

Closely connected with the problems just discussed is the third of the ways in which government policy may stifle enterprise, and I believe is in fact doing so in Egypt. This relates to the absence of incentive for managers of firms to exert themselves in any special way except for patriotic reasons—which can never be an adequate foundation for continuing long-term activity. There is even little or no incentive for firms to make special efforts to export, and sometimes positive disincentives in this respect. The lack of incentive is of two kinds, one relating to individuals and the other to the firm. It seems to me that the general system of incentives in industry, and in particular so far as it affects potential exports, is one of the single most serious handicaps which the industrialization of the Egyptian economy now faces.

I cannot take up the discussion of this problem in any detail at this point, which is in any event very well known, but so far as I can see little attempt has been made clearly to appraise its effects. Managers of aggressive and expanding firms must work very hard—there is no question of their working a 7 hour day or 5 day week, for they live with their job, which often absorbs the greater part of their waking hours in one way or another. But there must be some incentive for them to accept such a strenuous life, an incentive which is directly related to the effort they exert. Clearly an inefficient bonus system or a ceiling on income which can be reached as early as 40 years of age—which is just the age when a man can begin to make the greatest contribution to economic activity providing he is willing to accept responsibility and risk wholeheartedly—will call forth little such effort. There is little question that this leads to demoralization, lack of commitment and consequently a tendency to accept the routine tasks and refrain from exerting those special efforts with a consequent risk of ulcers! The potential genius of men is thus killed before it has had a chance even to develop, and the entire economy suffers. The problem is not confined, incidentally, to the managerial strata, but exists at all levels. Wage differentials are often such that the abler worker cannot be rewarded in accordance with his productivity, with the result that he finds no incentive to do his best since the lackadaisical worker is just as well off as he is.

The other type of incentive relates to the firm itself, quite apart from the individual incentives for management. As suggested above men have a tendency to identify themselves with the organization for which they work and in consequence they become discouraged and disgruntled when the organization is prevented from doing a job of which they can be proud, or when the system is such that the organization is faced with a conflict in the objectives imposed on it and must therefore fail in achieving one or more of them. This is partly the problem of multiple 'success criteria', which has received considerable attention in Eastern European literature, and partly the problem of what happens when the plan is wrong, which I discussed above. On the other hand, if it is desired to encourage firms to act in particular directions, they must be given particular incentives to do so. Although profit is an important criteria for judging the success of enterprises and an important incentive in almost all systems, the really effective incentive for an enterprising firm is the prospect not only of making profit but of retaining some of its earnings for expansion.

If the men in charge of a firm are enterprising, they will want to improve its factories, expand its activities, and have scope for putting their creative ideas into effect. This requires money, and in an underdeveloped country especially, it is likely to require foreign exchange. If the government appropriates the profits of a firm, and especially its foreign exchange earnings, while doling out investment funds in accordance with a plan, this incentive is almost completely destroyed. This problem is of especial importance for countries whose development depends heavily on their ability to produce a diversified range of manufactured exports, and if a government wants seriously to encourage firms to make special efforts in this direction it would do well to permit them to retain that part of their earnings which they could use in expanding and improving their plant.[5] In Egypt there seems to be little or no incentive for firms to make special efforts to export their products, incredible as this may seem to an outside observer. It is not surprising to me that, apart from special bilateral arrangements made with certain groups of countries, the growth in exports, and especially in the variety of exports, has been disappointing. In this connection too, it is important that firms be able to get into direct and constant touch with the foreign markets, and again in Egypt special obstacles are put in the way of their doing so. It is possible that bilateral trade agreements are to a large extent made necessary by the very policies of governments which restrict opportunities and deny incentives for enterprises to export in other ways.

*Conclusion*

This paper has been largely a critique. I do not deny the importance of planning for development and I have not tried to suggest how such planning should proceed if it is to be effective while at the same time giving freedom to enterprises to expand in directions that seem profitable to them. I do think that the centralized control of enterprises by higher governmental authorities is inimical to the development of a dynamic industrial economy, and the experience of socialist countries, especially the smaller ones for whom exports are important, seems to support this conclusion. More attention is everywhere being given to decentralization in planning and to the role of the 'market', since the market is an alternative method of coordinating economic activity and allocating resources. But very little attention is being given to the study of the nature of the growth of enterprises in relation to planning. I have tried here to expound some of the problems; I have not put forward many specific proposals regarding how the relationship between the plan and the enterprise might be worked out. Much can be learned from an examination of the methods of control and planning adopted in large diversified firms with many subsidiaries. But the analogy can be misleading and should be cautiously used, since there are great differences between the relation of a firm to its environment and that of a country to the outside world. The chief purpose of this paper is to point to an area where further research is needed, research that must be related to the specific conditions of specific countries. It seems highly probable that there are opportunities in Egypt that are being neglected, especially export opportunities, and that some of the misdirected effort in present policies and plans might well be avoided in the future if sufficient attention is given to the importance of the dynamic role of firms in the process of growth.

## NOTES

1 Miroslav Sokol, 'Changes in Economic Management in Czechoslovakia', *Czechoslovak Economic Papers*, No. 8, 1967. pp. 12, 13, 15.
2 It is partly for this reason that a closer association of 'the workers' with the control of the firm is often deemed desirable, but there is a difference in kind between activities that involve the acceptance of personal responsibility for a large variety of matters, the need to make important decisions affecting the firm as a whole, the organization and management of other men, and in particular the need to exercise imagination and creative judgement, and the routine activities undertaken by the majority of workers. It is sometimes feared that the latter will necessarily be 'exploited' or unjustly treated if they do not participate in management, or it is believed that since workers too are part of the firm they have a natural right to participate in the control of it. Thus, in some countries (and in some firms in all countries) arrangements are made to enable them to do so. Whether or not such arrangements exist makes no difference to the argument put forth above, which is not concerned with the composition of management but with the scope to be given it.
3 I have discussed this aspect of the growth of firms in some detail in my *Theory of the Growth of the Firm* (Oxford. Basil Blackwell. 1959).
4 This point too has been developed at greater length in my *Theory of the Growth of the Firm*.
5 Of course, if the exchange rate under values foreign currency, there might well be a tendency for managers to attempt to 'modernize' their plant by importing capital equipment which is more appropriately used in economies where labour is scarcer than it is in Egypt. Methods might have to be devised to discourage this unless it can be shown that increased capital intensiveness would increase the competitiveness of the firm in export markets enough to offset the increased import cost as well as the adverse effect on employment.

# SECTION IV

# ECONOMICS AND THE ASPIRATIONS OF
## LE TIERS MONDE\*

### I

IT is extremely difficult to keep scholars on a predetermined, straight, and narrow path. Academic life not only has an inner compulsion of its own, but scholars, like goats, tend to wander afield, and as they move from one object to another with somewhat indiscriminate though scientific curiosity, they are led further and further away from their original path. Originally the School of Oriental and African Studies was established strictly to be useful; its purpose as set forth in its 1916 Charter was to 'give instruction in the Languages of Eastern and African Peoples, Ancient and Modern, and in the Literature, History, Religion, and Customs of those peoples, especially with a view to the needs of persons about to proceed to the East or to Africa for the pursuit of study and research, commerce or a profession . . .'.

The Reay Committee, in recommending the establishment of the School, reported in 1908 that 'there now exists a very strong feeling that if the British are to maintain and improve their commercial position in the East and in the Far East, a knowledge of Oriental languages must be regarded as indispensable to the businessman doing business with Oriental peoples'.[1] In the Parliamentary debates on the Report the needs of business and of government officials were prominently set forth, and the high interest of Imperial purpose was stressed.

Evidently the early members of the School did not adhere very strictly to the assigned task of merely 'giving instruction . . .' Doubtless they served very nobly the needs of commerce and government, but they quite clearly went about their own business as well, and by 1932, in an act that I am sure must have been little more than a belated recognition of an existing state of affairs, the purposes of the School were widened to include, in addition to giving instruction, the prosecution of 'study and research', and in the additional subjects of Law and Art, still relating to Eastern and African peoples.[2] So far as Eastern peoples were concerned, most of the 'instruction' as well as the 'study and research' fell comfortably within the traditional categories of oriental studies.

\* An inaugural lecture delivered on 10 February 1965. Published by the School of Oriental and African Studies, University of London, 1965.

Next time the statement of the purposes of the School is formally revised it will again be to recognize the path already beaten by the irrepressible academic goats, and it will have to include instruction, study, and research in geography, politics and economics. Nevertheless, the wandering, though perhaps a bit outside the letter of the law, has been completely consistent with the spirit of those who supported and approved the establishment of the School during the first decade of this century.

Formal alterations of an academic structure are not made lightly. I have not carried out a careful investigation of this matter, but casual observation leads me to suggest that changes in the problems studied, in the syllabuses of instruction, and in the academic establishment, take place behind the developments in the outside world which called them forth with a lag of something like half a century. This School, for example, was the first Institute to specialize in teaching Oriental Languages in Britain, but it was founded only in 1916—surely some fifty to sixty years after Britain's increasing involvement in Oriental Affairs must have made the need obvious! The Reay Committee's Report included references to the desirability of the provision of teaching sociology, and political, economic, and commercial geography, and expressed the hope that 'before long such instruction may be provided directly or indirectly for the students of the School'.[3] That was, you will remember, in 1908. Last year—1964 —Professors of Sociology and Geography were appointed—the hopes of the Reay Committee are well on the way to realization—and the School will soon be celebrating its fiftieth anniversary.

But we have today gone even further. The pace of change in Asia and Africa has quickened, and these countries are increasingly adopting new and 'modern' ways; the techniques of examining, interpreting, and appraising contemporary social and economic conditions have become more ingenious, and the development of social accounting and other forms of economic statistics has increased the scope for useful statistical description (as well, incidentally, as the possibilities of government and business planning). As a result, it is abundantly clear that the range of studies required for an adequate interpretation of modern Asia must extend even to economics—the one discipline of all the social sciences that shares with linguistics an insistence that its principles are independent of place and period, and which has demonstrated its capacity to establish significant general relationships among phenomena.

Hence, the real significance of this occasion is not my own inauguration as a Professor of Economics, but rather the inauguration of Economics in the School of Oriental and African Studies. The presence of economists here today is the result of two parallel academic responses, both called forth by the same external stimulus, a stimulus which is, I suggest, also part of a more general movement toward closer integration of the world economy. The external stimulus is the economic and social modernization of Asia and Africa. This calls, on the one hand, for an extension of Oriental and African studies to include the modern social sciences, and, on the other, for an extension of economics—both in theory and application—to deal more effectively with the economic problems of most urgent interest to the Afro-Asian countries, along with other largely non-industrial economies. The School of Oriental and African Studies, in establishing a Department of Economics, has responded to the first call, economists joining the School to pursue their studies in an institution devoted to Asia and Africa have responded to the second call. I am convinced that today, perhaps more than ever before because of the very rapid economic changes that are taking place in the Afro-Asian world, both Asian and African studies and the Science of Economics have much to gain from a closer mutual association.

II

We are concerned with an extremely heterogeneous group of countries, each of which is different in one or more fundamental respects from the others. Yet all have one overriding characteristic in common—the great mass of their people are, in both aptitudes and attitudes, as yet only on the edge of the modern world in respect of their ability to make use of modern technology, modern scientific discoveries, and modern organization in industry and agriculture, and also in government. One must not forget that the severe lack of trained administrative cadres and of experienced political leaders is often one of the most serious aspects of the general state of under-development.

This clear political and sociological, as well as economic difference between the so-called developed countries (including Japan) and most of the rest of the world, vindicates, I think, my borrowing an expressive and evocative phrase from another language—*le tiers monde*—'the third world', which I chose deliberately, but not for precious effect. The English equivalents—'new' or 'emergent' (and many of these countries are neither new nor emergent), 'under-developed', 'developing', or even 'backward' and 'poor'—stress too specifically single aspects of an interrelated complex of motivations and economic and political circumstances, and in any event, some are often chosen more for their political acceptability than for their descriptive accuracy. Nevertheless, economic advance is central to the aspirations of all of these countries. Even political independence was often demanded partly in the belief that colonialism perpetuated economic backwardness, and today the political ambitions of the *tiers monde* are intimately connected with economic development: the realization of their nationalist aspirations depends on it.

I should perhaps make it clear from the beginning that I am adopting, without labouring to justify it, a reasonably optimistic approach with respect to the prospects of economic development of most of the poorer countries of the world. Provided that a major war is avoided and that some progress is made in controlling the rate of population increase, I see no conclusive reason to assume that the various problems of these countries will not in time be solvable, although the pace of progress will certainly not be as rapid as is hoped, and there will be many setbacks.

A technological revolution lies at the heart of this network of interrelated changes that we call economic development; such a revolution is not brought about by the mere importation of factories and machines of various kinds from abroad and the forced expansion of capital formation in industry. It is a much more difficult and much deeper process, for it requires the development, not only of technical and administrative skills, but of institutional arrangements and social and intellectual attitudes that will enable the local population to adopt, and also creatively to adapt, modern technology in appropriate measure for their particular economy. After all, it is not the machines in themselves, or even the savings-cum-investment, that are responsible for the high living standards now attained by the developed countries —it is the ability, skills, and economic behaviour of the people that have made possible the development and efficient use of these machines. A recent investigation into the sources of economic growth in the United States, for example, came up with estimates that education accounted for some 23% of the rate of growth of the aggregate national product between 1929 and 1957, or 42% of the rate of growth in output per person employed. Any such estimate is, of course, subject to a large margin of error, but it does perhaps indicate how very much importance must be attached to increasing the qualifications of the population.[4]

The successful adaptation of modern technology in industry and agriculture, in administration and management, in medicine and education, in finance and commerce, will inevitably impose a social and cultural transformation on the societies of the *tiers monde* that seems likely to be among the most far-reaching of any in their history. Development may be facilitated—even accelerated—from without, and economic relations with the outside world are of the utmost importance, but essentially it is an internal process, and it is beginning in a surprisingly large number of countries in a surprisingly large variety of ways. Where cultural and social attitudes are inconsistent with the exigencies of modern technological change, they are beginning to give at the margin, particularly among young people and students; anachronistic institutions, ranging from the extended family to forms of land tenure, are beginning to crack; the methods, content, and extent of education are being slowly adapted to the new needs. Wherever we look, the obstacles or combinations of obstacles blocking modernization are under attack, and one of the chief objectives of much of the attack is to create conditions in which advantage can be taken of modern technology, for this seems the one way to keep total output rising faster than the numbers of the people.

I am afraid it is probably true that the institutional and sociological—in short, cultural—changes required for an absorption of modern technology and for the efficient organization of economic activities will compel a fair degree of uniformity over wide areas of social life that will not appeal to those who have studied, loved, and admired what Lord Curzon called the 'genius of the East'. Nor does it appeal to the traditional elements in these societies themselves. As Lockwood has pointed out with respect to Japan, 'new modes of production are apt to encounter subtle and powerful resistance. And this may take the form not only of political disorder, or indifference to material progress, but a deeper hostility to the habits of mind and social arrangements which these modes require.' Nevertheless, he endorsed Sir George Sansom's remark that the notion of the 'unchanging East' is 'a very dubious dictum',[5] and no one reading Dr. Broadbridge's description of the industrial, organizational, and financial techniques used by the Japanese to make their country the world's leading shipbuilder could doubt the truth of this.[6] But few Oriental scholars will find solace in gazing at the involuted tubes of oil refineries, the rising smoke of factories, or even the majestic masses of concrete in the great dams and barrages to which their Asian (non-Orientalist) friends may often lead them in pride. Indeed many Western economists, too, will be distressed by some of the more grandiose manifestations of the desire for modernization, but most of these countries have now the responsibility of running their own affairs and they are setting their own objectives and choosing their own paths.

### III

The changes that are taking place, and which are fired by aspirations to obtain better standards of living and more modern economies, require that modern social scientists, and particularly economists, be drawn into Asian studies. These changes cannot be understood without reference to the relation between investment, saving, and consumption, the circular flow of income, the mechanism of inflation, the role of foreign trade, the significance of money, credit, and banking, the allocation of productive factors, social and private costs and benefits, and similar matters. Most of these affect the economic and social life of even the humblest peasant in some degree, and not only are they clearly the subject-matter of economics, but they also relate to the kinds of problems with which Western economists deal in their own societies. The familiarity of these problems, however, is at once an advantage and a danger for economists: it is an advantage because even the least-known society is not wholly unfamiliar; it is a danger because in this field a little knowledge is truly a dangerous thing, and economists have often made serious errors of judgement for lack of sufficient knowledge of the special or peculiar characteristics of the society with which they may be only momentarily concerned.

A Western economist, for example, may go as an adviser to an under-developed country and propose a tax system which seems ideally suited to meet the economic requirements of that country. But suppose that the ethnic divisions of the population coincide with its occupational divisions, and that the incidence of the proposed taxes on different occupations means that one of the important ethnic groups in the community will be harder hit than others. Plainly the new tax system may well end in disaster—even in revolution! In my opinion it is no valid defence of the economist that he was called in to advise *qua* economist and could therefore hardly be expected to know these ethnographic and sociological facts. In the first place, as an economist it was his business to know what kinds of ugly facts would be likely to murder his beautiful theory—which is only to say that he must know what assumptions and what theories are implicit in any advice he gives. And in the second place, as a Western adviser to a government with a civil service almost certainly extremely short of trained and experienced men, it was his responsibility to find out whether in the given environment it was safe to accept the assumptions.

Thus whether the economist is called upon merely to analyse, describe, and explain or also to advise (which essentially involves prediction) he has a great deal to learn from the orientalist and the historian, the anthropologist and the sociologist, for it is not enough that he should understand the nature of economic relationships 'in principle' if he is studying a particular economy; he must know a great deal about the particular economy. Responses to innovation, to changes in the flows of money and goods, to new ideas, incentives, and opportunities are very much influenced not only by a country's economic structure but also by the cultural characteristics of its people, their values, traditions, and customs. Economics is not just a technique of analysis; it is also the application of that technique to situations of the real world in which many considerations other than those admitted into theoretical economic models must be taken account of, and indeed are taken account of implicitly or explicitly by Western economists dealing with their own societies.

Nor is it enough that economists should simply recognize the differences between the economies of the under-developed world and those of the industrialized West; they must also know how to make allowance for these differences. This is a question of the range of problems to which modern economic analysis is in fact applicable, and of the appropriate use of economic concepts and analytical techniques. It seems to be widely believed, particularly by those who know African and Asian societies well, that economists (and others) make the mistakes they do *because* they apply Western concepts to Asian situations, and it is often argued that economic analysis is of very restricted applicability.[7] In my experience, however, the mistakes in diagnosis and prescription that are made are more commonly the consequence of an inadequate understanding of the society concerned and of a failure to select and use the most appropriate economic analysis, than they are mistakes resulting from the inapplicability of economic analysis itself. They may, of course, also arise from the same type of difficulty that creates disagreement among economists anywhere—differences of opinion over the relative desirability of different goals, over the appraisal of non-economic consequences, or over the likely reactions of important political groups to specific economic measures, &c. But I am never able to discuss any Asian economy without reference to standard economic relationships such as supply and demand, money and prices, consumption and investment, terms of trade, costs, inputs, outputs. And no one has ever been able to demonstrate to me that these concepts can be dispensed with. Most of them are needed even if one analyses a subsistence, non-monetary economy.

The fact that some concepts are applicable and some relationships important under one set of circumstances and not under others should disturb no one, even if these limitations should apply to some that play a leading role in modern theory. For example, I doubt very much whether the concept of full employment as we use it has much analytical value for the study of the economies of most of the under-developed countries. Moreover, many theories that were designed with an industrial economy in mind are irrelevant without very substantial alterations for the analysis of backward, largely agricultural economies—for example, the Keynesian multiplier. And finally there are many problems urgently needing study in the *tiers monde* that are just beginning to be examined in the modern economic literature; in other words, there are indeed many things of which economists are ignorant. But none of this seems to me to be relevant to the question whether economics has sufficient general applicability to be useful in the analysis of economic development and in the formation of policies to assist it. If one considers, for example, the industrialization of Egypt since 1952, as Mr. O'Brien has recently analysed it, the applicability of general economic principles is abundantly clear![8] In this, however, I am not denying that the problems economists have chosen to study, as well as the assumptions they have chosen for their theoretical investigations of them, have been often of more relevance to the articulated, largely industrial societies of the West than to the fragmented, largely agricultural societies of the *tiers monde*.[9]

As to relatively neglected subjects, consider the question of land reform. It is really quite astonishing how very little attention has been given to this subject by economists since the days of the Physiocrats, and yet in country after country it is a burning issue. Moreover, when economists do write about it with respect to any particular country, the value of their work is often seriously impaired by lack of knowledge, by a failure to use adequate analytical techniques, and by inappropriate use of economic analysis. Let me illustrate all of this very briefly with reference to land reform in Iraq.

Iraq is a country in which much of the cultivable land was in the hands of large, politically powerful landowners before the revolution of 1958. These landowners were not, by and large, noted for their progressive attitudes, to put it mildly, and there is no doubt that they did stand in the way of many legislative changes that were required to improve Iraqi agriculture, for their political power was excessively great. The popular demand was for expropriation of all large landlords and the redistribution of the land to small peasants. Most economists, Western as well as Iraqi (and these are largely Western-trained), who discussed this question endorsed the proposal, accepting the notion that redistribution of land would increase the productivity of agriculture and the incomes of the *fellahin*; '. . . land reform is worth doing,' wrote a Western analyst, 'even if it is not done with administrative efficiency, simply because redistribution of land can bring *immediate* improvements in the living standard. Reform need not wait on better farming; the agriculture of South Iraq is so primitive that there is *no* risk that a decline in production could follow a division of the big estates. There is no need to aim at perfection in equipping the farmers or in the organization of services.'[10]

Immediately after the revolution of July 1958, the expropriation of the large landowners and the redistribution of the land was decreed—without waiting for the government to organize the provision of services or anything else. This act may have been a political necessity, but it was economically disastrous in spite of the fact that it was moderate and reasonably liberal as land reforms go, with compensation being paid to the landlords. The psychological effect on both landlords and tenants and the absence of proper organization and planning resulted in near anarchy. The area planted to crops fell drastically—and, as usual, it was the peasants who really suffered. All observers, including the government officials charged with administering the reform, agree that the effects, which unfortunately were compounded by two years of bad weather, were very far reaching. Just last year—six years after the reform—the Agricultural Bank could report that finally some stability was beginning to appear in agriculture after the chaos in the years following the revolution and 'the utter confusion into which agricultural relationships had been thrown'.[11]

I cannot go into the sequence of events here: suffice it to say that the actual results were the opposite of those predicted, and this for reasons that could have been foreseen if economists had made serious attempts systematically to collect information, if they had used analytical techniques that took account of input-output relationships, and if adequate attention had been paid to the capacities of the cultivators. I will give only brief examples of each of these to illustrate what I mean.

Information about Iraqi agriculture was extremely poor in spite of numerous investigations and reports, including one by the International Bank for Reconstruction and Development in 1951. It has been improving in recent years, but nobody, not even economists, questioned some of the basic assumptions on which the economic case for land reform was built. To give just one example: it was assumed that tenant farmers were worse off than small farm-owners simply because the tenants had to pay a large proportion of their crop to the landlords as rent. I have never seen this assumption questioned in any of the general discussions about the need for land reform in Iraq. But no investigations were put forward to support it, no statistical surveys conducted—it was considered self-evident. Yet this kind of reasoning commits the elementary mistake of ignoring *ceteris paribus*, for other things were not equal when the position of tenant farmers and farm-owners was compared. In fact, sometimes tenants paying 50% of their crop as rent were better off than small owners, because the maintenance of yields in Iraqi agriculture, where techniques are very primitive, often requires that land be left fallow for a season. Farm-owners with small holdings have sometimes to use their land too intensively in order to scratch a bare livelihood and are unable to let it rest, with the result that the productivity of their land is reduced, and their income is often less than that of tenant farmers with larger areas at their disposal for rotation. The farms of tenants tended to be larger than those of farm-owners because landowners who let the farms to tenants 'could control the farm sizes, while the inheritance law of the Koran, which causes the splitting up of family-owned farms, did not apply to the farms of the tenants but does apply to the farms of the farm-owners'.[12] The problem of inheritance is a serious difficulty in the way of ensuring that farms will remain of economic size in the Middle East and there is need for research on the optimum sizes of farm in the light of expected changes in the rural population.

Not only was the necessary information woefully inadequate, and little attempt made to collect it, but there was also very little attempt to develop appropriate techniques to analyse what was available. The simplest input-output analysis would have made clear the extent to which agriculture depended on investments that were actually made by the landlords—the only important source of investment in agriculture—and would have shown the very great significance of livestock in the farming pattern. Yet even the International Bank had stated in its Report that livestock were only an 'adjunct' to agriculture and not an integral part of it.[13] How misleading this was became clear later from a detailed survey of six agricultural regions that was carried out in 1958/9 and showed that over two-thirds of the net income of farmers came from livestock in one of the regions, and never less than two-fifths in any of them.[14] That this was nothing new, but a long-standing characteristic of agriculture in Iraq has been confirmed by the national income estimates recently published by the Governor of the Central Bank showing that in the ten years covered by the estimates (1953–63) livestock contributed more to the value added in agriculture than did any other group of products, the proportion exceeding 50% in several years.[15]

Land reform was expected to increase productivity in agriculture because peasants, relieved of the necessity of paying rents, would work harder and invest in their own land. This simple notion of the effect of increased expectations of profit took no account of the resources of the Iraqi small farmers or of the environment within which they made their decisions, and was ludicrously unrealistic. This was indeed realized by many who insisted that extensive government organization of services, of finance, of advice, and of certain types of investment goods would be required. Some cash crops, cotton for example, simply will not be produced and marketed by uninstructed and unaided small farmers, and planting of cotton fell 35–45% in the first two years after the reform.

And finally, the subsequent repercussions throughout the economy were of a type that any economist, using the simple generalizations alleged to be impossible and so useless, would easily predict—rising prices, increased import demand, substitutions between crops, problems arising from increased tax burdens, &c.

Land reform may well have been required in Iraq for social, political, or even economic reasons, but the economists who examined the matter failed in their professional duty to bring out clearly the economic implication of various courses of action; indeed, they confused the issues and gave a false confidence to the politicians.

This extended illustration of faulty description, diagnosis, and prescription will serve sufficiently, I hope, to convince you that the remedy for the errors that have been committed by economists with respect to Asian and African countries is not necessarily the abolition of economists but the improvement of their work involving those countries. And this is one of the results I expect from the establishment in this School of a Department of Economics, the chief function of which is to teach economics with special reference to economic development and to carry out research into the characteristics, problems, and contemporary history of the African and Asian economies.

At the same time, however, I am not suggesting that economists should become general 'regional experts'; on the contrary, economists should stick to their last in their professional capacity. There is no contradiction between this and my insistence that if economists are going to be professionally interested in a country about whose history, outlook, manners, culture, geography, resources, language, and technology, they know little, they must remedy their lack of knowledge. Time was when many of the countries of the *tiers monde* were remote, unknown, exotic, and anyone who had 'been there' and had some first-hand experience, who knew the language or had read the literature, was widely accepted as an 'expert'—and indeed his knowledge often did have a high scarcity value at the very least. Today this will not any more do; knowledge has increased, standards of analysis have improved, especially in the study of the society and its culture with the rise of sociology as a social science in its own right, and in the study of the economy. It is as absurd to speak of an 'expert on India' generally as to speak of 'an expert' on Great Britain or the United States. True, there are people whose job it is to synthesize and explain contemporary events over a very wide area, for example journalists, broadcasters, diplomats reporting to their governments, those concerned to provide information for business, &c. These are important and necessary activities and thus there is scope for providing courses of study for students who want to acquire a reasonably broad knowledge of an area with such occupations in mind. But academic research and advances in knowledge will undoubtedly depend upon increasing specialization. After all, even in economics alone no respectable economist pretends that he is equally qualified in all branches of the subject.

At the same time most of the concrete problems faced by the individual countries attempting to promote their economic and social development will not be solved with reference to one type of analysis alone. This fact has caused a great deal of confusion, and added to already existing demands for an extension of what are called 'interdisciplinary' studies. Here we must make a distinction between the training of students and research; except for the purposes just mentioned, I think that students should be trained as specialists in one of the important disciplines. Interdisciplinary *research*, on the other hand, means that several people from different disciplines co-operate in the study of specific types of problems, and this calls for specialization on the part of the participants. A great difficulty, however, is to persuade the specialists from one discipline that the others have something significant to say: in particular, it often seems as if an almost genetic incompatibility exists between economists and some of the other social scientists. In any event, I am sure that progress can only be made if multi-disciplinary research is directed towards carefully defined problems.

## IV

In introducing economics formally into S.O.A.S., the School intended merely to broaden the range of its Asian and African studies. But this will be only one effect, for by formally linking the study of economics with that of Africa and Asia, the School will be participating in an aspect of economics that has been revived since the Second World War and is rapidly contributing to the further development of economic science itself—that aspect now called the economics of growth and development.

Economists do tend rather naturally to put a special emphasis on the issues of particular concern to the age and society in which they live. I have the impression, although in this I defer to our Chairman, whose fame rests among other things on his contribution to the history of economic thought, that towards the last quarter of the last century economists became less interested in problems of growth than they had been earlier, since by then rapid technological and other changes had removed any lingering doubts about immediate hindrances to economic expansion. With few exceptions, the interest in economic growth did not revive until the Great Depression of the 1930's, when men in the United States and Western Europe began to ask whether long-term growth was slowing down, and whether a 'secular stagnation', as it was then called, had set in. Then came the Second World War, but after the war the study of growth returned, this time stimulated partly by certain international political rivalries, partly by the changed nature of the economics of war, partly by the enhanced rate of population growth and by the demands of growing mass consumption, and partly by the plight of the poor countries of the world. Fairly quickly the field bifurcated into the 'economics of growth' and the 'economics of development'—the former referring largely to the industrial countries and the latter to the underdeveloped countries. And for the latter the analysis of the older classical economists is proving to be of considerable relevance.

Thus it has really been only since the war that, apart from a few mavericks, economists have taken special note of the economic problems of countries whose structure and characteristics are very unlike our own; but as they did so, research became increasingly directed to such problems as the changes in economic structure as economies grow richer (that is, the changes in the relative importance of the different sectors such as agriculture, industry, trade), the choice of technique that would maximize rates of economic growth, the role of money flows in a fragmented economy, changes in the distribution of income as total income grows, &c.

Of particular interest to me are the economic problems associated with the position of these countries in the world economy, and further theoretical as well as empirical work needs to be done on such questions as trade and foreign exchange policies, the implications of increasing amounts of private foreign investment, and the structure and operations of international firms such as those of the international petroleum industry. In addition, a host of practical questions such as the formation of technical and administrative cadres, the economics of education, the nature of rural unemployment, the organization of agriculture and, in particular, the problems of government planning, are receiving patient and detailed study in the specific context of particular countries.

All of this is not only making economics a more interesting and richer field of study, but is also contributing to the further development of analytical techniques, particularly in the field of linear programming or input-output analysis, cost-benefit analysis, and investment theory. Thus the applied analysis of agricultural systems is being sharpened, planning techniques are being refined, and ways of making cost-benefit analyses are being extended to cover a wide variety of situations. The tools with which the applied economist must work are constantly being improved and added to in order to increase his ability to deal with practical problems. Thus the emergence of an 'economics of development' does not imply, as some economists have apparently thought, a new and special 'economics' for the developing countries; it merely represents another category of problems, as does, for example, the economics of international trade, or of money and banking, and not another category of thought. After all, some of the problems of greatest interest to the developing countries are of similar interest to the industrialized countries as they consider their own problems of expansion—for example, the economics of education, i.e., the appropriate criteria for determining the amount and kind of investment in human beings, or the appropriate economic policies with respect to the social services, transportation, housing, &c.

On the other hand, the search for a general theory of growth has not yet got very far. In this respect, the theoretical investigations of some economists are strongly reminiscent of medieval scholasticism. Original work, highly abstract but nevertheless of considerable relevance to the functioning of a real economic system, often stimulates a chain of theoretical elaborations which seem to be controlled by a kind of 'law of diminishing relevance' as theorists of true scholastic temperament are drawn further and further into a fantasy world by the compelling intellectual fascination of increasingly trivial complications.[16]

But by and large, economic theorists are very much concerned that their theories be verifiable and their theoretical concepts empirically usable, and here a certain division of labour tends to arise, which is of particular importance in the field of economic development.

The testing of economic theories requires that information of the right kind be available in the right form for the relevant periods of time. Such information is scarce enough in the developed countries and almost non-existent in the developing ones. The economists who construct economic theories, or models, are not often the same economists who either collect or organize the data, but the activities of both groups of economists will be more fruitful if each works with the problems of the other in mind. This constant interplay between those who compile the data and the theorists who use them to test their theories is evident in the development of national income statistics which have rapidly become more and more useful for economic analysis. It may well be that economists who have spent some time assiduously learning an Asian language, Arabic or Chinese for example, and extending their knowledge of an Asian society in order better to study its economy, may find such study an absorbing pursuit, fascinating for its own sake, and may themselves be uninterested in contributing to theoretical advances, even if their intellectual proclivities lean in that direction. Nevertheless, the work they do, the research they publish, should make additional information available to those whose interest lies in the creation and formal testing of theoretical structures—information that is essential if theories of growth and development are to be kept from empty formalism. On the other hand, all applied economics is, in a sense, a test of the applicability of some theory, and therefore it is an essential part of the evolutionary development of theoretical analysis. The difficulty is that such 'tests' by their very nature are usually without controls, and the interpretation of the results thus depends very heavily on the judgement of the analyst. This is one reason why an economist should have a sound knowledge of the countries he studies.

V

At the beginning of this lecture I stated that I was reasonably optimistic about the prospects for the economic development of the poor countries of the world. I have also suggested that considerable progress is being made in economic understanding. But is there any particular reason to think that the countries most in need of sound economic policies will adopt them? It can hardly be denied that in many countries it is very difficult sometimes to discern the direction of the tide, so complex are the cross-currents and undertows of ideological dissension, sectional strife, imperialism under the guise of nationalism, and personal political ambition. Any or all of these may for long periods so dominate the affairs of a particular country that its economic position stagnates or even regresses.

In the best of circumstances the task of raising the standard of living of the people of the *tiers monde* will require sustained effort over a long period of time, and obviously the policies adopted should be such that every available resource makes its contribution. Some of the problems of the poor countries can be seen as a series of vicious circles, and certainly one of these is the relation between the state of backwardness and the supply, or the use, of the qualified leaders and officials that are required for economic development. The countries most in need of sound economic policy are often those to whom it is most denied, either because trained and experienced nationals are simply not available, or because qualified people are not given power, or even made use of, by the country's political or military leaders, who may themselves hold incredibly naïve views on economic affairs.

It must be admitted that there is great scope for disagreement over what constitutes 'sound' economic policy, particularly with regard to such matters as the optimum rate of industrialization, foreign trade and exchange policy, the choice of technique, and the appropriate functions of public and private enterprise in an economic plan. (It is interesting to note, incidentally, that even in China there is controversy over the role of the private sector in agriculture, as Dr. Walker's studies of Chinese agricultural policy show.[17]) Mistakes, serious mistakes, will undoubtedly be made everywhere, but mistakes are unavoidable consequences of the limitations of our knowledge at the present time—for we really must ignore attempts of any particular economist to insist that mistakes could be avoided if only his advice were taken! If impartial, professional, and calm judgement is brought to bear on the formulation of economic policy, this is all we can ask. But when economic judgement is completely ignored, and economics made the handmaiden of politics, then it is indeed tragic for the ordinary people, regardless of the type of system under which they live.

The inadequacies of our understanding, and of the inevitable, perhaps inherent, uncertainties of our so-called science must not lead to an underestimation of what we do know and are agreed upon. Man does not have complete power over economic possibilities; there are predictable limitations to what he can do. It may be that we economists are not always agreed on the *best* ways of achieving given ends, yet we certainly can point out many of the ways most likely to fail to achieve them!

But it is important to recognize that the developing countries can today as never before look to two distinctly different models of successful economic development—the western industrialized countries and those the United Nations calls the 'centrally planned economies'. The very existence of this choice creates severe internal tensions and controversies within the under-developed countries, which will inevitably become entangled with almost all other divisive issues. Even non-communist observers sometimes consider the communist model the more efficient means of promoting economic development.[18] The practical demonstration of the possibilities of government running an economy, the moral appeal of the socialist emphasis on human equality in societies where most people live in appalling poverty, and the lack of responsible experience of the difficulties of economic and social administration, and of the complexities of economic and social organization, provide fertile ground for seeds of controversy; and combined with political irresponsibility these have also led a large number of governments to adopt economic policies which seriously jeopardize economic progress.

Our task as economists is to study the economies of the countries of Asia and Africa as objectively as is humanly possible, irrespective of the type of economic system, and to make hard-headed independent judgements of our own, but with great sympathy, in full awareness of the real difficulty of the often unpalatable choices the fallible human beings who exercise the functions of governments in these countries have been forced to make; and with humility in the light of our own sometimes disastrous mistakes, and of our own ignorance. Our task is important academically for all the reasons I have given; it is urgent practically because the rapidly increasing population of the world will create extremely dangerous problems if the efficiency with which the world's resources are used does not correspondingly increase. Our own future may well be determined by events in Africa and Asia; it is in our mutual interest that our studies of the economic conditions and problems of the Afro-Asian world should be as extensive and competent as we can possibly make them.

NOTES

1 *Report of the Committee appointed to consider the organization of Oriental Studies in London . . .*, Cd. 4560 (1909), p. 14.
2 See the Charter and Notes thereto in the Calendar of the School.
3 Ibid., p. 18.
4 See E. F. Denison, *The Sources of Economic Growth in the United States and the Alternatives before Us*, Supplementary Paper No. 13, New York, Committee for Economic Development, 1962. See also the review article by Moses Abramovitz, 'Economic Growth in the United States', *American Economic Review*, vol. lii, no. 4 (Sept. 1962), pp. 762–82.
5 W. W. Lockwood, *The Economic Development of Japan* (Princeton, 1954), p. 581.
6 S. A. Broadbridge, 'State Support and Technological Progress in the Japanese Shipbuilding Industry', *Journal of Development Studies*, vol. I, no. 2 (Jan. 1965).
7 See, for example, the argument of Dudley Seers, in 'The Limitations of the Special Case' in the *Bulletin of the Oxford Institute of Economics and Statistics*, vol. xxv, no. 2 (May 1963).
8 P. K. O'Brien, *The Revolution in Egypt's Economic System*, published in *Middle Eastern Monographs* of the Royal Institute of International Affairs, London, 1966.
9 Compare Marshall commenting on Ricardo and his followers: 'The same bent of mind that led our lawyers to impose English civil law on the Hindoos, led our economists to work out their theories on the tacit supposition that the world was made up of city men.' Alfred Marshall, *Principles of Economics*, 8th ed., p. 762.
10 Doreen Warriner, *Land Reform and Development in the Middle East* (London, 1962, 2nd ed.), p. 169. (Italics mine.)
11 *Annual Report on the Operations of the Agricultural Bank, 1963/4* (Baghdad 1964), p. 5.
12 A. P. G. Poyck, *Farm Studies in Iraq* (Wageningen, Netherlands, 1962), p. 58.
13 I.B.R.D., *Economic Development of Iraq* (Baltimore, 1952), pp. 246–7.
14 Poyck, op. cit., p. 59.
15 K. Haseeb, *The National Income of Iraq* (London, 1964), p. 34, for the years

1953–61. Data for 1962–3 were presented in his lecture to a Seminar of the Economic Research Institute of the American University of Beirut, 29 May 1964. (Mimeo.)

16 I stress the word 'trivial', for I am not taking issue with the degree of abstraction—an altogether different question. See, for example, the concluding remarks of F. H. Hahn and R. C. O. Matthews, in their article 'The Theory of Economic Growth: A Survey', *Economic Journal,* vol. lxxiv, no. 296 (Dec. 1964), pp. 779 and 891.

17 K. R. Walker, *Planning in Chinese Agriculture* (London: Frank Cass, 1965).

18 For example, Andrew Shonfield, Director of Studies of Chatham House, after examining the various ways in which ruthless centralized direction has an advantage in promoting development over the more democratic type of system, wrote: 'Considering all the advantages of communism, it might be thought that the wisest and perhaps the most generous course would be to let the underdeveloped countries get on with it and go communist. Certainly it is true that if two poor nations, equally endowed by nature, were starting from scratch, the odds would be strongly in favour of the one which adopted a communist system.' He rejected the 'communist path' on political grounds: the 'moral and political damage done to a nation by being pushed through the communist mangle is of a lasting character'. See his *The Attack on World Poverty* (London, 1960), pp. 16–17.